Evolutionary Computation 1

Basic Algorithms and Operators

Evolutionary Computation 1

Basic Algorithms and Operators

Edited by

Thomas Bäck, David B Fogel
and Zbigniew Michalewicz

Taylor & Francis
Taylor & Francis Group
New York London

Taylor & Francis is an imprint of the
Taylor & Francis Group, an **informa** business

Published in 2000 by
Taylor & Francis Group
270 Madison Avenue
New York, NY 10016

Published in Great Britain by
Taylor & Francis Group
2 Park Square
Milton Park, Abingdon
Oxon OX14 4RN

International Standard Book Number-10: 0-7503-0664-5 (Softcover)
International Standard Book Number-13: 978-0-7503-0664-5 (Softcover)

This book contains information obtained from authentic and highly regarded sources. Reprinted material is quoted with permission, and sources are indicated. A wide variety of references are listed. Reasonable efforts have been made to publish reliable data and information, but the author and the publisher cannot assume responsibility for the validity of all materials or for the consequences of their use.

Library of Congress Cataloging-in-Publication Data

Catalog record is available from the Library of Congress

Taylor & Francis Group
is the Academic Division of Informa plc.

Visit the Taylor & Francis Web site at
http://www.taylorandfrancis.com

Contents

Preface   **xiii**

List of contributors   **xvii**

Glossary   **xxi**

PART 1   WHY EVOLUTIONARY COMPUTATION?

1   Introduction to evolutionary computation   **1**
David B Fogel
   1.1   Introductory remarks   1
   1.2   Optimization   1
   1.3   Robust adaptation   2
   1.4   Machine intelligence   2
   1.5   Biology   2
   1.6   Discussion   3
        References   3

2   Possible applications of evolutionary computation   **4**
David Beasley
   2.1   Introduction   4
   2.2   Applications in planning   4
   2.3   Applications in design   6
   2.4   Applications in simulation and identification   7
   2.5   Applications in control   8
   2.6   Applications in classification   9
   2.7   Summary   10
        References   10
        Further reading   18

**3   Advantages (and disadvantages) of evolutionary computation over
other approaches**   **20**
Hans-Paul Schwefel
   3.1   No-free-lunch theorem   20
   3.2   Conclusions   21
        References   22

PART 2   EVOLUTIONARY COMPUTATION: THE BACKGROUND

4   Principles of evolutionary processes   **23**
    David B Fogel
    4.1   Overview   23
          References   26

5   Principles of genetics   **27**
    Raymond C Paton
    5.1   Introduction   27
    5.2   Some fundamental concepts in genetics   27
    5.3   The gene in more detail   33
    5.4   Options for change   35
    5.5   Population thinking   35
          References   38

6   A history of evolutionary computation   **40**
    Kenneth De Jong, David B Fogel and Hans-Paul Schwefel
    6.1   Introduction   40
    6.2   Evolutionary programming   41
    6.3   Genetic algorithms   44
    6.4   Evolution strategies   48
          References   51

**PART 3   EVOLUTIONARY ALGORITHMS AND THEIR
         STANDARD INSTANCES**

7   Introduction to evolutionary algorithms   **59**
    Thomas Bäck
    7.1   General outline of evolutionary algorithms   59
          References   62
          Further reading   62

8   Genetic algorithms   **64**
    Larry J Eshelman
    8.1   Introduction   64
    8.2   Genetic algorithm basics and some variations   65
    8.3   Mutation and crossover   68
    8.4   Representation   75
    8.5   Parallel genetic algorithms   77
    8.6   Conclusion   78
          References   78

9   Evolution strategies   **81**
    Günter Rudolph
    9.1   The archetype of evolution strategies   81
    9.2   Contemporary evolution strategies   83

| | 9.3 | Nested evolution strategies | 86 |
| | | References | 87 |

10 Evolutionary programming — **89**
V William Porto
	10.1	Introduction	89
	10.2	History	90
	10.3	Current directions	97
	10.4	Future research	100
		References	100
		Further reading	102

11 Derivative methods in genetic programming — **103**
Kenneth E Kinnear, Jr
	11.1	Introduction	103
	11.2	Genetic programming defined and explained	103
	11.3	The development of genetic programming	108
	11.4	The value of genetic programming	109
		References	111
		Further reading	112

12 Learning classifier systems — **114**
Robert E Smith
	12.1	Introduction	114
	12.2	Types of learning problem	114
	12.3	Learning classifier system introduction	117
	12.4	'Michigan' and 'Pitt' style learning classifier systems	118
	12.5	The bucket brigade algorithm (implicit form)	119
	12.6	Internal messages	120
	12.7	Parasites	120
	12.8	Variations of the learning classification system	121
	12.9	Final comments	122
		References	122

13 Hybrid methods — **124**
Zbigniew Michalewicz
| | | References | 126 |

PART 4 REPRESENTATIONS

14 Introduction to representations — **127**
Kalyanmoy Deb
	14.1	Solutions and representations	127
	14.2	Important representations	128
	14.3	Combined representations	130
		References	131

15 Binary strings   **132**
Thomas Bäck
    References   135

16 Real-valued vectors   **136**
David B Fogel
    16.1 Object variables   136
    16.2 Object variables and strategy parameters   137
    References   138

17 Permutations   **139**
Darrell Whitley
    17.1 Introduction   139
    17.2 Mapping integers to permutations   141
    17.3 The inverse of a permutation   141
    17.4 The mapping function   142
    17.5 Matrix representations   143
    17.6 Alternative representations   145
    17.7 Ordering schemata and other metrics   146
    17.8 Operator descriptions and local search   149
    References   149

18 Finite-state representations   **151**
David B Fogel
    18.1 Introduction   151
    18.2 Applications   152
    References   154

19 Parse trees   **155**
Peter J Angeline
    References   158

20 Guidelines for a suitable encoding   **160**
David B Fogel and Peter J Angeline
    References   162

21 Other representations   **163**
Peter J Angeline and David B Fogel
    21.1 Mixed-integer structures   163
    21.2 Introns   163
    21.3 Diploid representations   164
    References   164

PART 5   SELECTION

22 Introduction to selection   **166**
Kalyanmoy Deb

22.1 Working mechanisms   166
22.2 Pseudocode   167
22.3 Theory of selective pressure   170
      References   171

23 Proportional selection and sampling algorithms   172
John Grefenstette
23.1 Introduction   172
23.2 Fitness functions   172
23.3 Selection probabilities   175
23.4 Sampling   175
23.5 Theory   176
      References   180

24 Tournament selection   181
Tobias Blickle
24.1 Working mechanism   181
24.2 Parameter settings   182
24.3 Formal description   182
24.4 Properties   183
      References   185

25 Rank-based selection   187
John Grefenstette
25.1 Introduction   187
25.2 Linear ranking   188
25.3 Nonlinear ranking   188
25.4 (μ, λ), $(\mu + \lambda)$ and threshold selection   189
25.5 Theory   190
      References   194

26 Boltzmann selection   195
Samir W Mahfoud
26.1 Introduction   195
26.2 Simulated annealing   196
26.3 Working mechanism for parallel recombinative simulated
      annealing   196
26.4 Pseudocode for a common variation of parallel recombinative
      simulated annealing   197
26.5 Parameters and their settings   197
26.6 Global convergence theory and proofs   199
      References   200

27 Other selection methods   201
David B Fogel
27.1 Introduction   201
27.2 Tournament selection   201

27.3 Soft brood selection   202
27.4 Disruptive selection   202
27.5 Boltzmann selection   202
27.6 Nonlinear ranking selection   202
27.7 Competitive selection   203
27.8 Variable lifespan   203
      References   204

28 Generation gap methods   **205**
Jayshree Sarma and Kenneth De Jong
28.1 Introduction   205
28.2 Historical perspective   206
28.3 Steady state and generational evolutionary algorithms   207
28.4 Elitist strategies   210
      References   211

29 A comparison of selection mechanisms   **212**
Peter J B Hancock
29.1 Introduction   212
29.2 Simulations   213
29.3 Population models   214
29.4 Equivalence: expectations and reality   214
29.5 Simulation results   218
29.6 The effects of evaluation noise   222
29.7 Analytic comparison   224
29.8 Conclusions   225
      References   225

30 Interactive evolution   **228**
Wolfgang Banzhaf
30.1 Introduction   228
30.2 History   228
30.3 The problem   229
30.4 The interactive evolution approach   229
30.5 Difficulties   231
30.6 Application areas   231
30.7 Further developments and perspectives   232
      References   233
      Further reading   234

PART 6   SEARCH OPERATORS

31 Introduction to search operators   **235**
Zbigniew Michalewicz
      References   236

32 Mutation operators   **237**

Thomas Bäck, David B Fogel, Darrell Whitley and Peter J Angeline

32.1 Binary strings   237
32.2 Real-valued vectors   239
32.3 Permutations   243
32.4 Finite-state machines   246
32.5 Parse trees   248
32.6 Other representations   250
      References   252

33 Recombination   **256**

Lashon B Booker , David B Fogel, Darrell Whitley, Peter J Angeline
and A E Eiben

33.1 Binary strings   256
33.2 Real-valued vectors   270
33.3 Permutations   274
33.4 Finite-state machines   284
33.5 Crossover: parse trees   286
33.6 Other representations   289
33.7 Multiparent recombination   289
      References   302

34 Other operators   **308**

Russell W Anderson, David B Fogel and Martin Schütz

34.1 The Baldwin effect   308
34.2 Knowledge-augmented operators   317
34.3 Gene duplication and deletion   319
      References   326
      Further reading   329

Index   **331**

Preface

The original *Handbook of Evolutionary Computation* (Bäck *et al* 1997) was designed to fulfill the need for a broad-based reference book reflecting the important role that evolutionary computation plays in a variety of disciplines—ranging from the natural sciences and engineering to evolutionary biology and computer sciences. The basic idea of evolutionary computation, which came onto the scene in the 1950s, has been to make use of the powerful process of natural evolution as a problem-solving paradigm, either by simulating it ('by hand' or automatically) in a laboratory, or by simulating it on a computer. As the history of evolutionary computation is the topic of one of the introductory sections of the *Handbook*, we will not go into the details here but simply mention that genetic algorithms, evolution strategies, and evolutionary programming are the three independently developed mainstream representatives of evolutionary computation techniques, and genetic programming and classifier systems are the most prominent derivative methods.

In the 1960s, visionary researchers developed these mainstream methods of evolutionary computation, namely J H Holland (1962) at Ann Arbor, Michigan, H J Bremermann (1962) at Berkeley, California, and A S Fraser (1957) at Canberra, Australia, for genetic algorithms, L J Fogel (1962) at San Diego, California, for evolutionary programming, and I Rechenberg (1965) and H P Schwefel (1965) at Berlin, Germany, for evolution strategies. The first generation of books on the topic of evolutionary compuation, written by several of the pioneers themselves, still gives an impressive demonstration of the capabilities of evolutionary algorithms, especially if one takes account of the limited hardware capacity available at that time (see Fogel *et al* (1966), Rechenberg (1973), Holland (1975), and Schwefel (1977)).

Similar in some ways to other early efforts towards imitating nature's powerful problem-solving tools, such as artificial neural networks and fuzzy systems, evolutionary algorithms also had to go through a long period of ignorance and rejection before receiving recognition. The great success that these methods have had, in extremely complex optimization problems from various disciplines, has facilitated the undeniable breakthrough of evolutionary computation as an accepted problem-solving methodology. This breakthrough is reflected by an exponentially growing number of publications in the field, and an increasing interest in corresponding conferences and journals. With these activities, the field now has its own archivable high-quality publications in which the actual research results are published. The publication of a considerable amount of application-specific work is, however, widely scattered over different disciplines and their specific conferences and journals, thus reflecting the general applicability and success of evolutionary computation methods.

The progress in the theory of evolutionary computation methods since 1990 impressively confirms the strengths of these algorithms as well as their limitations. Research in this field has reached maturity, concerning theoretical and application aspects, so it becomes important to provide a complete reference for practitioners, theorists, and teachers in a variety of disciplines. The original *Handbook of Evolutionary Computation* was designed to provide such a reference work. It included complete, clear, and accessible information, thoroughly describing state-of-the-art evolutionary computation research and application in a comprehensive style.

These new volumes, based in the original *Handbook*, but updated, are designed to provide the material in units suitable for coursework as well as for individual researchers. The first volume, *Evolutionary Computation 1: Basic Algorithms and Operators*, provides the basic information on evolutionary algorithms. In addition to covering all paradigms of evolutionary computation in detail and giving an overview of the rationale of evolutionary computation and of its biological background, this volume also offers an in-depth presentation of basic elements of evolutionary computation models according to the types of representations used for typical problem classes (e.g. binary, real-valued, permutations, finite-state machines, parse trees). Choosing this classification based on representation, the search operators mutation and recombination (and others) are straightforwardly grouped according to the semantics of the data they manipulate. The second volume, *Evolutionary Computation 2: Advanced Algorithms and Operators*, provides information on additional topics of major importance for the design of an evolutionary algorithm, such as the fitness evaluation, constraint-handling issues, and population structures (including all aspects of the parallelization of evolutionary algorithms). This volume also covers some advanced techniques (e.g. parameter control, meta-evolutionary approaches, coevolutionary algorithms, etc) and discusses the efficient implementation of evolutionary algorithms.

Organizational support provided by Institute of Physics Publishing makes it possible to prepare this second version of the *Handbook*. In particular, we would like to express our gratitude to our project editor, Robin Rees, who worked with us on editorial and organizational issues.

Thomas Bäck, David B Fogel and Zbigniew Michalewicz
August 1999

References

Bäck T, Fogel D B and Michalewicz Z 1997 *Handbook of Evolutionary Computation* (Bristol: Institute of Physics Publishing and New York: Oxford University Press)

Bezdek J C 1994 What is computational intelligence ? *Computational Intelligence: Imitating Life* ed J M Zurada, R J Marks II and C J Robinson (New York: IEEE Press) pp 1–12

Bremermann H J 1962   Optimization through evolution and recombination *Self-Organizing Systems* ed M C Yovits, G T Jacobi and G D Goldstine (Washington, DC: Spartan Book) pp 93–106

Fogel L J 1962 Autonomous automata *Industrial Research* **4** 14–9

Fogel L J, Owens A J and Walsh M J 1966 *Artificial Intelligence through Simulated Evolution* (New York: Wiley)

Fraser A S 1957 Simulation of genetic systems by automatic digital computers: I. Introduction *Austral. J. Biol. Sci.* **10** pp 484–91

Holland J H 1962 Outline for a logical theory of adaptive systems *J. ACM* **3** 297–314

——1975 *Adaptation in Natural and Artificial Systems* (Ann Arbor, MI: University of Michigan Press)

Rechenberg I 1965 Cybernetic solution path of an experimental problem *Royal Aircraft Establishment Library Translation No 1122* (Farnborough, UK)

Rechenberg I 1973 *Evolutionsstrategie: Optimierung technischer Systeme nach Prinzipien der biologischen Evolution* (Stuttgart: Frommann-Holzboog)

Schwefel H-P 1965   Kybernetische Evolution als Strategie der experimentellen Forschung in der Strömungstechnik *Diplomarbeit* Hermann Föttinger Institut für Strömungstechnik, Technische Universität, Berlin

——1977 *Numerische Optimierung von Computer-Modellen mittels der Evolutionsstrategie* Interdisciplinary Systems Research vol 26 (Basel: Birkhäuser)

List of Contributors

Peter J Angeline (Chapters 19–21, 32, 33)

Senior Scientist, Natural Selection, Inc., Vestal, NY, USA
e-mail: angeline@natural-selection.com

Thomas Bäck (Chapters 7, 15, 32, Glossary)

Associate Professor of Computer Science, Leiden University, The Netherlands; and Managing Director and Senior Research Fellow, Center for Applied Systems Analysis, Informatik Centrum Dortmund, Germany
e-mail: baeck@ls11.informatik.uni-dortmund.de

Wolfgang Banzhaf (Chapter 30)

Professor of Computer Science, University of Dortmund, Germany
e-mail: banzhaf@ls11.informatik.uni-dortmund.de

David Beasley (Chapter 2)

Software Engineer, Praxis PLC, Deloitte and Touche Consulting Group, Bath, United Kingdom
e-mail: dabley@praxis.co.uk

Tobias Blickle (Chapter 24)

Electrical Engineer, Institute TIK, ETH Zurich, Switzerland
e-mail: blickle@tik.ee.ethz.ch

Lashon B Booker (Chapter 33)

Principal Scientist, Artificial Intelligence Technical Center, The MITRE Corporation, McLean, VA, USA
e-mail: booker@mitre.org

Kalyanmoy Deb (Chapters 14, 22)

Associate Professor of Mechanical Engineering, Indian Institute of Technology, Kanpur, India
e-mail: deb@iitk.ernet.in

Kenneth De Jong (Chapters 6, 28)

Professor of Computer Science, George Mason University, Fairfax, VA, USA
e-mail: kdejong@gmu.edu

A E Eiben (Chapter 33)

Leiden Institute of Advanced Computer Science, Leiden University, The Netherlands; and Faculty of Sciences, Vrije Universiteit Amsterdam, The Netherlands
e-mail: gusz@cs.leidenuniv.nl and gusz@cs.vu.nl

Larry J Eshelman (Chapter 8)

Principal Member of Research Staff, Philips Research, Briarcliff Manor, NY, USA
e-mail: lje@philabs.philips.com

David B Fogel (Chapters 1, 4, 6, 16, 18, 20, 21, 27, 32–34, Glossary)

Executive Vice President and Chief Scientist, Natural Selection Inc., La Jolla, CA, USA
e-mail: dfogel@natural-selection.com

John Grefenstette (Chapters 23, 25)

Head of the Machine Learning Section, Navy Center for Applied Research in Artificial Intelligence, Naval Research Laboratory, Washington, DC, USA
e-mail: gref@aic.nrl.navy.mil

Peter J B Hancock (Chapter 29)

Lecturer in Psychology, University of Stirling, United Kingdom
e-mail: pjh@psych.stir.ac.uk

Kenneth E Kinnear Jr (Chapter 11)

Chief Technical Officer, Adaptive Computing Technology, Boxboro, MA, USA
e-mail: kinnear@adapt.com

Samir W Mahfoud (Chapter 26)

Vice President of Research and Software Engineering, Advanced Investment Technology, Clearwater, FL, USA
e-mail: sam@ait-tech.com

Zbigniew Michalewicz (Chapters 13, 31)

Professor of Computer Science, University of North Carolina, Charlotte, USA; and Institute of Computer Science, Polish Academy of Sciences, Warsaw, Poland
e-mail: zbyszek@uncc.edu

Raymond C Paton (Chapter 5)

Lecturer in Computer Science, University of Liverpool, United Kingdom
e-mail: r.c.paton@csc.liv.ac.uk

V William Porto (Chapter 10)

Senior Staff Scientist, Natural Selection Inc., La Jolla, CA, USA
e-mail: bporto@natural-selection.com

Günter Rudolph (Chapter 9)

Senior Research Fellow, Center for Applied Systems Analysis, Informatik Centrum Dortmund, Germany
e-mail: rudolph@icd.de

Jayshree Sarma (Chapter 28)

Computer Science Researcher, George Mason University, Fairfax, VA, USA
e-mail: jsarma@gmu.edu

Martin Schütz (Chapter 34)

Computer Scientist, Systems Analysis Research Group, University of Dortmund,
Germany
e-mail: schuetz@ls11.informatik.uni-dortmund.de

Hans-Paul Schwefel (Chapters 3, 6)

Chair of Systems Analysis, and Professor of Computer Science, University of
Dortmund, Germany; and Director, Center for Applied Systems Analysis,
Informatik Centrum Dortmund, Germany
e-mail: schwefel@ls11.informatik.uni-dortmund.de

Robert E Smith (Chapter 12)

Senior Research Fellow, Intelligent Computing Systems Centre, Computer Studies
and Mathematics Faculty, University of the West of England, Bristol, United
Kingdom
rsmith@btc.uwe.ac.uk

Darrell Whitley (Chapters 17, 32, 33)

Professor of Computer Science, Colorado State University, Fort Collins, CO, USA
e-mail: whitley@cs.colostate.edu

Glossary

Thomas Bäck and David B Fogel

Bold text within definitions indicates terms that are also listed elsewhere in this glossary.

Adaptation: This denotes the general advantage in ecological or physiological efficiency of an **individual** in contrast to other members of the **population**, and it also denotes the process of attaining this state.

Adaptive behavior: The underlying mechanisms to allow living organisms, and, potentially, robots, to adapt and survive in uncertain environments (cf **adaptation**).

Adaptive surface: Possible biological trait combinations in a **population** of **individuals** define points in a high-dimensional sequence space, where each coordinate axis corresponds to one of these traits. An additional dimension characterizes the **fitness** values for each possible trait combination, resulting in a highly multimodal fitness landscape, the so-called adaptive surface or adaptive topography.

Allele: An alternative form of a **gene** that occurs at a specified chromosomal position (**locus**).

Artificial life: A terminology coined by C G Langton to denote the '...study of simple computer generated hypothetical life forms, i.e. life-as-it-could-be.' Artificial life and **evolutionary computation** have a close relationship because **evolutionary algorithms** are often used in artificial life research to breed the survival strategies of **individuals** in a **population** of artificial life forms.

Automatic programming: The task of finding a program which calculates a certain input–output function. This task has to be performed in automatic programming by another computer program (cf **genetic programming**).

Baldwin effect: Baldwin theorized that **individual** learning allows an organism to exploit genetic variations that only partially determine a physiological structure. Consequently, the ability to learn can guide evolutionary processes by rewarding partial genetic successes. Over evolutionary time, learning can guide evolution because individuals with useful genetic variations are maintained by learning, such that useful **genes** are utilized more widely in the subsequent generation. Over time, abilities that previously required learning are replaced by genetically determinant systems. The guiding effect of learning on evolution is referred to as the Baldwin effect. (*See also Section 34.1.*)

Behavior: The response of an organism to the present environmental stimulus. The collection of behaviors of an organism defines the **fitness** of the organism to its present environment.

Boltzmann selection: The Boltzmann selection method transfers the probabilistic acceptance criterion of **simulated annealing** to **evolutionary algorithms**. The method operates by creating an offspring **individual** from two parents and accepting improvements (with respect to the parent's fitness) in any case and deteriorations according to an exponentially decreasing function of an exogeneous 'temperature' parameter. (*See also Chapter 26.*)

Building block: Certain forms of recombination in **evolutionary algorithms** attempt to bring together building blocks, shorter pieces of an overall solution, in the hope that together these blocks will lead to increased performance. (*See also Section 26.3.*)

Central dogma: The fact that, by means of **translation** and **transcription** processes, the genetic information is passed from the **genotype** to the **phenotype** (i.e. from **DNA** to **RNA** and to the **proteins**). The dogma implies that behaviorally acquired characteristics of an **individual** are not inherited to its offspring (cf **Lamarckism**).

Chromatids: The two identical parts of a duplicated **chromosome**.

Chromosome: Rod-shaped bodies in the nucleus of **eukaryotic cells**, which contain the hereditary units or **genes**.

Classifier systems: Dynamic, rule-based systems capable of learning by examples and induction. Classifier systems evolve a **population** of production rules (in the so-called Michigan approach, where an **individual** corresponds to a single rule) or a population of production rule bases (in the so-called Pittsburgh approach, where an individual represents a complete rule base) by means of an **evolutionary algorithm**. The rules are often encoded by a ternary alphabet, which contains a 'don't care' symbol facilitating a generalization capability of condition or action parts of a rule, thus allowing for an inductive learning of concepts. In the Michigan approach, the rule **fitness** (its strength) is incrementally updated at each generation by the 'bucket brigade' credit assignment algorithm based on the reward the system obtains from the environment, while in the Pittsburgh approach the fitness of a complete rule base can be calculated by testing the behavior of the individual within its environment.

Codon: A group of three nucleotide bases within the **DNA** that encodes a single amino acid or start and stop information for the **transcription** process.

Coevolutionary system: In coevolutionary systems, different **populations** interact with each other in a way such that the evaluation function of one population may depend on the state of the evolution process in the other population(s).

Comma strategy: The notation (μ, λ) strategy describes a selection method introduced in **evolution strategies** and indicates that a parent **population** of μ **individuals** generates $\lambda > \mu$ offspring and the best out of these λ offspring are deterministically selected as parents of the next generation. (*See also Section 25.4.*)

Computational intelligence: The field of computational intelligence is currently seen to include subsymbolic approaches to artificial intelligence, such as **neural networks, fuzzy systems,** and **evolutionary computation,** which are gleaned from the model of information processing in natural systems. Following a commonly accepted characterization, a system is computationally intelligent if it deals only with numerical data, does not use knowledge in the classical expert system sense, and exhibits computational adaptivity, fault tolerance, and speed and error rates approaching human performance.

Convergence reliability: Informally, the convergence reliability of an **evolutionary algorithm** means its capability to yield reasonably good solutions in the case of highly multimodal topologies of the objective function. Mathematically, this is closely related to the property of global convergence with probability one, which states that, given infinite running time, the algorithm finds a global optimum point with probability one. From a theoretical point of view, this is an important property to justify the feasibility of evolutionary algorithms as **global optimization** methods.

Convergence velocity: In the theory of **evolutionary algorithms,** the convergence velocity is defined either as the expectation of the change of the distance towards the optimum between two subsequent generations, or as the expectation of the change of the objective function value between two subsequent generations. Typically, the best **individual** of a **population** is used to define the convergence velocity.

Crossover: A process of information exchange of genetic material that occurs between adjacent **chromatids** during **meiosis.**

Cultural algorithm: Cultural algorithms are special variants of **evolutionary algorithms** which support two models of inheritance, one at the microevolutionary level in terms of traits, and the other at the macroevolutionary level in terms of beliefs. The two models interact via a communication channel that enables the behavior of individuals to alter the belief structure and allows the belief structure to constrain the ways in which individuals can behave. The belief structure represents 'cultural' knowledge about a certain problem and therefore helps in solving the problem on the level of traits.

Cycle crossover: A **crossover** operator used in **order-based genetic algorithms** to manipulate permutations in a permutation preserving way. Cycle crossover performs recombination under the constraint that each element must come from one parent or the other by transferring element cycles between the mates. The cycle crossover operator preserves absolute positions of the elements of permutations. (*See also Section 33.3.*)

Darwinism: The theory of evolution, proposed by Darwin, that evolution comes about through random variation (**mutation**) of heritable characteristics, coupled with **natural selection**, which favors those species for further survival and evolution that are best adapted to their environmental conditions. (*See also Chapter 4.*)

Deception: Objective functions are called deceptive if the combination of good **building blocks** by means of **recombination** leads to a reduction of fitness rather than an increase.

Deficiency: A form of **mutation** that involves a terminal segment loss of **chromosome** regions.

Defining length: The defining length of a **schema** is the maximum distance between specified positions within the schema. The larger the defining length of a schema, the higher becomes its disruption probability by **crossover**.

Deletion: A form of **mutation** that involves an internal segment loss of a **chromosome** region.

Deme: An independent subpopulation in the **migration model** of parallel **evolutionary algorithms**.

Diffusion model: The diffusion model denotes a massively parallel implementation of **evolutionary algorithms**, where each **individual** is realized as a single process being connected to neighboring individuals, such that a spatial individual structure is assumed. **Recombination** and **selection** are restricted to the neighborhood of an individual, such that information is locally preserved and spreads only slowly over the **population**.

Diploid: In diploid organisms, each body cell carries two sets of **chromosomes**; that is, each chromosome exists in two **homologous** forms, one of which is **phenotypically** realized.

Discrete recombination: Discrete recombination works on two vectors of **object variables** by performing an exchange of the corresponding object variables with probability one half (other settings of the exchange probability are in principle possible) (cf **uniform crossover**). (*See also Section 33.2.*)

DNA: Deoxyribonucleic acid, a double-stranded macromolecule of helical structure (comparable to a spiral staircase). Both single strands are linear, unbranched nucleic acid molecules built up from alternating deoxyribose (sugar) and phosphate molecules. Each deoxyribose part is coupled to a nucleotide base, which is responsible for establishing the connection to the other strand of the DNA. The four nucleotide bases adenine (A), thymine (T), cytosine (C) and guanine (G) are the alphabet of the genetic information. The sequences of these bases in the DNA molecule determines the building plan of any organism.

Duplication: A form of **mutation** that involves the doubling of a certain region of a **chromosome** at the expense of a corresponding **deficiency** on the other of two **homologous** chromosomes.

Elitism: Elitism is a feature of some **evolutionary algorithms** ensuring that the maximum objective function value within a **population** can never reduce from one generation to the next. This can be assured by simply copying the best **individual** of a population to the next generation, if none of the selected offspring constitutes an improvement of the best value.

Eukaryotic cell: A cell with a membrane-enclosed nucleus and organelles found in animals, fungi, plants, and protists.

Evolutionary algorithm: *See* **evolutionary computation**.

Evolutionary computation: This encompasses methods of simulating evolution, most often on a computer. The field encompasses methods that comprise a population-based approach that relies on random variation and selection. Instances of algorithms that rely on evolutionary principles are called **evolutionary algorithms**. Certain historical subsets of evolutionary algorithms include **evolution strategies**, **evolutionary programming**, and **genetic algorithms**.

Evolutionary operation (EVOP): An industrial management technique presented by G E P Box in the late fifties, which provides a systematic way to test alternative production processes that result from small modifications of the standard parameter settings. From an abstract point of view, the method resembles a $(1 + \lambda)$ strategy with a typical setting of $\lambda = 4$ and $\lambda = 8$ (the so-called 2^2 and 2^3 factorial design), and can be interpreted as one of the earliest **evolutionary algorithms**.

Evolutionary programming: An **evolutionary algorithm** developed by L J Fogel at San Diego, CA, in the 1960s and further refined by D B Fogel and others in the 1990s. Evolutionary programming was originally developed as a method to evolve **finite-state machines** for solving time series prediction tasks and was later extended to parameter optimization problems. Evolutionary programming typically relies on variation operators that are tailored to the problem, and these often are based on a single parent; however, the earliest versions of evolutionary programming considered the possibility for recombining three or more **finite-state machines**. **Selection** is a stochastic **tournament selection** that determines μ **individuals** to survive out of the μ parents and the μ (or other number of) offspring generated by mutation. Evolutionary programming also uses the **self-adaptation** principle to evolve **strategy parameters** on-line during the search (cf **evolution strategy**). (*See also Chapter 10.*)

Evolution strategy: An **evolutionary algorithm** developed by I Rechenberg and H-P Schwefel at the Technical University of Berlin in the 1960s. The evolution strategy typically employs real-valued parameters, though it has also been used for discrete problems. Its basic features are the distinction between a parent **population** (of size μ) and an offspring population (of size $\lambda \geq \mu$), the explicit emphasis on normally distributed **mutations**, the utilization of different forms of **recombination**, and the incorporation of the **self-adaptation** principle for **strategy parameters**; that is, those parameters that determine the mutation probability density function are evolved on-line, by the same principles which are used to evolve the **object variables**. (*See also Chapter 9.*)

Exon: A region of **codons** within a **gene** that is expressed for the **phenotype** of an organism.

Finite-state machine: A transducer that can be stimulated by a finite alphabet of input symbols, responds in a finite alphabet of output symbols, and possesses some finite number of different internal states. The behavior of the finite-state machine is specified by the corresponding input–output symbol pairs and next-state transitions for each input symbol, taken over every state. In **evolutionary programming**, finite-state machines are historically the first structures that were evolved to find optimal predictors of the environmental behavior. (*See also Chapter 18.*)

Fitness: The propensity of an **individual** to survive and reproduce in a particular environment. In **evolutionary algorithms**, the fitness value of an individual is closely related (and sometimes identical) to the objective function value of the solution represented by the individual, but especially when using **proportional selection** a **scaling function** is typically necessary to map objective function values to positive values such that the best-performing individual receives maximum fitness.

Fuzzy system: Fuzzy systems try to model the the fact that real-world circumstances are typically not precise but 'fuzzy'. This is achieved by generalizing the idea of a crisp membership function of sets by allowing for an arbitrary degree of membership in the unit interval. A fuzzy set is then described by such a generalized membership function. Based on membership functions, linguistic variables are defined that capture real-world concepts such as 'low temperature'. Fuzzy rule-based systems then allow for knowledge processing by means of fuzzification, fuzzy inference, and defuzzification operators which often enable a more realistic modeling of real-world situations than expert systems do.

Gamete: A **haploid** germ cell that fuses with another in fertilization to form a **zygote**.

Gene: A unit of **codons** on the **DNA** that encodes the synthesis for a **protein**.

Generation gap: The generation gap characterizes the percentage of the **population** to be replaced during each generation. The remainder of the population is chosen (at random) to survive intact. The generation gap allows for gradually shifting from the generation-based working scheme towards the extreme of just generating one new **individual** per 'generation', the so-called **steady-state selection** algorithm. (*See also Chapter 28.*)

Genetic algorithm: An **evolutionary algorithm** developed by J H Holland and his students at Ann Arbor, MI, in the 1960s. Fundamentally equivalent procedures were also offered earlier by H J Bremermann at UC Berkeley and A S Fraser at the University of Canberra, Australia in the 1960s and 1950s. Originally, the genetic algorithm or adaptive plan was designed as a formal system for **adaptation** rather than an optimization system. Its basic features are the strong emphasis on **recombination** (crossover), use of a probabilistic **selection** operator (**proportional selection**), and the interpretation of **mutation** as a background operator, playing a minor role for the algorithm. While the original form of genetic algorithms (the canonical genetic algorithm) represents solutions by binary strings, a number of variants including real-coded genetic algorithms and order-based genetic algorithms have also been developed to make the algorithm applicable to other than binary search spaces. (*See also Chapter 8.*)

Genetic code: The **translation** process performed by the ribosomes essentially maps triplets of nucleotide bases to single amino acids. This (redundant) mapping between the $4^3 = 64$ possible **codons** and the 20 amino acids is the so-called genetic code.

Genetic drift: A random decrease or increase of biological trait frequencies within the **gene** pool of a **population**.

Genetic programming: Derived from genetic algorithms, the genetic programming paradigm characterizes a class of **evolutionary algorithms** aiming at the automatic generation of computer programs. To achieve this, each **individual** of a **population** represents a complete computer program in a suitable programming language. Most commonly, symbolic expressions representing parse trees in (a subset of) the LISP language are used to represent these programs, but also other representations (including binary representation) and other programming languages (including machine code) are successfully employed. (*See also Chapter 11.*)

Genome: The total genetic information of an organism.

Genotype: The sum of inherited characters maintained within the entire reproducing population. Often also the genetic constitution underlying a single trait or set of traits.

Global optimization: Given a function $f : M \rightarrow \mathbb{R}$, the problem of determining a point $x^* \in M$ such that $f(x^*)$ is minimal (i.e. $f(x^*) \leq f(x) \ \forall x \in M$) is called the global optimization problem.

Global recombination: In **evolution strategies**, **recombination** operators are sometimes used which potentially might take all **individuals** of a **population** into account for the creation of an offspring individual. Such recombination operators are called global recombination (i.e. global **discrete recombination** or global **intermediate recombination**).

Gradient method: Local optimization algorithms for continuous parameter optimization problems that orient their choice of search directions according to the first partial derivatives of the objective function (its gradient) are called gradient strategies (cf **hillclimbing strategy**).

Gray code: A binary code for integer values which ensures that adjacent integers are encoded by binary strings with **Hamming distance** one. Gray codes play an important role in the application of canonical **genetic algorithms** to parameter optimization problems, because there are certain situations in which the use of Gray codes may improve the performance of an **evolutionary algorithm**.

Hamming distance: For two binary vectors, the Hamming distance is the number of different positions.

Haploid: Haploid organisms carry one set of genetic information.

Heterozygous: **Diploid** organisms having different **alleles** for a given trait.

Hillclimbing strategy: Hillclimbing methods owe their name to the analogy of their way of searching for a maximum with the intuitive way a sightless climber might feel his way from a valley up to the peak of a mountain by steadily moving upwards. These strategies follow a nondecreasing path to an optimum by a sequence of neighborhood moves. In the case of multimodal landscapes, hillclimbing locates the optimum closest to the starting point of its search.

Homologues: **Chromosomes** of identical structure, but with possibly different genetic information contents.

Homozygous: **Diploid** organisms having identical **alleles** for a given trait.

Hybrid method: **Evolutionary algorithms** are often combined with classical optimization techniques such as **gradient methods** to facilitate an efficient local search in the final stage of the evolutionary optimization. The resulting combinations of algorithms are often summarized by the term hybrid methods.

Implicit parallelism: The concept that each individual solution offers partial information about sampling from other solutions that contain similar subsections. Although it was once believed that maximizing implicit parallelism would increase the efficiency of an **evolutionary algorithm**, this notion has been proved false in several different mathematical developments (See **no-free-lunch theorem**).

Individual: A single member of a **population**. In **evolutionary algorithms**, an individual contains a **chromosome** or **genome**, that usually contains at least a representation of a possible solution to the problem being tackled (a single point in the search space). Other information such as certain **strategy parameters** and the individual's **fitness** value are usually also stored in each individual.

Intelligence: The definition of the term intelligence for the purpose of clarifying what the essential properties of artificial or computational intelligence should be turns out to be rather complicated. Rather than taking the usual anthropocentric view on this, we adopt a definition by D Fogel which states that intelligence is the capability of a system to adapt its behavior to meet its goals in a range of environments. This definition also implies that

evolutionary algorithms provide one possible way to evolve intelligent systems.

Interactive evolution: The interactive evolution approach involves the human user of the **evolutionary algorithm** on-line into the variation–**selection** loop. By means of this method, subjective judgment relying on human intuition, esthetical values, or taste can be utilized for an evolutionary algorithm if a **fitness** criterion can not be defined explicitly. Furthermore, human problem knowledge can be utilized by interactive evolution to support the search process by preventing unnecessary, obvious detours from the global optimization goal. (*See also Chapter 30.*)

Intermediate recombination: Intermediate recombination performs an averaging operation on the components of the two parent vectors. (*See also Section 33.2.*)

Intron: A region of **codons** within a **gene** that do not bear genetic information that is expressed for the **phenotype** of an organism.

Inversion: A form of **mutation** that changes a **chromosome** by rotating an internal segment by $180°$ and refitting the segment into the chromosome.

Lamarckism: A theory of evolution which preceded Darwin's. Lamarck believed that acquired characteristics of an **individual** could be passed to its offspring. Although Lamarckian inheritance does not take place in nature, the idea has been usefully applied within some **evolutionary algorithms**.

Locus: A particular location on a **chromosome**.

Markov chain: A **Markov process** with a finite or countable finite number of states.

Markov process: A stochastic process (a family of random variables) such that the probability of the process being in a certain state at time k depends on the state at time $k - 1$, not on any states the process has passed earlier. Because the offspring **population** of an **evolutionary algorithm** typically depends only on the actual population, Markov processes are an appropriate mathematical tool for the analysis of evolutionary algorithms.

Meiosis: The process of cell division in **diploid** organisms through which germ cells (**gametes**) are created.

Metaevolution: The problem of finding optimal settings of the exogeneous parameters of an **evolutionary algorithm** can itself be interpreted as an optimization problem. Consequently, the attempt has been made to use an evolutionary algorithm on the higher level to evolve optimal **strategy parameter** settings for evolutionary algorithms, thus hopefully finding a best-performing parameter set that can be used for a variety of objective functions. The corresponding technique is often called a metaevolutionary algorithm. An alternative approach involves the **self-adaptation** of strategy parameters by evolutionary learning.

Migration: The transfer of an **individual** from one subpopulation to another.

Migration model: The migration model (often also referred to as the island model) is one of the basic models of parallelism exploited by **evolutionary algorithm** implementations. The population is no longer **panmictic**, but distributed into several independent subpopulations (so-called **demes**), which coexist (typically on different processors, with one subpopulation per processor) and may mutually exchange information by interdeme **migration**. Each of the subpopulations corresponds to a conventional (i.e. sequential) evolutionary algorithm. Since **selection** takes place only locally inside a **population**, every deme is able to concentrate on different promising regions of the search space, such that the global search capabilities of migration models often exceed those of panmictic populations. The fundamental parameters introduced by the migration principle are the exchange frequency of information, the number of **individuals** to exchange, the selection strategy for the emigrants, and the replacement strategy for the immigrants.

Monte Carlo algorithm: *See* **uniform random search**.

(μ, λ) strategy: *See* **comma strategy**.

($\mu + \lambda$) strategy: *See* **plus strategy**.

Multiarmed bandit: Classical analysis of **schema** processing relied on an analogy to sampling from a number of slot machines (one-armed bandits) in order to minimize expected losses.

Multimembered evolution strategy: All variants of **evolution strategies** that use a parent **population** size of $\mu > 1$ and therefore facilitate the utilization of **recombination** are summarized under the term multimembered evolution strategy.

Multiobjective optimization: In multiobjective optimization, the simultaneous optimization of several, possibly competing, objective functions is required. The family of solutions to a multiobjective optimization problem is composed of all those elements of the search space sharing the property that the corresponding objective vectors cannot be all simultaneously improved. These solutions are called Pareto optimal.

Multipoint crossover: A **crossover** operator which uses a predefined number of uniformly distributed crossover points and exchanges alternating segments between pairs of crossover points between the parent **individuals** (cf **one-point crossover**).

Mutation: A change of the genetic material, either occurring in the germ path or in the **gametes** (generative) or in body cells (somatic). Only generative mutations affect the offspring. A typical classification of mutations distinguishes **gene** mutations (a particular gene is changed), **chromosome** mutations (the gene order is changed by **translocation** or **inversion**, or the chromosome number is changed by **deficiencies**, **deletions**, or **duplications**), and **genome** mutations (the number of chromosomes or genomes is changed). In **evolutionary algorithms**, mutations are either modeled on the **phenotypic** level (e.g. by using normally distributed variations with expectation zero for continuous traits) or on the **genotypic** level (e.g. by using bit inversions with small probability as an equivalent for nucleotide base changes). (*See also Chapter 32.*)

Mutation rate: The probability of the occurrence of a **mutation** during **DNA** replication.

Natural selection: The result of competitive exclusion as organisms fill the available finite resource space.

Neural network: Artificial neural networks try to implement the data processing capabilities of brains on a computer. To achieve this (at least in a very simplified form regarding the number of processing units and their interconnectivity), simple units (corresponding to neurons) are arranged in a number of layers and allowed to communicate via weighted connections (corresponding to axons and dendrites). Working (at least principally) in parallel, each unit of the network typically calculates a weighted sum of its inputs, performs some internal mapping of the result, and eventually propagates a nonzero value to its output connection. Though the artificial models are strong simplifications of the natural model, impressive results have been achieved in a variety of application fields.

Niche: **Adaptation** of a **species** occurs with respect to any major kind of environment, the adaptive zone of this species. The set of possible environments that permit survival of a species is called its (ecological) niche.

Niching methods: In **evolutionary algorithms**, niching methods aim at the formation and maintenance of stable subpopulations (**niches**) within a single **population**. One typical way to achieve this proceeds by means of **fitness sharing** techniques.

No-free-lunch theorem: This theorem proves that when applied across all possible problems, all algorithms that do not resample points from the search space perform exactly the same on average. This result implies that it is necessary to tune the operators of an **evolutionary algorithm** to the problem at hand in order to perform optimally, or even better than random search. The no-free-lunch theorem has been extended to apply to certain subsets of all possible problems. Related theorems have been developed indicating that

Object variables: The parameters that are directly involved in the calculation of the objective function value of an **individual**.

Off-line performance: A performance measure for **genetic algorithms**, giving the average of the best **fitness** values found in a **population** over the course of the search.

1/5 success rule: A theoretically derived rule for the deterministic adjustment of the standard deviation of the **mutation** operator in a $(1 + 1)$ **evolution strategy**. The 1/5 success rule reflects the theoretical result that, in order to maximize the **convergence velocity**, on average one out of five mutations should cause an improvement with respect to the objective function value. (*See also Chapter 9.*)

One-point crossover: A **crossover** operator using exactly one crossover point on the **genome**.

On-line performance: A performance measure giving the average **fitness** over all tested search points over the course of the search.

Ontogenesis: The development of an organism from the fertilized **zygote** until its death.

Order: The order of a **schema** is given by the number of specified positions within the schema. The larger the order of a schema, the higher becomes its probability of disruption by **mutation**.

Order-based problems: A class of optimization problems that can be characterized by the search for an optimal permutation of specific items. Representative examples of this class are the traveling salesman problem or scheduling problems. In principle, any of the existing **evolutionary algorithms** can be reformulated for order-based problems, but the first permutation applications were handled by so-called order-based **genetic algorithms**, which typically use **mutation** and **recombination** operators that ensure that the result of the application of an operator to a permutation is again a permutation.

Order crossover: A **crossover** operator used in **order-based genetic algorithms** to manipulate permutations in a permutation preserving way. The order crossover (OX) starts in a way similar to **partially matched crossover** by picking two crossing sites uniformly at random along the permutations and mapping each string to constituents of the matching section of its mate. Then, however, order crossover uses a sliding motion to fill the holes left by transferring the mapped positions. This way, order crossover preserves the relative positions of elements within the permutation. (*See also Section 33.3.*)

Order statistics: Given λ independent random variables with a common probability density function, their arrangement in nondecreasing order is called the order statistics of these random variables. The theory of order statistics provides many useful results regarding the moments (and other properties) of the members of the order statistics. In the theory of **evolutionary algorithms**, the order statistics are widely utilized to describe deterministic **selection** schemes such as the **comma strategy** and **tournament selection**.

Panmictic population: A mixed **population**, in which any **individual** may be mated with any other individual with a probability that depends only on **fitness**. Most conventional **evolutionary algorithms** have panmictic populations.

Parse tree: The syntactic structure of any program in computer programming languages can be represented by a so-called parse tree, where the internal nodes of the tree correspond to operators and leaves of the tree correspond to constants. Parse trees (or, equivalently, S-expressions) are the fundamental data structure in **genetic programming**, where **recombination** is usually implemented as a subtree exchange between two different parse trees. (*See also Chapter 19.*)

Partially matched crossover: A **crossover** operator used to manipulate permutations in a permutation preserving way. The partially matched crossover (PMX) picks two crossing sites uniformly at random along the permutations, thus defining a matching section used to effect a cross through position-by-position exchange operations. (*See also Section 33.3.*)

Penalty function: For constraint optimization problems, the penalty function method provides one possible way to try to achieve feasible solutions: the unconstrained objective function is extended by a penalty function that penalizes infeasible solutions and vanishes for feasible solutions. The penalty function is also typically graded in the sense that the closer a solution is to feasibility, the smaller is the value of the penalty term for that solution. By means of this property, an **evolutionary algorithm** is often able to approach the feasible region although initially all members of the population might be infeasible.

Phenotype: The behavioral expression of the **genotype** in a specific environment.

Phylogeny: The evolutionary relationships among any group of organisms.

Pleiotropy: The influence of a single **gene** on several **phenotypic** features of an organism.

Plus strategy: The notation $(\mu + \lambda)$ strategy describes a **selection** method introduced in **evolution strategies** and indicates that a parent population of μ **individuals** generates $\lambda \geq \mu$ offspring and all $\mu + \lambda$ individuals compete directly, such that the μ best out of parents and offspring are deterministically selected as parents of the next generation.

Polygeny: The combined influence of several **genes** on a single **phenotypical** characteristic.

Population: A group of **individuals** that may interact with each other, for example, by mating and offspring production. The typical population sizes in **evolutionary algorithms** range from one (for $(1 + 1)$ **evolution strategies**) to several thousands (for **genetic programming**).

Prokaryotic cell: A cell lacking a membrane-enclosed nucleus and organelles.

Proportional selection: A **selection** mechanism that assigns selection probabilities in proportion to the relative **fitness** of an individual. (*See also Chapter 23.*)

Protein: A multiply folded biological macromolecule consisting of a long chain of amino acids. The metabolic effects of proteins are basically caused by their three-dimensional folded structure (the tertiary structure) as well as their symmetrical structure components (secondary structure), which result from the amino acid order in the chain (primary structure).

Punctuated crossover: A **crossover** operator to explore the potential for **self-adaptation** of the number of crossover points and their positions. To achieve this, the vector of **object variables** is extended by a crossover mask, where a one bit indicates the position of a crossover point in the object variable part of the **individual**. The crossover mask itself is subject to recombination and mutation to allow for a self-adaptation of the crossover operator.

Rank-based selection: In rank-based selection methods, the selection probability of an **individual** does not depend on its absolute **fitness** as in case of **proportional selection**, but only on its relative fitness in comparison with the other population members: its rank when all individuals are ordered in increasing (or decreasing) order of fitness values. (*See also Chapter 25.*)

Recombination: *See* **crossover**.

RNA: Ribonucleic acid. The **transcription** process in the cell nucleus generates a copy of the nucleotide sequence on the coding strand of the **DNA**. The resulting copy is an RNA molecule, a single-stranded molecule which carries information by means of the necleotide bases adenine, cytosine, guanine, and uracil (U) (replacing the thymine in the DNA). The RNA molecule acts as a messenger that transfers information from the cell nucleus to the ribosomes, where the **protein** synthesis takes place.

Scaling function: A scaling function is often used when applying **proportional selection**, particularly when needing to treat individuals with non-positive evaluations. Scaling functions typically employ a linear, logarithmic, or exponential mapping. (*See also Chapter 23.*)

Schema: A schema describes a subset of all binary vectors of fixed length that have similarities at certain positions. A schema is typically specified by a vector over the alphabet {0, 1, #}, where the # denotes a 'wildcard' matching both zero and one.

Schema theorem: A theorem offered to describe the expected number of instances of a **schema** that are represented in the next generation of an **evolutionary algorithm** when **proportional selection** is used. Although once considered to be a 'fundamental' theorem, mathematical results show that the theorem does not hold in general when iterated over more than one generation and that it may not hold when individual solutions have noisy **fitness** evaluations. Furthermore, the theorem cannot be used to determine which schemata should be recombined in future generations and has little or no predictive power.

Segmented crossover: A **crossover** operator which works similarly to **multipoint crossover**, except that the number of crossover points is not fixed but may vary around an expectation value. This is achieved by a segment switch rate that specifies the probability that a segment will end at any point in the string.

Selection: The operator of **evolutionary algorithms**, modeled after the principle of **natural selection**, which is used to direct the search process towards better regions of the search space by giving preference to **individuals** of higher **fitness** for mating and reproduction. The most widely used selection methods include the **comma** and **plus strategies**, **ranking selection**, **proportional selection**, and **tournament selection**. (*See also Chapters 22–30.*)

Self-adaptation: The principle of self-adaptation facilitates evolutionary algorithms learning their own **strategy parameters** on-line during the search, without any deterministic exogeneous control, by means of evolutionary processes in the same way as the **object variables** are modified. More precisely, the strategy parameters (such as **mutation** rates, variances, or covariances of normally distributed variations) are part of the individual and undergo mutation (**recombination**) and **selection** as the object variables do. The biological analogy consists in the fact that some portions of the **DNA** code for mutator **genes** or repair enzymes; that is, some partial control over the **DNA's** mutation rate is encoded in the **DNA**.

Sharing: Sharing (short for **fitness** sharing) is a **niching method** that derates the fitnesses of **population** elements according to the number of **individuals** in a **niche**, so that the population ends up distributed across multiple niches.

Simulated annealing: An optimization strategy gleaned from the model of thermodynamic evolution, modeling an annealing process in order to reach a state of minimal energy (where energy is the analogue of **fitness** in **evolutionary algorithms**). The strategy works with one trial solution and generates a new solution by means of a variation (or **mutation**) operator. The new solution is always accepted if it represents a decrease of energy, and it is also accepted with a certain parameter-controlled probability if it represents an increase of energy. The control parameter (or **strategy parameter**) is commonly called temperature and makes the thermodynamic origin of the strategy obvious.

Speciation: The process whereby a new **species** comes about. The most common cause of speciation is that of geographical isolation. If a subpopulation of a single species is separated geographically from the main **population** for a sufficiently long time, its **genes** will diverge (either due to differences in **selection** pressures in different locations, or simply due to **genetic drift**). Eventually, genetic differences will be so great that members of the subpopulation must be considered as belonging to a different (and new) species.

Species: A **population** of similarly constructed organisms, capable of producing fertile offspring. Members of one species occupy the same ecological **niche**.

Steady-state selection: A **selection** scheme which does not use a generation-wise replacement of the **population**, but rather replaces one **individual** per iteration of the main **recombine–mutate–select** loop of the algorithm.

Usually, the worst population member is replaced by the result of recombination and mutation, if the resulting individual represents a **fitness** improvement compared to the worst population member. The mechanism corresponds to a $(\mu + 1)$ selection method in **evolution strategies** (cf **plus strategy**).

Strategy parameter: The control parameters of an **evolutionary algorithm** are often referred to as strategy parameters. The particular setting of strategy parameters is often critical to gain good performance of an evolutionary algorithm, and the usual technique of empirically searching for an appropriate set of parameters is not generally satisfying. Alternatively, some researchers try techniques of **metaevolution** to optimize the strategy parameters, while in **evolution strategies** and **evolutionary programming** the technique of **self-adaptation** is successfully used to evolve strategy parameters in the same sense as **object variables** are evolved.

Takeover time: A characteristic value to measure the selective pressure of **selection** methods utilized in **evolutionary algorithms**. It gives the expected number of generations until, under repeated application of selection as the only operator acting on a **population**, the population is completely filled with copies of the initially best **individual**. The smaller the takeover time of a selection mechanism, the higher is its emphasis on reproduction of the best individual, i.e. its selective pressure.

Tournament selection: Tournament selection methods share the principle of holding tournaments between a number of **individuals** and selecting the best member of a tournament group for survival to the next generation. The tournament members are typically chosen uniformly at random, and the tournament sizes (number of individuals involved per tournament) are typically small, ranging from two to ten individuals. The tournament process is repeated μ times in order to select a population of μ members. (*See also Chapter 24.*)

Transcription: The process of synthesis of a messenger **RNA** (mRNA) reflecting the structure of a part of the **DNA**. The synthesis is performed in the cell nucleus.

Translation: The process of synthesis of a **protein** as a sequence of amino acids according to the information contained in the messenger **RNA** and the **genetic code** between triplets of nucleotide bases and amino acids. The synthesis is performed by the ribosomes under utilization of transfer **RNA** molecules.

Two-membered evolution strategy: The two-membered or $(1 + 1)$ **evolution strategy** is an **evolutionary algorithm** working with just one ancestor individual. A descendant is created by means of **mutation**, and **selection** selects the better of ancestor and descendant to survive to the next generation (cf **plus strategy**).

Uniform crossover: A **crossover** operator which was originally defined to work on binary strings. The uniform crossover operator exchanges each bit with a certain probability between the two parent individuals. The exchange probability typically has a value of one half, but other settings are possible (cf **discrete recombination**). (*See also Section 33.3.*)

Uniform random search: A random search algorithm which samples the search space by drawing points from a uniform distribution over the search space. In contrast to **evolutionary algorithms**, uniform random search does not update its sampling distribution according to the information gained from past samples, i.e. it is not a **Markov process**.

Zygote: A fertilized egg that is always **diploid**.

1

Introduction to evolutionary computation

David B Fogel

1.1 Introductory remarks

As a recognized field, evolutionary computation is quite young. The term itself was invented as recently as 1991, and it represents an effort to bring together researchers who have been following different approaches to simulating various aspects of evolution. These techniques of genetic algorithms (Chapter 7), evolution strategies (Chapter 8), and evolutionary programming (Chapter 9) have one fundamental commonality: they each involve the reproduction, random variation, competition, and selection of contending individuals in a population. These form the essential essence of evolution, and once these four processes are in place, whether in nature or in a computer, evolution is the inevitable outcome (Atmar 1994). The impetus to simulate evolution on a computer comes from at least four directions.

1.2 Optimization

Evolution is an optimization process (Mayr 1988, p 104). Darwin (1859, ch 6) was struck with the 'organs of extreme perfection' that have been evolved, one such example being the image-forming eye (Atmar 1976). Optimization does not imply perfection, yet evolution can discover highly precise functional solutions to particular problems posed by an organism's environment, and even though the mechanisms that are evolved are often overly elaborate from an engineering perspective, function is the sole quality that is exposed to natural selection, and functionality is what is optimized by iterative selection and mutation.

It is quite natural, therefore, to seek to describe evolution in terms of an algorithm that can be used to solve difficult engineering optimization problems. The classic techniques of gradient descent, deterministic hill climbing, and purely random search (with no heredity) have been generally unsatisfactory when applied to nonlinear optimization problems, especially those with stochastic, temporal, or chaotic components. But these are the problems that nature has seemingly solved so very well. Evolution provides inspiration for computing the solutions to problems that have previously appeared intractable. This was a key foundation for the efforts in evolution strategies (Rechenberg 1965, 1994, Schwefel 1965, 1995).

1.3   Robust adaptation

The real world is never static, and the problems of temporal optimization are some of the most challenging. They require changing behavioral strategies in light of the most recent feedback concerning the success or failure of the current strategy. Holland (1975), under the framework of genetic algorithms (formerly called reproductive plans), described a procedure that can evolve strategies, either in the form of coded strings or as explicit behavioral rule bases called classifier systems (Chapter 12), by exploiting the potential to recombine successful pieces of competing strategies, bootstrapping the knowledge gained by independent individuals. The result is a robust procedure that has the potential to adjust performance based on feedback from the environment.

1.4   Machine intelligence

Intelligence may be defined as the capability of a system to adapt its behavior to meet desired goals in a range of environments (Fogel 1995, p xiii). Intelligent behavior then requires prediction, for adaptation to future circumstances requires predicting those circumstances and taking appropriate action. Evolution has created creatures of increasing intelligence over time. Rather than seek to generate machine intelligence by replicating humans, either in the rules they may follow or in their neural connections, an alternative approach to generating machine intelligence is to simulate evolution on a class of predictive algorithms. This was the foundation for the evolutionary programming research of Fogel (1962, Fogel *et al* 1966).

1.5   Biology

Rather than attempt to use evolution as a tool to solve a particular engineering problem, there is a desire to capture the essence of evolution in a computer simulation and use the simulation to gain new insight into the physics of natural evolutionary processes (Ray 1991) (see also Chapter 4). Success raises the possibility of studying alternative biological systems that are merely plausible images of what life might be like in some way. It also raises the question of what properties such imagined systems might have in common with life as evolved on Earth (Langton 1987). Although every model is incomplete, and assessing what life might be like in other instantiations lies in the realm of pure speculation, computer simulations under the rubric of artificial life have generated some patterns that appear to correspond with naturally occurring phenomena.

1.6 Discussion

The ultimate answer to the question 'why simulate evolution?' lies in the lack of good alternatives. We cannot easily germinate another planet, wait several millions of years, and assess how life might develop elsewhere. We cannot easily use classic optimization methods to find global minima in functions when they are surrounded by local minima. We find that expert systems and other attempts to mimic human intelligence are often brittle: they are not robust to changes in the domain of application and are incapable of correctly predicting future circumstances so as to take appropriate action. In contrast, by successfully exploiting the use of randomness, or in other words *the useful use of uncertainty*, 'all possible pathways are open' for evolutionary computation (Hofstadter 1995, p 115). Our challenge is, at least in some important respects, to not allow our own biases to constrain the potential for evolutionary computation to discover new solutions to new problems in fascinating and unpredictable ways. However, as always, the ultimate advancement of the field will come from the careful abstraction and interpretation of the natural processes that inspire it.

References

Atmar J W 1976 *Speculation on the Evolution of Intelligence and its Possible Realization in Machine Form* Doctoral Dissertation, New Mexico State University

Atmar W 1994 Notes on the simulation of evolution *IEEE Trans. Neural Networks* **NN-5** 130–47

Darwin C R 1859 *On the Origin of Species by Means of Natural Selection or the Preservation of Favoured Races in the Struggle for Life* (London: Murray)

Fogel D B 1995 *Evolutionary Computation: Toward a New Philosophy of Machine Intelligence* (Piscataway, NJ: IEEE)

Fogel L J 1962 Autonomous automata *Industr. Res.* **4** 14–9

Fogel L J, Owens A J and Walsh M J 1966 *Artificial Intelligence through Simulated Evolution* (New York: Wiley)

Hofstadter D 1995 *Fluid Concepts and Creative Analogies: Computer Models of the Fundamental Mechanisms of Thought* (New York: Basic Books)

Holland J H 1975 *Adaptation in Natural and Artificial Systems* (Ann Arbor, MI: University of Michigan Press)

Langton C G 1987 Artificial life *Artificial Life* ed C G Langton (Reading, MA: Addison-Wesley) pp 1–47

Mayr E 1988 *Toward a New Philosophy of Biology: Observations of an Evolutionist* (Cambridge, MA: Belknap)

Ray T 1991 An approach to the synthesis of life *Artificial Life II* ed C G Langton, C Taylor, J D Farmer and S Rasmussen (Reading, MA: Addison-Wesley) pp 371–408

Rechenberg I 1965 *Cybernetic Solution Path of an Experimental Problem* Royal Aircraft Establishment Library Translation 1122, Farnborough, UK

——1994 *Evolutionsstrategies '94* (Stuttgart: Frommann-Holzboog)

Schwefel H-P 1965 *Kybernetische Evolution als Strategie der Experimentellen Forschung in der Strömungstechnik* Diploma Thesis, Technical University of Berlin

——1995 *Evolution and Optimum Seeking* (New York: Wiley)

2

Possible applications of evolutionary computation

David Beasley

2.1 Introduction

Applications of evolutionary computation (EC) fall into a wide continuum of areas. For convenience, in this chapter they have been split into five broad categories:

- planning
- design
- simulation and identification
- control
- classification.

These categories are by no means meant to be absolute or definitive. They all overlap to some extent, and many applications could rightly appear in more than one of the categories.

A number of bibliographies where more extensive information on EC applications can be found are listed after the references at the end of this chapter.

2.2 Applications in planning

2.2.1 Routing

Perhaps one of the best known combinatorial optimization problems is the traveling salesman problem or TSP (Goldberg and Lingle 1985, Grefenstette 1987, Fogel 1988, Oliver *et al* 1987, Mühlenbein 1989, Whitley *et al* 1989, Fogel 1993a, Homaifar *et al* 1993). A salesman must visit a number of cities, and then return home. In which order should the cities be visited to minimize the distance traveled? Optimizing the tradeoff between speed and accuracy of solution has been one aim (Verhoeven *et al* 1992).

A generalization of the TSP occurs when there is more than one salesman (Fogel 1990). The *vehicle routing problem* is similar. There is a fleet of vehicles, all based at the same depot. A set of customers must each receive one delivery. Which route should each vehicle take for minimum cost? There are constraints, for example, on vehicle capacity and delivery times (Blanton and Wainwright 1993, Thangia *et al* 1993).

Closely related to this is the transportation problem, in which a single commodity must be distributed to a number of customers from a number of depots. Each customer may receive deliveries from one or more depots. What is the minimum-cost solution? (Michalewicz 1992, 1993).

Planning the path which a robot should take is another route planning problem. The path must be feasible and safe (i.e. it must be achievable within the operational constraints of the robot) and there must be no collisions. Examples include determining the joint motions required to move the gripper of a robot arm between locations (Parker *et al* 1989, Davidor 1991, McDonnell *et al* 1992), and autonomous vehicle routing (Jakob *et al* 1992, Page *et al* 1992). In unknown areas or nonstatic environments, on-line planning/navigating is required, in which the robot revises its plans as it travels.

2.2.2 *Scheduling*

Scheduling involves devising a plan to carry out a number of activities over a period of time, where the activities require resources which are limited, there are various constraints and there are one or more objectives to be optimized.

Job shop scheduling is a widely studied NP-complete problem (Davis 1985, Biegel and Davern 1990, Syswerda 1991, Yamada and Nakano 1992). The scenario is a manufacturing plant, with machines of different types. There are a number of jobs to be completed, each comprising a set of tasks. Each task requires a particular type of machine for a particular length of time, and the tasks for each job must be completed in a given order. What schedule allows all tasks to be completed with minimum cost? Husbands (1993) has used the additional biological metaphor of an ecosystem. His method optimizes the sequence of tasks in each job at the same time as it builds the schedule. In real job shops the requirements may change while the jobs are being carried out, requiring that the schedule be replanned (Fang *et al* 1993). In the limit, the manufacturing process runs continuously, so all scheduling must be carried out on-line, as in a chemical flowshop (Cartwright and Tuson 1994).

Another scheduling problem is to devise a timetable for a set of examinations (Corne *et al* 1994), university lectures (Ling 1992), a staff rota (Easton and Mansour 1993) or suchlike.

In computing, scheduling problems include efficiently allocating tasks to processors in a multiprocessor system (Van Driessche and Piessens 1992, Kidwell 1993, Fogel and Fogel 1996), and devising memory cache replacement policies (Altman *et al* 1993).

2.2.3 Packing

Evolutionary algorithms (EAs) have been applied to many packing problems, the simplest of which is the one-dimensional zero–one knapsack problem. Given a knapsack of a certain capacity, and a set of items, each with a particular size and value, find the set of items with maximum value which can be accommodated in the knapsack. Various real-world problems are of this type: for example, the allocation of communication channels to customers who are charged at different rates.

There are various examples of two-dimensional packing problems. When manufacturing items are cut from sheet materials (e.g. metal or cloth), it is desirable to find the most compact arrangement of pieces, so as to minimize the amount of scrap (Smith 1985, Fujita *et al* 1993). A similar problem arises in the design of layouts for integrated circuits—how should the subcircuits be arranged to minimize the total chip area required (Fourman 1985, Cohoon and Paris 1987, Chan *et al* 1991)?

In three dimensions, there are obvious applications in which the best way of packing objects into a restricted space is required. Juliff (1993) has considered the problem of packing goods into a truck for delivery.

2.3 Applications in design

The design of filters has received considerable attention. EAs have been used to design electronic or digital systems which implement a desired frequency response. Both finite impulse response (FIR) and infinite impulse response (IIR) filter structures have been employed (Etter *et al* 1982, Suckley 1991, Fogel 1991, Fonseca *et al* 1993, Ifeachor and Harris 1993, Namibar and Mars 1993, Roberts and Wade 1993, Schaffer and Eshelman 1993, White and Flockton 1993, Wicks and Lawson 1993, Wilson and Macleod 1993). EAs have also been used to optimize the design of signal processing systems (San Martin and Knight 1993) and in integrated circuit design (Louis and Rawlins 1991, Rahmani and Ono 1993). The *unequal-area facility layout problem* (Smith and Tate 1993) is similar to integrated circuit design. It involves finding a two-dimensional arrangement of 'departments' such that the distance which information has to travel between departments is minimized.

EC techniques have been widely applied to artificial neural networks, both in the design of network topologies and in the search for optimum sets of weights (Miller *et al* 1989, Fogel *et al* 1990, Harp and Samad 1991, Baba 1992, Hancock 1992, Feldman 1993, Gruau 1993, Polani and Uthmann 1993, Romaniuk 1993, Spittle and Horrocks 1993, Zhang and Mühlenbein 1993, Porto *et al* 1995). They have also been applied to Kohonen feature map design (Polani and Uthmann 1992). Other types of network design problems have also been approached, for example, in telecommunications (Cox *et al* 1991, Davis and Cox 1993).

There have been many engineering applications of EC: structure design, both two-dimensional, such as a plane truss (Lohmann 1992, Watabe and Okino 1993), and three-dimensional, such as aircraft design (Bramlette and Bouchard 1991), actuator placement on space structures (Furuya and Haftka 1993), linear accelerator design, gearbox design, and chemical reactor design (Powell and Skolnick 1993). In relation to high-energy physics, the design of Monte Carlo generators has been tackled.

In order to perform parallel computations requiring global coordination, EC has been used to design cellular automata with appropriate communication mechanisms.

There have also been applications in testing and fault diagnosis. For example, an EA can be used to search for challenging fault scenarios for an autonomous vehicle controller.

2.4 Applications in simulation and identification

Simulation involves taking a design or model for a system, and determining how the system will behave. In some cases this is done because we are unsure about the behavior (e.g. when designing a new aircraft). In other cases, the behavior is known, but we wish to test the accuracy of the model.

EC has been applied to difficult problems in chemistry and biology. Roosen and Meyer (1992) used an evolution strategy to determine the equilibrium of chemically reactive systems, by determining the minimum free enthalpy of the compounds involved. The determination of the three-dimensional structure of a protein, given its amino acid sequence, has been tackled (Lucasius et al 1991). Lucasius and Kateman (1992) approached this as a sequenced subset selection problem, using two-dimensional nuclear magnetic resonance spectrum data as a starting point. Others have searched for energetically favorable protein conformations (Schulze-Kremer 1992, Unger and Moult 1993), and used EC to assist with drug design (Gehlhaar et al 1995). EC has been used to simulate how the nervous system learns in order to test an existing theory. Similarly, EC has been used in order to help develop models of biological evolution.

In the field of economics, EC has been used to model economic interaction of competing firms in a market.

Identification is the inverse of simulation. It involves determining the design of a system given its behavior.

Many systems can be represented by a model which produces a single-valued output in response to one or more input signals. Given a number of observations of input and output values, system identification is the task of deducing the details of the model. Flockton and White (1993) concern themselves with determining the poles and zeros of the system.

One reason for wanting to identify systems is so that we can predict the output in response to a given set of inputs. EC may also employed to fit equations to noisy, chaotic medical data, in order to predict future values.

Janikow and Cai (1992) similarly used EC to estimate statistical functions for survival analysis in clinical trials. In a similar area, Manela *et al* (1993) used EC to fit spline functions to noisy pharmaceutical fermentation process data.

EC may also be used to identify the sources of airborne pollution, given data from a number of monitoring points in an urban area—the source apportionment problem. In electromagnetics, Tanaka *et al* (1993) have applied EC to determining the two-dimensional current distribution in a conductor, given its external magnetic field. Away from conventional system identification, an EC approach has been used to help with identifying criminal suspects. This system helps witnesses to create a likeness of the suspect, without the need to give an explicit description.

2.5   Applications in control

There are two distinct approaches to the use of EC in control: off-line and on-line. The off-line approach uses an EA to design a controller, which is then used to control the system. The on-line approach uses an EA as an active part of the control process. Therefore, with the off-line approach there is nothing evolutionary about the control process itself, only about the design of the controller.

Some researchers (Fogel *et al* 1966, DeJong 1980) have sought to use the adaptive qualities of EAs in order to build on-line controllers for dynamic systems. The advantage of an evolutionary controller is that it can adapt to cope with systems whose characteristics change over time, whether the change is gradual or sudden. Most researchers, however, have taken the off-line approach to the control of relatively unchanging systems.

Fonseca and Fleming (1993) used an EA to design a controller for a gas turbine engine, to optimize its step response, and a control system has been used to optimize combustion in multiple-burner furnaces and boiler plants. EC has also been applied to the control of guidance and navigation systems (Krishnakumar and Goldberg 1990, 1992).

Hunt (1992b) has tackled the problem of synthesizing LQG (linear–quadratic–Gaussian) and H_∞ (H-infinity) optimal controllers. He has also considered the frequency domain optimization of controllers with fixed structures (Hunt 1992a).

Two control problems which have been well studied are balancing a pole on a movable cart (Fogel 1995), and backing up a trailer truck to a loading bay from an arbitrary starting point (Abu Zitar and Hassoun 1993). In robotics, EAs have been developed which can evolve control systems for visually guided behaviors. They can also learn how to control mobile robots (Kim and Shim 1995), for example, controlling the legs of a six-legged 'insect' to make it crawl or walk (Spencer 1993). Almássy and Verschure (1992) modeled the interaction between natural selection and the adaptation of individuals during their lifetimes to develop an agent with a distributed adaptive control framework which learns to avoid obstacles and locate food sources.

2.6 Applications in classification

As described in Chapter 12, a significant amount of EC research has concerned the theory and practice of classifier systems (CFS) (Booker 1985, Holland 1985, 1987, Holland *et al* 1987, Robertson 1987, Wilson 1987, Fogarty 1994). Classifier systems are at the heart of many other types of system. For example, many control systems rely on being able to classify the characteristics of their environment before an appropriate control decision can be made. This is true in many robotics applications of EC, for example, learning to control robot arm motion (Patel and Dorigo 1994) and learning to solve mazes (Pipe and Carse 1994).

An important aspect of a classifier system, especially in a control application, is how the state space is partitioned. Many applications take for granted a particular partitioning of the state space, while in others, the appropriate partitioning of the state space is itself part of the problem (Melhuish and Fogarty 1994).

Game playing is another application for which classification plays a key role. Although EC is often applied to rather simple games (e.g. the prisoner's dilemma (Axelrod 1987, Fogel 1993b)), this is sometimes motivated by more serious applications, such as military ones (e.g. the two-tanks game (Fairley and Yates 1994) and air combat maneuvering.

EC has been hybridized with feature partitioning and applied to a range of tasks (Güvenir and Şirin 1993), including classification of iris flowers, prediction of survival for heart attack victims from echocardiogram data, diagnosis of heart disease, and classification of glass samples. In linguistics, EC has been applied to the classification of Swedish words.

In economics, Oliver (1993) has found rules to reflect the way in which consumers choose one brand rather than another, when there are multiple criteria on which to judge a product. A fuzzy hybrid system has been used for financial decision making, with applications to credit evaluation, risk assessment, and insurance underwriting.

In biology, EC has been applied to the difficult task of protein secondary-structure determination, for example, classifying the locations of particular protein segments (Handley 1993). It has also been applied to the classification of soil samples (Punch *et al* 1993).

In image processing, there have been further military applications, classifying features in images as targets (Bala and Wechsler 1993, Tackett 1993), and also non-military applications, such as optical character recognition.

Of increasing importance is the efficient storage and retrieval of information, including the generation of equifrequency distributions of material, to improve that efficiency. EC has also been employed to assist with the representation and storage of chemical structures, and the retrieval from databases of molecules containing certain substructures (Jones *et al* 1993). The retrieval of documents which match certain characteristics is becoming increasingly important as more and more information is held on-line. Tools to retrieve documents which contain specified words have been available for many years, but they have the limitation that constructing an appropriate search query can be difficult. Researchers are now using EAs to help with query construction (Yang and Korfhage 1993).

2.7   Summary

EC has been applied in a vast number of application areas. In some cases it has advantages over existing computerized techniques. More interestingly, perhaps, it is being applied to an increasing number of areas in which computers have not been used before. We can expect to see the number of applications grow considerably in the future. Comprehensive bibliographies in many different application areas are listed after the References.

References

Abu Zitar R A and Hassoun M H 1993 Regulator control via genetic search and assisted reinforcement *Proc. 5th Int. Conf. on Genetic Algorithms (Urbana-Champaign, IL, July 1993)* ed S Forrest (San Mateo, CA: Morgan Kaufmann) pp 254–62

Almássy N and Verschure P 1992 Optimizing self-organising control architectures with genetic algorithms: the interaction between natural selection and ontogenesis *Parallel Problem Solving from Nature, 2 (Proc. 2nd Int. Conf. on Parallel Problem Solving from Nature, Brussels, 1992)* ed R Männer and B Manderick (Amsterdam: Elsevier) pp 451–60

Altman E R, Agarwal V K and Gao G R 1993 A novel methodology using genetic algorithms for the design of caches and cache replacement policy *Proc. 5th Int. Conf. on Genetic Algorithms (Urbana-Champaign, IL, July 1993)* ed S Forrest (San Mateo, CA: Morgan Kaufmann) pp 392–9

Axelrod R 1987 The evolution of strategies in the iterated prisoner's dilemma *Genetic Algorithms and Simulated Annealing* ed L Davis (Boston, MA: Pitman) ch 3, pp 32–41

Baba N 1992 Utilization of stochastic automata and genetic algorithms for neural network learning *Parallel Problem Solving from Nature, 2 (Proc. 2nd Int. Conf. on Parallel Problem Solving from Nature, Brussels, 1992)* ed R Männer and B Manderick (Amsterdam: Elsevier) pp 431–40

Bagchi S, Uckun S, Miyabe Y and Kawamura K 1991 Exploring problem-specific recombination operators for job shop scheduling *Proc. 4th Int. Conf. on Genetic Algorithms (San Diego, CA, July 1991)* ed R Belew and L Booker (San Mateo, CA: Morgan Kaufmann) pp 10–7

Bala J W and Wechsler H 1993 Learning to detect targets using scale-space and genetic search *Proc. 5th Int. Conf. on Genetic Algorithms (Urbana-Champaign, IL, July 1993)* ed S Forrest (San Mateo, CA: Morgan Kaufmann) pp 516–22

Biegel J E and Davern J J 1990 Genetic algorithms and job shop scheduling *Comput. Indust. Eng.* **19** 81–91

Blanton J L and Wainwright R L 1993 Multiple vehicle routing with time and capacity constraints *Proc. 5th Int. Conf. on Genetic Algorithms (Urbana-Champaign, IL, July 1993)* ed S Forrest (San Mateo, CA: Morgan Kaufmann) pp 452–9

Booker L 1985 Improving the performance of genetic algorithms in classifier systems *Proc. 1st Int. Conf. on Genetic Algorithms (Pittsburgh, PA, July 1985)* ed J J Grefenstette (Hillsdale, NJ: Lawrence Erlbaum Associates) pp 80–92

Bramlette M F and Bouchard E E 1991 Genetic algorithms in parametric design of aircraft *Handbook of Genetic Algorithms* ed L Davis (New York: Van Nostrand Reinhold) ch 10, pp 109–23

Cartwright H M and Tuson A L 1994 Genetic algorithms and flowshop scheduling: towards the development of a real-time process control system *Evolutionary Computing (AISB Workshop, Leeds, 1994, Selected Papers) (Lecture Notes in Computer Science 865)* ed T C Fogarty (Berlin: Springer) pp 277–90

Chan H, Mazumder P and Shahookar K 1991 Macro-cell and module placement by genetic adaptive search with bitmap-represented chromosome *Integration VLSI J.* **12** 49–77

Cohoon J P and Paris W D 1987 Genetic placement *IEEE Trans. Computer-Aided Design* **CAD-6** 956–64

Corne D, Ross P and Fang H-L 1994 Fast practical evolutionary timetabling *Evolutionary Computing (AISB Workshop, Leeds, 1994, Selected Papers) (Lecture Notes in Computer Science 865)* ed T C Fogarty (Berlin: Springer) pp 250–63

Cox L A, Davis L and Qiu Y 1991 Dynamic anticipatory routing in circuit-switched telecommunications networks *Handbook of Genetic Algorithms* ed L Davis (New York: Van Nostrand Reinhold) ch 11, pp 124–43

Davidor Y 1991 A genetic algorithm applied to robot trajectory generation *Handbook of Genetic Algorithms* ed L Davis (New York: Van Nostrand Reinhold) ch 12, pp 144–65

Davis L 1985 Job shop scheduling with genetic algorithms *Proc. 1st Int. Conf. on Genetic Algorithms (Pittsburgh, PA, July 1985)* ed J J Grefenstette (Hillsdale, NJ: Lawrence Erlbaum Associates) pp 136–40

Davis L and Cox A 1993 A genetic algorithm for survivable network design *Proc. 5th Int. Conf. on Genetic Algorithms (Urbana-Champaign, IL, July 1993)* ed S Forrest (San Mateo, CA: Morgan Kaufmann) pp 408–15

DeJong K 1980 Adaptive system design: a genetic approach *IEEE Trans. Systems, Man Cybern.* **SMC-10** 566–74

Easton F F and Mansour N 1993 A distributed genetic algorithm for employee staffing and scheduling problems *Proc. 5th Int. Conf. on Genetic Algorithms (Urbana-Champaign, IL, July 1993)* ed S Forrest (San Mateo, CA: Morgan Kaufmann) pp 360–67

Etter D M, Hicks M J and Cho K H 1982 Recursive adaptive filter design using an adaptive genetic algorithm *IEEE Int. Conf. on Acoutics, Speech and Signal Processing* (Piscataway, NJ: IEEE) pp 635–8

Fairley A and Yates D F 1994 Inductive operators and rule repair in a hybrid genetic learning system: some initial results *Evolutionary Computing (AISB Workshop, Leeds, 1994, Selected Papers) (Lecture Notes in Computer Science 865)* ed T C Fogarty (Berlin: Springer) pp 166–79

Fang H-L, Ross P and Corne D 1993 A promising genetic algorithm approach to job-shop scheduling, rescheduling and open-shop scheduling problems *Proc. 5th Int. Conf. on Genetic Algorithms (Urbana-Champaign, IL, July 1993)* ed S Forrest (San Mateo, CA: Morgan Kaufmann) pp 375–82

Feldman D S 1993 Fuzzy network synthesis with genetic algorithms *Proc. 5th Int. Conf. on Genetic Algorithms (Urbana-Champaign, IL, July 1993)* ed S Forrest (San Mateo, CA: Morgan Kaufmann) pp 312–7

Flockton S J and White M 1993 Pole-zero system identification using genetic algorithms *Proc. 5th Int. Conf. on Genetic Algorithms (Urbana-Champaign, IL, July 1993)* ed S Forrest (San Mateo, CA: Morgan Kaufmann) pp 531–5

Fogarty T C 1994 Co-evolving co-operative populations of rules in learning control systems *Evolutionary Computing (AISB Workshop, Leeds, 1994, Selected Papers) (Lecture Notes in Computer Science 865)* ed T C Fogarty (Berlin: Springer) pp 195–209

Fogel D B 1988 An evolutionary approach to the traveling salesman problem *Biol. Cybernet.* **6** 139–44

——1990 A parallel processing approach to a multiple traveling salesman problem using evolutionary programming *Proc. 4th Ann. Symp. on Parallel Processing* (Piscataway, NJ: IEEE) pp 318–26

——1991 *System Identification through Simulated Evolution* (Needham, MA: Ginn)

——1993a Applying evolutionary programming to selected traveling salesman problems *Cybernet. Syst.* **24** 27–36

——1993b Evolving behaviors in the iterated prisoner's dilemma *Evolut. Comput.* **1** 77–97

——1995 *Evolutionary Computation: Toward a New Philosophy of Machine Intelligence* (Piscataway, NJ: IEEE)

Fogel D B and Fogel L J 1996 Using evolutionary programming to schedule tasks on a suite of heterogeneous computers *Comput. Operat. Res.* **23** 527–34

Fogel D B, Fogel L J and Porto V W 1990 Evolving neural networks *Biol. Cybern.* **63** 487–93

Fogel L J, Owens A J and Walsh M J 1966 *Artificial intelligence Through Simulated Evolution* (New York: Wiley)

Fonseca C M and Fleming P J 1993 Genetic algorithms for multiobjective optimization: formulation, discussion and generalization *Proc. 5th Int. Conf. on Genetic Algorithms (Urbana-Champaign, IL, July 1993)* ed S Forrest (San Mateo, CA: Morgan Kaufmann) pp 416–23

Fonseca C M, Mendes E M, Fleming P J and Billings S A 1993 Non-linear model term selection with genetic algorithms *Natural Algorithms in Signal Processing (Workshop, Chelmsford, UK, November 1993)* vol 2 (London: IEE) pp 27/1–27/8

Fourman M P 1985 Compaction of symbolic layout using genetic algorithms *Proc. 1st Int. Conf. on Genetic Algorithms (Pittsburgh, PA, July 1985)* ed J J Grefenstette (Hillsdale, NJ: Lawrence Erlbaum Associates) pp 141–53

Fujita K, Akagi S and Hirokawa N 1993 Hybrid approach for optimal nesting using genetic algorithm and a local minimization algorithm *Advances in Design Automation* vol 1, DE-65-1 (ASME) pp 477–84

Furuya H and Haftka R T 1993 Genetic algorithms for placing actuators on space structures *Proc. 5th Int. Conf. on Genetic Algorithms (Urbana-Champaign, IL, July 1993)* ed S Forrest (San Mateo, CA: Morgan Kaufmann) pp 536–42

Gehlhaar D K, Verkhivker G M, Rejto P A, Sherman C J, Fogel D B, Fogel L J and Freer S T 1995 Molecular recognition of the inhibitor AG-1343 by HIV-1 protease: conformationally flexible docking by evolutionary programming *Chem. Biol.* **2** 317–24

Goldberg D E and Lingle R 1985 Alleles, loci and the travelling salesman problem *Proc. 1st Int. Conf. on Genetic Algorithms (Pittsburgh, PA, July 1985)* ed J J Grefenstette (Hillsdale, NJ: Lawrence Erlbaum Associates) pp 154–9

Grefenstette J J 1987 Incorporating problem specific knowledge into genetic algorithms *Genetic Algorithms and Simulated Annealing* ed L Davis (Boston, MA: Pitman) ch 4, pp 42–60

Gruau F 1993 Genetic synthesis of modular neural networks *Proc. 5th Int. Conf. on Genetic Algorithms (Urbana-Champaign, IL, July 1993)* ed S Forrest (San Mateo, CA: Morgan Kaufmann) pp 318–25

Güvenir H A and Şirin I 1993 A genetic algorithm for classification by feature recognition *Proc. 5th Int. Conf. on Genetic Algorithms (Urbana-Champaign, IL, July 1993)* ed S Forrest (San Mateo, CA: Morgan Kaufmann) pp 543–8

Hancock P 1992 Recombination operators for the design of neural nets by genetic algorithm *Parallel Problem Solving from Nature, 2 (Proc. 2nd Int. Conf. on Parallel Problem Solving from Nature, Brussels, 1992)* ed R Männer and B Manderick (Amsterdam: Elsevier) pp 441–50

Handley S 1993 Automated learning of a detector for α-helices in protein sequences via genetic programming *Proc. 5th Int. Conf. on Genetic Algorithms (Urbana-Champaign, IL, July 1993)* ed S Forrest (San Mateo, CA: Morgan Kaufmann) pp 271–8

Harp S A and Samad T 1991 Genetic synthesis of neural network architecture *Handbook of Genetic Algorithms* ed L Davis (New York: Van Nostrand Reinhold) ch 15, pp 202–21

Holland J H 1985 Properties of the bucket-brigade algorithm *Proc. 1st Int. Conf. on Genetic Algorithms (Pittsburgh, PA, July 1985)* ed J J Grefenstette (Hillsdale, NJ: Lawrence Erlbaum Associates) pp 1–7

——1987 Genetic algorithms and classifier systems: foundations and future directions *Proc. 2nd Int. Conf. on Genetic Algorithms (Pittsburgh, PA, July 1985)* ed J J Grefenstette (Hillsdale, NJ: Lawrence Erlbaum Associates) pp 82–9

Holland J H, Holyoak K J, Nisbett R E and Thagard P R 1987 Classifier systems, Q-morphisms and induction *Genetic Algorithms and Simulated Annealing* ed L Davis (Boston, MA: Pitman) ch 9, pp 116–28

Homaifar, A., Guan S and Liepins G 1993 A new approach to the travelling salesman problem by genetic algorithms *Proc. 5th Int. Conf. on Genetic Algorithms (Urbana-Champaign, IL, July 1993)* ed S Forrest (San Mateo, CA: Morgan Kaufmann) pp 460–6

Hunt K J 1992a Optimal control system synthesis with genetic algorithms *Parallel Problem Solving from Nature, 2 (Proc. 2nd Int. Conf. on Parallel Problem Solving from Nature, Brussels, 1992)* ed R Männer and B Manderick (Amsterdam: Elsevier) pp 381–9

——1992b Polynomial LQG and H_∞ controller synthesis: a genetic algorithm solution *Proc. IEEE Conf. on Decision and Control (Tuscon, AZ)* (Picataway, NJ: IEEE)

Husbands P 1993 An ecosystems model for integrated production planning *Int. J. Comput. Integrated Manufacturing* **6** 74–86

Ifeachor E C and Harris S P 1993 A new approach to frequency sampling filter design using genetic algorithms *Natural Algorithms in Signal Processing (Workshop, Chelmsford, UK, November 1993)* vol 1 (London: IEE) pp 5/1–5/8

Jakob W, Gorges-Schleuter M and Blume C 1992 Application of genetic algorithms to task planning and learning *Parallel Problem Solving from Nature, 2 (Proc. 2nd Int. Conf. on Parallel Problem Solving from Nature, Brussels, 1992)* ed R Männer and B Manderick (Amsterdam: Elsevier) pp 291–300

Janikow C Z and Cai H 1992 A genetic algorithm application in nonparametric functional estimation *Parallel Problem Solving from Nature, 2 (Proc. 2nd Int. Conf. on Parallel Problem Solving from Nature, Brussels, 1992)* ed R Männer and B Manderick (Amsterdam: Elsevier) pp 249–58

Jones G, Brown R D, Clark D E, Willett P and Glen R C 1993 Searching databases of two-dimensional and three dimensional chemical structures using genetic algorithms *Proc. 5th Int. Conf. on Genetic Algorithms (Urbana-Champaign, IL, July 1993)* ed S Forrest (San Mateo, CA: Morgan Kaufmann) pp 597–602

Juliff K 1993 A multi-chromosome genetic algorithm for pallet loading *Proc. 5th Int. Conf. on Genetic Algorithms (Urbana-Champaign, IL, July 1993)* ed S Forrest (San Mateo, CA: Morgan Kaufmann) pp 467–73

Kidwell M D 1993 Using genetic algorithms to schedule distributed tasks on a bus-based system *Proc. 5th Int. Conf. on Genetic Algorithms (Urbana-Champaign, IL, July 1993)* ed S Forrest (San Mateo, CA: Morgan Kaufmann) pp 368–74

Kim J-H and Shim H-S 1995 Evolutionary programming-based optimal robust locomotion control of autonomous mobile robots *Proc. 4th Ann. Conf. on Evolutionary Programming* ed J R McDonnell, R G Reynolds and D B Fogel (Cambridge, MA: MIT Press) pp 631–44

Krishnakumar K and Goldberg D E 1990 Genetic algorithms in control system optimization. *Proc. AIAA Guidance, Navigation, and Control Conf. (Portland, OR)* pp 1568–77

——1992 Control system optimization using genetic algorithms *J. Guidance Control Dynam.* **15** 735–40

Ling S-E 1992 Integrating genetic algorithms with a prolog assignment program as a hybrid solution for a polytechnic timetable problem *Parallel Problem Solving from Nature, 2 (Proc. 2nd Int. Conf. on Parallel Problem Solving from Nature, Brussels, 1992)* ed R Männer and B Manderick (Amsterdam: Elsevier) pp 321–9

Lohmann R 1992 Structure evolution and incomplete induction *Parallel Problem Solving from Nature, 2 (Proc. 2nd Int. Conf. on Parallel Problem Solving from Nature, Brussels, 1992)* ed R Männer and B Manderick (Amsterdam: Elsevier) pp 175–85

Louis S J and Rawlins G J E 1991 Designer genetic algorithms: genetic algorithms in structure design *Proc. 4th Int. Conf. on Genetic Algorithms (San Diego, CA, July 1991)* ed R Belew and L Booker (San Mateo, CA: Morgan Kaufmann) pp 53–60

Lucasius C B, Blommers M J, Buydens L M and Kateman G 1991 A genetic algorithm for conformational analysis of DNA *Handbook of Genetic Algorithms* ed L Davis (New York: Van Nostrand Reinhold) ch 18, pp 251–81

Lucasius C B and Kateman G 1992 Towards solving subset selection problems with the aid of the genetic algorithm *Parallel Problem Solving from Nature, 2 (Proc. 2nd Int. Conf. on Parallel Problem Solving from Nature, Brussels, 1992)* ed R Männer and B Manderick (Amsterdam: Elsevier) pp 239–47

Manela, M., Thornhill N and Campbell J A 1993 Fitting spline functions to noisy data using a genetic algorithm *Proc. 5th Int. Conf. on Genetic Algorithms (Urbana-Champaign, IL, July 1993)* ed S Forrest (San Mateo, CA: Morgan Kaufmann) pp 549–56

McDonnell J R, Andersen B L, Page W C and Pin F G 1992 Mobile manipulator configuration optimization using evolutionary programming *Proc. 1st Ann. Conf. on Evolutionary Programming* ed D B Fogel and W Atmar (La Jolla, CA: Evolutionary Programming Society) pp 52–62

Melhuish C and Fogarty T C 1994 Applying a restricted mating policy to determine state space niches using immediate and delayed reinforcement *Evolutionary Computing (AISB Workshop, Leeds, 1994, Selected Papers) (Lecture Notes in Computer Science 865)* ed T C Fogarty (Berlin: Springer) pp 224–37

Michalewicz Z 1992 *Genetic Algorithms + Data Structures = Evolution Programs* (Berlin: Springer)

——1993 A hierarchy of evolution programs: an experimental study *Evolut. Comput.* **1** 51–76

Miller G F, Todd P M and Hegde S U 1989 Designing neural networks using genetic algorithms. *Proc. 3rd Int. Conf. on Genetic Algorithms (Fairfax, VA, June 1989)* ed J D Schaffer (San Mateo, CA: Morgan Kaufmann) pp 379–84

Mühlenbein H 1989 Parallel genetic algorithms, population genetics and combinatorial optimization *Proc. 3rd Int. Conf. on Genetic Algorithms (Fairfax, VA, June 1989)* ed J D Schaffer (San Mateo, CA: Morgan Kaufmann) pp 416–21

Namibar R and Mars P 1993 Adaptive IIR filtering using natural algorithms *Natural Algorithms in Signal Processing (Workshop, Chelmsford, UK, November 1993)* vol 2 (London: IEE) pp 20/1–20/10

Oliver J R 1993 Discovering individual decision rules: an application of genetic algorithms *Proc. 5th Int. Conf. on Genetic Algorithms (Urbana-Champaign, IL, July 1993)* ed S Forrest (San Mateo, CA: Morgan Kaufmann) pp 216–22

Oliver I M, Smith D J and Holland J R C 1987 A study of permutation crossover operators on the travelling salesman problem *Proc. 2nd Int. Conf. on Genetic Algorithms (Cambridge, MA, 1987)* ed J J Grefenstette (Hillsdale, NJ: Erlbaum) pp 224–30

Page W C, McDonnell J R and Anderson B 1992 An evolutionary programming approach to multi-dimensional path planning *Proc. 1st Ann. Conf. on Evolutionary Programming* ed D B Fogel and W Atmar (La Jolla, CA: Evolutionary Programming Society) pp 63–70

Parker J K, Goldberg D E and Khoogar A R 1989 Inverse kinematics of redundant robots using genetic algorithms *Proc. Int. Conf. on Robotics and Automation (Scottsdale, AZ, 1989)* vol 1 (Los Alamitos: IEEE Computer Society Press) pp 271–6

Patel M J and Dorigo M 1994 Adaptive learning of a robot arm *Evolutionary Computing (AISB Workshop, Leeds, 1994, Selected Papers) (Lecture Notes in Computer Science 865)* ed T C Fogarty (Berlin: Springer) pp 180–94

Pipe A G and Carse B 1994 A comparison between two architectures for searching and learning in maze problems *Evolutionary Computing (AISB Workshop, Leeds, 1994, Selected Papers) (Lecture Notes in Computer Science 865)* ed T C Fogarty (Berlin: Springer) pp 238–49

Polani D and Uthmann T 1992 Adaptation of Kohonen feature map topologies by genetic algorithms *Parallel Problem Solving from Nature, 2 (Proc. 2nd Int. Conf. on Parallel*

Problem Solving from Nature, Brussels, 1992) ed R Männer and B Manderick (Amsterdam: Elsevier) pp 421–9

——1993 Training Kohonen feature maps in different topologies: an analysis using genetic algorithms *Proc. 5th Int. Conf. on Genetic Algorithms (Urbana-Champaign, IL, July 1993)* ed S Forrest (San Mateo, CA: Morgan Kaufmann) pp 326–33

Porto V W, Fogel D B and Fogel L J 1995 Alternative neural network training methods *IEEE Expert* **10** 16–22

Powell D and Skolnick M M 1993 Using genetic algorithms in engineering design optimization with non-linear constraints *Proc. 5th Int. Conf. on Genetic Algorithms (Urbana-Champaign, IL, July 1993)* ed S Forrest (San Mateo, CA: Morgan Kaufmann) pp 424–31

Punch W F, Goodman E D, Pei, M, Chia-Shun L, Hovland P and Enbody R 1993 Further research on feature selection and classification using genetic algorithms *Proc. 5th Int. Conf. on Genetic Algorithms (Urbana-Champaign, IL, July 1993)* ed S Forrest (San Mateo, CA: Morgan Kaufmann) pp 557–64

Rahmani A T and Ono N 1993 A genetic algorithm for channel routing problem *Proc. 5th Int. Conf. on Genetic Algorithms (Urbana-Champaign, IL, July 1993)* ed S Forrest (San Mateo, CA: Morgan Kaufmann) pp 494–8

Roberts A and Wade G 1993 A structured GA for FIR filter design *Natural Algorithms in Signal Processing (Workshop, Chelmsford, UK, November 1993)* vol 1 (London: IEE) pp 16/1–16/8

Robertson G 1987 Parallel implementation of genetic algorithms in a classifier system *Genetic Algorithms and Simulated Annealing* ed L Davis (Boston, MA: Pitman) ch 10, pp 129–40

Romaniuk S G 1993 Evolutionary growth perceptrons *Proc. 5th Int. Conf. on Genetic Algorithms (Urbana-Champaign, IL, July 1993)* ed S Forrest (San Mateo, CA: Morgan Kaufmann) pp 334–41

Roosen P and Meyer F 1992 Determination of chemical equilibria by means of an evolution strategy *Parallel Problem Solving from Nature, 2 (Proc. 2nd Int. Conf. on Parallel Problem Solving from Nature, Brussels, 1992)* ed R Männer and B Manderick (Amsterdam: Elsevier) pp 411–20

San Martin R and Knight J P 1993 Genetic algorithms for optimization of integrated circuit synthesis *Proc. 5th Int. Conf. on Genetic Algorithms (Urbana-Champaign, IL, July 1993)* ed S Forrest (San Mateo, CA: Morgan Kaufmann) pp 432–8

Schaffer J D and Eshelman L J 1993 Designing multiplierless digital filters using genetic algorithms *Proc. 5th Int. Conf. on Genetic Algorithms (Urbana-Champaign, IL, July 1993)* ed S Forrest (San Mateo, CA: Morgan Kaufmann) pp 439–44

Schulze-Kremer S 1992 Genetic algorithms for protein tertiary structure prediction *Parallel Problem Solving from Nature, 2 (Proc. 2nd Int. Conf. on Parallel Problem Solving from Nature, Brussels, 1992)* ed R Männer and B Manderick (Amsterdam: Elsevier) pp 391–400

Smith A E and Tate D M 1993 Genetic optimization using a penalty function *Proc. 5th Int. Conf. on Genetic Algorithms (Urbana-Champaign, IL, July 1993)* ed S Forrest (San Mateo, CA: Morgan Kaufmann) pp 499–505

Smith D 1985 Bin packing with adaptive search *Proc. 1st Int. Conf. on Genetic Algorithms (Pittsburgh, PA, July 1985)* ed J J Grefenstette (Hillsdale, NJ: Lawrence Erlbaum Associates)

Spencer G F 1993 Automatic generation of programs for crawling and walking *Proc. 5th Int. Conf. on Genetic Algorithms (Urbana-Champaign, IL, July 1993)* ed S Forrest (San Mateo, CA: Morgan Kaufmann) p 654

Spittle M C and Horrocks D H 1993 Genetic algorithms and reduced complexity artificial neural networks *Natural Algorithms in Signal Processing (Workshop, Chelmsford, UK, November 1993)* vol 1 (London: IEE) pp 8/1–8/9

Suckley D 1991 Genetic algorithm in the design of FIR filters *IEE Proc.* G **138** 234–8

Syswerda G 1991 Schedule optimization using genetic algorithms *Handbook of Genetic Algorithms* ed L Davis(New York: Van Nostrand Reinhold) ch 21, pp 332–49

Tackett W A 1993 Genetic programming for feature discovery and image discrimination *Proc. 5th Int. Conf. on Genetic Algorithms (Urbana-Champaign, IL, July 1993)* ed S Forrest (San Mateo, CA: Morgan Kaufmann) pp 303–9

Tanaka, Y., Ishiguro A and Uchikawa Y 1993 A genetic algorithms application to inverse problems in electromagnetics *Proc. 5th Int. Conf. on Genetic Algorithms (Urbana-Champaign, IL, July 1993)* ed S Forrest (San Mateo, CA: Morgan Kaufmann) p 656

Thangia S R, Vinayagamoorthy R and Gubbi A V 1993 Vehicle routing with time deadlines using genetic and local algorithms *Proc. 5th Int. Conf. on Genetic Algorithms (Urbana-Champaign, IL, July 1993)* ed S Forrest (San Mateo, CA: Morgan Kaufmann) pp 506–13

Unger R and Moult J 1993 A genetic algorithm for 3D protein folding simulations *Proc. 5th Int. Conf. on Genetic Algorithms (Urbana-Champaign, IL, July 1993)* ed S Forrest (San Mateo, CA: Morgan Kaufmann) pp 581–8

Van Driessche R and Piessens R 1992 Load balancing with genetic algorithms *Parallel Problem Solving from Nature, 2 (Proc. 2nd Int. Conf. on Parallel Problem Solving from Nature, Brussels, 1992)* ed R Männer and B Manderick (Amsterdam: Elsevier) pp 341–50

Verhoeven M G A, Aarts E H L, van de Sluis E and Vaessens R J M 1992 Parallel local search and the travelling salesman problem *Parallel Problem Solving from Nature, 2 (Proc. 2nd Int. Conf. on Parallel Problem Solving from Nature, Brussels, 1992)* ed R Männer and B Manderick (Amsterdam: Elsevier) pp 543–52

Watabe H and Okino N 1993 A study on genetic shape design *Proc. 5th Int. Conf. on Genetic Algorithms (Urbana-Champaign, IL, July 1993)* ed S Forrest (San Mateo, CA: Morgan Kaufmann) pp 445–50

White M and Flockton S 1993 A comparative study of natural algorithms for adaptive IIR filtering *Natural Algorithms in Signal Processing (Workshop, Chelmsford, UK, November 1993)* vol 2 (London: IEE) pp 22/1–22/8

Whitley D, Starkweather T and Fuquay D 1989 Scheduling problems and travelling salesmen: the genetic edge recombination operator *Proc. 3rd Int. Conf. on Genetic Algorithms (Fairfax, VA, June 1989)* ed J D Schaffer (San Mateo, CA: Morgan Kaufmann) pp 133–40

Wicks T and Lawson S 1993 Genetic algorithm design of wave digital filters with a restricted coefficient set *Natural Algorithms in Signal Processing (Workshop, Chelmsford, UK, November 1993)* vol 1 (London: IEE) pp 17/1–17/7

Wilson P B and Macleod M D 1993 Low implementation cost IIR digital filter design using genetic algorithms *Natural Algorithms in Signal Processing (Workshop, Chelmsford, UK, November 1993)* vol 1 (London: IEE) pp 4/1–4/8

Wilson S W 1987 Hierarchical credit allocation in a classifier system *Genetic Algorithms and Simulated Annealing* ed L Davis (Boston, MA: Pitman) ch 8, pp 104–15

Yamada T and Nakano R 1992 A genetic algorithm applicable to large-scale job-shop problems *Parallel Problem Solving from Nature, 2 (Proc. 2nd Int. Conf. on Parallel Problem Solving from Nature, Brussels, 1992)* ed R Männer and B Manderick (Amsterdam: Elsevier) pp 281–90

Yang J-J and Korfhage R R 1993 Query optimization in information retrieval using genetic algorithms *Proc. 5th Int. Conf. on Genetic Algorithms (Urbana-Champaign, IL, July 1993)* ed S Forrest (San Mateo, CA: Morgan Kaufmann) pp 603–11

Zhang B-T and Mühlenbein H 1993 Genetic programming of minimal neural nets using Occam's razor *Proc. 5th Int. Conf. on Genetic Algorithms (Urbana-Champaign, IL, July 1993)* ed S Forrest (San Mateo, CA: Morgan Kaufmann) pp 342–9

Further reading

This article has provided only a glimpse into the range of applications for evolutionary computing. A series of comprehensive bibliographies has been produced by J T Alander of the Department of Information Technology and Production Economics, University of Vaasa, as listed below.

Art and Music: *Indexed Bibliography of Genetic Algorithms in Art and Music* Report 94-1-ART (ftp.uwasa.fi/cs/report94-1/gaARTbib.ps.Z)

Chemistry and Physics: *Indexed Bibliography of Genetic Algorithms in Chemistry and Physics* Report 94-1-CHEMPHYS (ftp.uwasa.fi/cs/report94-1/gaCHEMPHYSbib.ps.Z)

Control: *Indexed Bibliography of Genetic Algorithms in Control.* Report 94-1-CONTROL (ftp.uwasa.fi/cs/report94-1/gaCONTROLbib.ps.Z)

1. **Computer Aided Design:** *Indexed Bibliography of Genetic Algorithms in Computer Aided Design* Report 94-1-CAD (ftp.uwasa.fi/cs/report94-1/gaCADbib.ps.Z)

2. **Computer Science:** *Indexed Bibliography of Genetic Algorithms in Computer Science* Report 94-1-CS (ftp.uwasa.fi/cs/report94-1/gaCSbib.ps.Z)

3. **Economics:** *Indexed Bibliography of Genetic Algorithms in Economics* Report 94-1-ECO (ftp.uwasa.fi/cs/report94-1/gaECObib.ps.Z)

4. **Electronics and VLSI Design and Testing:** *Indexed Bibliography of Genetic Algorithms in Electronics and VLSI Design and Testing* Report 94-1-VLSI (ftp.uwasa.fi/cs/report94-1/gaVLSIbib.ps.Z)

5. **Engineering:** *Indexed Bibliography of Genetic Algorithms in Engineering* Report 94-1-ENG (ftp.uwasa.fi/cs/report94-1/gaENGbib.ps.Z)

6. **Fuzzy Systems:** *Indexed Bibliography of Genetic Algorithms and Fuzzy Systems* Report 94-1-FUZZY (ftp.uwasa.fi/cs/report94-1/gaFUZZYbib.ps.Z)

7. **Logistics:** *Indexed Bibliography of Genetic Algorithms in Logistics* Report 94-1-LOGISTICS (ftp.uwasa.fi/cs/report94-1/gaLOGISTICSbib.ps.Z)

8. **Manufacturing:** *Indexed Bibliography of Genetic Algorithms in Manufacturing* Report 94-1-MANU (ftp.uwasa.fi/cs/report94-1/gaMANUbib.ps.Z)

9. **Neural Networks:** *Indexed Bibliography of Genetic Algorithms and Neural Networks* Report 94-1-NN (ftp.uwasa.fi/cs/report94-1/gaNNbib.ps.Z)

10. **Optimization:** *Indexed Bibliography of Genetic Algorithms and Optimization* Report 94-1-OPTIMI (ftp.uwasa.fi/cs/report94-1/gaOPTIMIbib.ps.Z)

11. **Operations Research:** *Indexed Bibliography of Genetic Algorithms in Operations Research* Report 94-1-OR (ftp.uwasa.fi/cs/report94-1/gaORbib.ps.Z)

12. **Power Engineering:** *Indexed Bibliography of Genetic Algorithms in Power Engineering* Report 94-1-POWER (ftp.uwasa.fi/cs/report94-1/gaPOWERbib.ps.Z)

13. **Robotics:** *Indexed Bibliography of Genetic Algorithms in Robotics* Report 94-1-ROBO (ftp.uwasa.fi/cs/report94-1/gaROBObib.ps.Z)

14. **Signal and Image Processing:** *Indexed Bibliography of Genetic Algorithms in Signal and Image Processing* Report 94-1-SIGNAL (ftp.uwasa.fi/cs/report94-1/gaSIGNALbib.ps.Z)

3

Advantages (and disadvantages) of evolutionary computation over other approaches

Hans-Paul Schwefel

3.1 No-free-lunch theorem

Since, according to the no-free-lunch (NFL) theorem (Wolpert and Macready 1996), there cannot exist any algorithm for solving *all* (e.g. optimization) problems that is generally (on average) superior to any competitor, the question of whether evolutionary algorithms (EAs) are inferior/superior to any alternative approach is senseless. What could be claimed solely is that EAs behave better than other methods with respect to solving a specific class of problems—with the consequence that they behave worse for other problem classes.

The NFL theorem can be corroborated in the case of EAs versus many classical optimization methods insofar as the latter are more efficient in solving linear, quadratic, strongly convex, unimodal, separable, and many other special problems. On the other hand, EAs do not give up so early when discontinuous, nondifferentiable, multimodal, noisy, and otherwise unconventional response surfaces are involved. Their effectiveness (or robustness) thus extends to a broader field of applications, of course with a corresponding loss in efficiency when applied to the classes of simple problems classical procedures have been specifically devised for.

Looking into the historical record of procedures devised to solve optimization problems, especially around the 1960s (see the book by Schwefel (1995)), when a couple of direct optimum-seeking algorithms were published, for example, in the *Computer Journal*, a certain pattern of development emerges. Author A publishes a procedure and demonstrates its suitability by means of tests using some test functions. Next, author B comes along with a counterexample showing weak performance of A's algorithm in the case of a certain test problem. Of course, he also presents a new or modified technique that outperforms the older one(s) with respect to the additional test problem. This game could in principle be played *ad infinitum*.

A better means of clarifying the scene ought to result from theory. This should clearly define the domain of applicability of each algorithm by presenting convergence proofs and efficiency results. Unfortunately, however, it is possible to prove abilities of algorithms only by simplifying them as well as the situations to which they are confronted. The huge remainder of questions must be answered by means of (always limited) test series, and even that cannot tell much about an actual real-world problem-solving situation with yet unanalyzed features, that is, the normal case in applications.

Again unfortunately, there does not exist an agreed-upon test problem catalogue to evaluate old as well as new algorithms in a concise way. It is doubtful whether such a test bed will ever be agreed upon, but efforts in that direction would be worthwhile.

3.2 Conclusions

Finally, what are the truths and consequences? First, there will always remain a dichotomy between efficiency and general applicability, between reliability and effort of problem-solving, especially optimum-seeking, algorithms. Any specific knowledge about the situation at hand may be used to specify an adequate specific solution algorithm, the optimal situation being that one knows the solution in advance. On the other hand, there cannot exist one method that solves all problems effectively as well as efficiently. These goals are contradictory.

If there is already a traditional method that solves a given problem, EAs should not be used. They cannot do it better or with less computational effort. In particular, they do not offer an escape from the curse of dimensionality—the often quadratic, cubic, or otherwise polynomial increase in instructions used as the number of decision variables is increased, arising, for example, from matrix manipulation.

To develop a new solution method suitable for a problem at hand may be a nice challenge to a theoretician, who will afterwards get some merit for his effort, but from the application point of view the time for developing the new technique has to be added to the computer time invested. In that respect, a nonspecialized, robust procedure (and EAs belong to this class) may be, and often proves to be, worthwhile.

A warning should be given about a common practice—the linearization or other decomplexification of the situation in order to make a traditional method applicable. Even a guaranteed globally optimal solution for the simplified task may be a long way off and thus greatly inferior to an approximate solution to the real problem.

The best one can say about EAs, therefore, is that they present a methodological framework that is easy to understand and handle, and is either usable as a black-box method or open to the incorporation of new or old recipes for further sophistication, specialization or hybridization. They are applicable even in dynamic situations where the goal or constraints are moving over time or changing, either exogenously or self-induced, where parameter adjustments and fitness measurements are disturbed, and where the landscape is rough, discontinuous, multimodal, even fractal or cannot otherwise be handled by traditional methods, especially those that need global prediction from local surface analysis.

There exist EA versions for multiple criterion decision making (MCDM) and many different parallel computing architectures. Almost forgotten today is their applicability in experimental (non-computing) situations.

Sometimes striking is the fact that even obviously wrong parameter settings do not prevent fairly good results: this certainly can be described as robustness. Not yet well understood, but nevertheless very successful are those EAs which self-adapt some of their internal parameters, a feature that can be described as collective learning of the environmental conditions. Nevertheless, even self-adaptation does not circumvent the NFL theorem.

In this sense, and only in this sense, EAs always present an intermediate compromise; the enthusiasm of their inventors is not yet taken into account here, nor the insights available from the analysis of the algorithms for natural evolutionary processes which they try to mimic.

References

Schwefel H-P 1995 *Evolution and Optimum Seeking* (New York: Wiley)
Wolpert D H and Macready W G 1996 *No Free Lunch Theorems for Search* Technical Report SFI-TR-95-02-010 Santa Fe Institute

4

Principles of evolutionary processes

David B Fogel

4.1 Overview

The most widely accepted collection of evolutionary theories is the neo-Darwinian paradigm. These arguments assert that the vast majority of the history of life can be fully accounted for by physical processes operating on and within populations and species (Hoffman 1989, p 39). These processes are reproduction, mutation, competition, and selection. Reproduction is an obvious property of extant species. Further, species have such great reproductive potential that their population size would increase at an exponential rate if all individuals of the species were to reproduce successfully (Malthus 1826, Mayr 1982, p 479). Reproduction is accomplished through the transfer of an individual's genetic program (either asexually or sexually) to progeny. Mutation, in a positively entropic system, is guaranteed, in that replication errors during information transfer will necessarily occur. Competition is a consequence of expanding populations in a finite resource space. Selection is the inevitable result of competitive replication as species fill the available space. Evolution becomes the inescapable result of interacting basic physical statistical processes (Huxley 1963, Wooldridge 1968, Atmar 1979).

Individuals and species can be viewed as a duality of their genetic program, the genotype (Section 5.2), and their expressed behavioral traits, the *phenotype*. The genotype provides a mechanism for the storage of experiential evidence, of historically acquired information. Unfortunately, the results of genetic variations are generally unpredictable due to the universal effects of pleiotropy and polygeny (figure 4.1) (Mayr 1959, 1963, 1982, 1988, Wright 1931, 1960, Simpson 1949, p 224, Dobzhansky 1970, Stanley 1975, Dawkins 1986). Pleiotropy is the effect that a single gene may simultaneously affect several phenotypic traits. Polygeny is the effect that a single phenotypic characteristic may be determined by the simultaneous interaction of many genes. There are no one-gene, one-trait relationships in naturally evolved systems. The phenotype varies as a complex, nonlinear function of the interaction between underlying genetic structures and current environmental conditions. Very different genetic

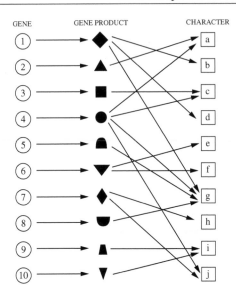

Figure 4.1. Pleiotropy is the effect that a single gene may simultaneously affect several phenotypic traits. Polygeny is the effect that a single phenotypic characteristic may be determined by the simultaneous interaction of many genes. These one-to-many and many-to-one mappings are pervasive in natural systems. As a result, even small changes to a single gene may induce a raft of behavioral changes in the individual (after Mayr 1963).

structures may code for equivalent behaviors, just as diverse computer programs can generate similar functions.

Selection directly acts only on the expressed behaviors of individuals and species (Mayr 1988, pp 477–8). Wright (1932) offered the concept of adaptive topography to describe the fitness of individuals and species (minimally, isolated reproductive populations termed demes). A population of genotypes maps to respective phenotypes (*sensu* Lewontin 1974), which are in turn mapped onto the adaptive topography (figure 4.2). Each peak corresponds to an optimized collection of phenotypes, and thus to one of more sets of optimized genotypes. Evolution probabilistically proceeds up the slopes of the topography toward peaks as selection culls inappropriate phenotypic variants.

Others (Atmar 1979, Raven and Johnson 1986, pp 400–1) have suggested that it is more appropriate to view the adaptive landscape from an inverted position. The peaks become troughs, 'minimized prediction error entropy wells' (Atmar 1979). Searching for peaks depicts evolution as a slowly advancing, tedious, uncertain process. Moreover, there appears to be a certain fragility to an evolving phyletic line; an optimized population might be expected to quickly fall of the peak under slight perturbations. The inverted topography leaves an altogether different impression. Populations advance rapidly down the walls of the error troughs until their cohesive set of interrelated behaviors is optimized,

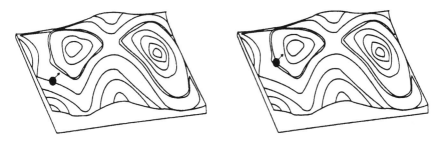

Figure 4.2. Wright's adaptive topography, inverted. An adaptive topography, or adaptive landscape, is defined to represent the fitness of all possible phenotypes (generated by the interaction between the genotypes and the environment). Wright (1932) proposed that as selection culls the last appropriate existing behaviors relative to others in the population, the population advances to areas of higher fitness on the landscape. Atmar (1979) and others have suggested viewing the topography from an inverted perspective. Populations advance to areas of lower behavioral error.

at which point stagnation occurs. If the topography is generally static, rapid descents will be followed by long periods of stasis. If, however, the topography is in continual flux, stagnation may never set in.

Viewed in this manner, evolution is an obvious optimizing problem-solving process (not to be confused with a process that leads to perfection). Selection drives phenotypes as close to the optimum as possible, given initial conditions and environment constraints. However the environment is continually changing. Species lag behind, constantly evolving toward a new optimum. No organism should be viewed as being perfectly adapted to its environment. The suboptimality of behavior is to be expected in any dynamic environment that mandates tradeoffs between behavioral requirements. However selection never ceases to operate, regardless of the population's position on the topography.

Mayr (1988, p 532) has summarized some of the more salient characteristics of the neo-Darwinian paradigm. These include:

(i)   The individual is the primary target of selection.
(ii) Genetic variation is largely a chance phenomenon. Stochastic processes play a significant role in evolution.
(iii) Genotypic variation is largely a product of recombination and 'only ultimately of mutation'.
(iv) 'Gradual' evolution may incorporate phenotypic discontinuities.
(v)   Not all phenotypic changes are necessarily consequences of *ad hoc* natural selection.
(vi) Evolution is a change in adaptation and diversity, not merely a change in gene frequencies.
(vii) Selection is probabilistic, not deterministic.

These characteristics form a framework for evolutionary computation.

References

Atmar W 1979 The inevitability of evolutionary invention, unpublished manuscript

Dawkins R 1986 *The Blind Watchmaker* (Oxford: Clarendon)

Dobzhansky T 1970 *Genetics of the Evolutionary Processes* (New York: Columbia University Press)

Hoffman A 1989 *Arguments on Evolution: a Paleontologist's Perspective* (New York: Oxford University Press)

Huxley J 1963 The evolutionary process *Evolution as a Process* ed J Huxley, A C Hardy and E B Ford (New York: Collier) pp 9–33

Lewontin R C 1974 *The Genetic Basis of Evolutionary Change* (New York: Columbia University Press)

Malthus T R 1826 *An Essay on the Principle of Population, as it Affects the Future Improvement of Society* 6th edn (London: Murray)

Mayr E 1959 Where are we? *Cold Spring Harbor Symp. Quant. Biol.* **24** 409–40

——1963 *Animal Species and Evolution* (Cambridge, MA: Belknap)

——1982 *The Growth of Biological Thought: Diversity, Evolution and Inheritance* (Cambridge, MA: Belknap)

——1988 *Toward a New Philosophy of Biology: Observations of an Evolutionist* (Cambridge, MA: Belknap)

Raven P H and Johnson G B 1986 *Biology* (St Louis, MO: Times Mirror)

Simpson G G 1949 *The Meaning of Evolution: a Study of the History of Life and its Significance for Man* (New Haven, CT: Yale University Press)

Stanley S M 1975 A theory of evolution above the species level *Proc. Natl Acad. Sci. USA* **72** 646–50

Wooldridge D E 1968 *The Mechanical Man: the Physical Basis of Intelligent Life* (New York: McGraw-Hill)

Wright S 1931 Evolution in Mendelian populations *Genetics* **16** 97–159

——1932 The roles of mutation, inbreeding, crossbreeding, and selection in evolution *Proc. 6th Int. Congr. on Genetics (Ithaca, NY)* vol 1, pp 356–66

——1960 The evolution of life, panel discussion *Evolution After Darwin: Issues in Evolution* vol 3, ed S Tax and C Callender (Chicago, IL: University of Chicago Press)

5

Principles of genetics

Raymond C Paton

5.1 Introduction

The material covers a number of key areas which are necessary to understanding the nature of the evolutionary process. We begin by looking at some basic ideas of heredity and how variation occurs in interbreeding populations. From here we look at the gene in more detail and then consider how it can undergo change. The next section looks at aspects of population thinking needed to appreciate selection. This is crucial to an appreciation of Darwinian mechanisms of evolution. The chapter concludes with selected references to further information. In order to keep this contribution within its size limits, the material is primarily about the biology of higher plants and animals.

5.2 Some fundamental concepts in genetics

Many plants and animals are produced through sexual means by which the nucleus of a male sperm cell fuses with a female egg cell (ovum). Sperm and ovum nuclei each contain a single complement of nuclear material arranged as ribbon-like structures called chromosomes. When a sperm fuses with an egg the resulting cell, called a zygote, has a double complement of chromosomes together with the cytoplasm of the ovum. We say that a single complement of chromosomes constitutes a haploid set (abbreviated as n) and a double complement is called the diploid set ($2n$). Gametes (sex cells) are haploid whereas most other cells are diploid. The formation of gametes (gametogenesis) requires the number of chromosomes in the gamete-forming cells to be halved (see figure 5.1).

Gametogenesis is achieved through a special type of cell division called meiosis (also called reduction division). The intricate mechanics of meiosis ensures that gametes contain only one copy of each chromosome.

A genotype is the genetic constitution that an organism inherits from its parents. In a diploid organism, half the genotype is inherited from one parent and half from the other. Diploid cells contain two copies of each chromosome. This

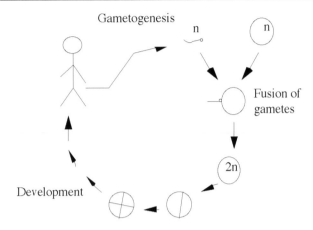

Figure 5.1. A common life cycle model.

rule is not universally true when it comes to the distribution of sex chromosomes. Human diploid cells contain 46 chromosomes of which there are 22 pairs and an additional two sex chromosomes. Sex is determined by one pair (called the sex chromosomes); female is X and male is Y. A female human has the sex chromosome genotype of XX and a male is XY. The inheritance of sex is summarized in figure 5.2. The members of a pair of nonsex chromosomes are said to be homologous (this is also true for XX genotypes whereas XY are not homologous).

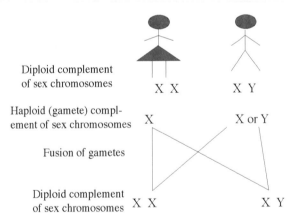

Figure 5.2. Inheritance of sex chromosomes.

Although humans have been selectively breeding domestic animals and plants for a long time, the modern study of genetics began in the mid-19th century with the work of Gregor Mendel. Mendel investigated the inheritance of particular traits in peas. For example, he took plants that had wrinkled seeds and plants that had round seeds and bred them with plants of the same phenotype (i.e. observable appearance), so wrinkled were bred with wrinkled and round were bred with round. He continued this over a number of generations until round always produced round offspring and wrinkled, wrinkled. These are called pure breeding plants. He then cross-fertilized the plants by breeding rounds with wrinkles. The subsequent generation (called the F1 hybrids) was all round. Then Mendel crossed the F1 hybrids with each other and found that the next generation, the F2 hybrids, had round and wrinkled plants in the ratio of 3 (round) : 1 (wrinkled).

Mendel did this kind of experiment with a number of pea characteristics such as:

color of cotyledons	yellow or green
color of flowers	red or white
color of seeds	gray/brown or white
length of stem	tall or dwarf.

In each case he found that the the F1 hybrids were always of one form and the two forms reappeared in the F2. Mendel called the form which appeared in the F1 generation dominant and the form which reappeared in the F2 recessive (for the full text of Mendel's experiments see an older genetics book, such as that by Sinnott *et al* (1958)).

A modern interpretation of inheritance depends upon a proper understanding of the nature of a gene and how the gene is expressed in the phenotype. The nature of a gene is quite complex as we shall see later (see also Alberts *et al* 1989, Lewin 1990, Futuyma 1986). For now we shall take it to be the functional unit of inheritance. An allele (allelomorph) is one of several forms of a gene occupying a given locus (location) on a chromosome. Originally related to pairs of contrasting characteristics (see examples above), the idea of observable unit characters was introduced to genetics around the turn of this century by such workers as Bateson, de Vries, and Correns (see Darden 1991). The concept of a gene has tended to replace allele in general usage although the two terms are not the same.

How can the results of Mendel's experiments be interpreted? We know that each parent plant provides half the chromosome complement found in its offspring and that chromosomes in the diploid cells are in pairs of homologues. In the pea experiments pure breeding parents had homologous chromosomes which were identical for a particular gene; we say they are homozygous for a particular gene. The pure breeding plants were produced through self-fertilization and by selecting those offspring of the desired phenotype. As round was dominant to wrinkled we say that the round form of the gene is R ('big r') and the wrinkled r ('little r'). Figure 5.3 summarizes the cross of a pure breeding round (RR) with a pure breeding wrinkled (rr).

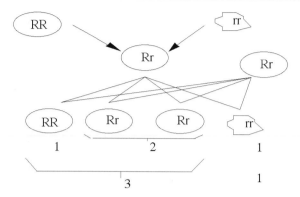

Figure 5.3. A simple Mendelian experiment.

We see the appearance of the heterozygote (in this case Rr) in the F1 generation. This is phenotypically the same as the dominant phenotype but genotypically contains both a dominant and a recessive form of the particular gene under study. Thus when the heterozygotes are randomly crossed with each other the phenotype ratio is three dominant : one recessive. This is called the monohybrid ratio (i.e. for one allele). We see in Mendel's experiments the independent segregation of alleles during breeding and their subsequent independent assortment in offspring.

In the case of two genes we find more phenotypes and genotypes appearing. Consider what happens when pure breeding homozygotes for round yellow seeds (RRYY) are bred with pure breeding homozygotes for wrinkled green seeds (rryy). On being crossed we end up with heterozygotes with a genotype of RrYy and phenotype of round yellow seeds. We have seen that the genes segregate independently during meiosis so we have the combinations shown in figure 5.4.

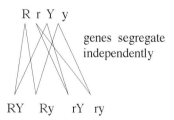

Figure 5.4. Genes segregating independently.

Thus the gametes of the heterozygote can be of four kinds though we assume that each form can occur with equal frequency. We may examine the possible combinations of gametes for the next generation by producing a contingency table for possible gamete combinations. These are shown in figure 5.5.

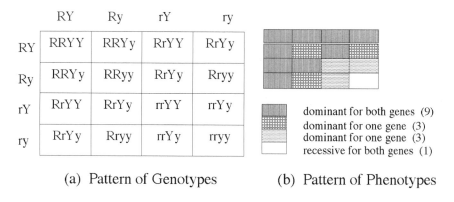

(a) Pattern of Genotypes   (b) Pattern of Phenotypes

Figure 5.5. Genotype and phenotype patterns in F2.

We summarize this set of genotype combinations in the phenotype table
(figure 5.5(*b*)). The resulting ratio of phenotypes is called the dihybrid ratio
(9:3:3:1). We shall consider one final example in this very brief summary.
When pure breeding red-flowered snapdragons were crossed with pure breeding
white-flowered plants the F1 plants were all pink. When these were selfed the
population of offspring was in the ratio of one red : two pink : one white. This
is a case of incomplete dominance in the heterozygote.

It has been found that the Mendelian ratios do not always apply in breeding
experiments. In some cases this is because certain genes interact with each
other. Epistasis occurs when the expression of one gene masks the phenotypic
effects of another. For example, certain genotypes (cyanogenics) of clover can
resist grazing because they produce low doses of cyanide which makes them
unpalatable. Two genes are involved in cyanide production, one which produces
an enzyme which converts a precursor molecule into a glycoside and another
gene which produces an enzyme which converts the glycoside into hydrogen
cyanide (figure 5.6(*a*)). If two pure breeding acyanogenic strains are crossed
the heterozygote is cyanogenic (figure 5.6(*b*)).

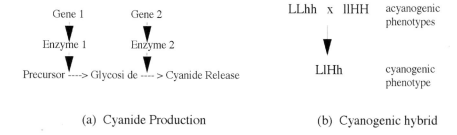

(a) Cyanide Production   (b) Cyanogenic hybrid

Figure 5.6. Cyanogenic clover: cyanide production and cyanogenic hybrid.

	LH	Lh	lH	lh
LH	LLHH	LLHh	LlHH	LlHh
Lh	LLHh	LLhh	LlHh	Llhh
lH	LlHH	LlHh	llHH	llHh
lh	LlHH	Llhh	llHh	llhh

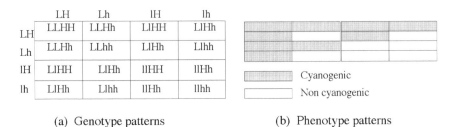

▨ Cyanogenic
☐ Non cyanogenic

(a) Genotype patterns   (b) Phenotype patterns

Figure 5.7. Epistasis in clover: genotype and phenotype patterns.

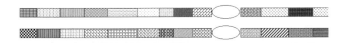

Figure 5.8. Pairing of homologous chromosomes.

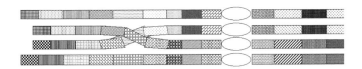

Figure 5.9. Crossing over in a tetrad.

When the cyanogenic strain is selfed the genotypes are as summarized in figure 5.7(a). There are only two phenotypes produced, cyanogenic and acyanogenic, as summarized in figure 5.7(b).

So far we have followed Mendel's laws regarding the independent segregation of genes. This independent segregation does not occur when genes are located on the same chromosome. During meiosis homologous chromosomes (i.e. matched pairs one from each parental gamete) move together and are seen to be joined at the centromere (the clear oval region in figure 5.8).

In this simplified diagram we show a set of genes (rectangles) in which those on the top are of the opposite form to those on the bottom. As the chromosomes are juxtaposed they each are doubled up so that four strands (usually called chromatids) are aligned. The close proximity of the inner two chromatids and the presence of enzymes in the cellular environment can result in breakages and recombinations of these strands as summarized in figure 5.9.

The result is that of the four juxtaposed strands two are the same as the parental chromosomes and two, called the recombinants, are different. This crossover process mixes up the genes with respect to original parental chromosomes. The chromosomes which make up a haploid gamete will be a random mixture of parental and recombinant forms. This increases the variability between parents and offspring and reduces the chance of harmful recessives becoming homozygous.

5.3   The gene in more detail

Genes are located on chromosomes. Chromosomes segregate independently during meiosis whereas genes can be linked on the same chromosome. The conceptual reasons why there has been confusion are the differences in understanding about gene and chromosome such as which is the unit of heredity (see Darden 1991). The discovery of the physicochemical nature of hereditary material culminated in the Watson–Crick model in 1953 (see figure 5.10). The coding parts of the deoxyribonucleic acid (DNA) are called bases; there are four types (adenine, thymine, cytosine, and guanine). They are strung along a sugar-and-phosphate string, which is arranged as a helix. Two intertwined strings then form the double helix. The functional unit of this code is a triplet of bases which can code for a single amino acid. The genes are located along the DNA strand.

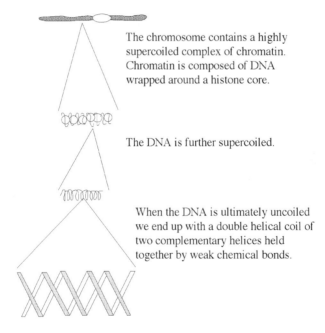

The chromosome contains a highly supercoiled complex of chromatin. Chromatin is composed of DNA wrapped around a histone core.

The DNA is further supercoiled.

When the DNA is ultimately uncoiled we end up with a double helical coil of two complementary helices held together by weak chemical bonds.

Figure 5.10. Idealization of the organization of chromosomes in a eukaryotic cell. (A eukaryotic cell has an organized nucleus and cytoplasmic organelles.)

Transcription is the synthesis of ribonucleic acid (RNA) using the DNA template. It is a preliminary step in the ultimate synthesis of protein. A gene can be transcribed through the action of enzymes and a chain of transcript is formed as a polymer called messenger RNA (mRNA). This mRNA can then be translated into protein. The translation process converts the mRNA code into a protein sequence via another form of RNA called transfer RNA (tRNA). In this way, genes are transcribed so that mRNA may be produced, from which protein molecules (typically the 'workhorses' and structural molecules of a cell) can be formed. This flow of information is generally unidirectional. (For more details on this topic the reader should consult a molecular biology text and look at the central dogma of molecular biology, see e.g. Lewin 1990, Alberts *et al* 1989.)

Figure 5.11 provides a simplified view of the anatomy of a structural gene, that is, one which codes for a protein or RNA.

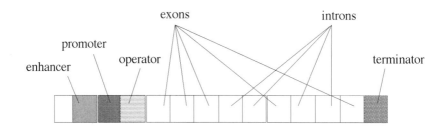

Figure 5.11. A simplified diagram of a structural gene.

That part of the gene which ultimately codes for protein or RNA is preceded upstream by three stretches of code. The enhancer facilitates the operation of the promoter region, which is where RNA polymerase is bound to the gene in order to initiate transcription. The operator is the site where transcription can be halted by the presence of a repressor protein. Exons are expressed in the final gene product (e.g. the protein molecule) whereas introns are transcribed but are removed from the transcript leaving the fragments of exon material to be spliced. One stretch of DNA may consist of several overlapping genes. For example, the introns in one gene may be the exons in another (Lewin 1990). The terminator is the postexon region of the gene which causes transcription to be terminated. Thus a biological gene contains not only code to be read but also coded instructions on how it should be read and what should be read. Genes are highly organized. An operon system is located on one chromosome and consists of a regulator gene and a number of contiguous structural genes which share the same promoter and terminator and code for enzymes which are involved in specific metabolic pathways (the classical example is the Lac operon, see figure 5.12).

Operons can be grouped together into higher-order (hierarchical) regulatory genetic systems (Neidhart *et al* 1990). For example, a number of operons from different chromosomes may be regulated by a single gene known as a regulon. These higher-order systems provide a great challenge for change in a genome. Modification of the higher-order gene can have profound effects on the expression of structural genes that are under its influence.

Figure 5.12. A visualization of an operon.

5.4 Options for change

We have already seen how sexual reproduction can mix up the genes which are incorporated in a gamete through the random reassortment of paternal and maternal chromosomes and through crossing over and recombination. Effectively though, the gamete acquires a subset of the same genes as the diploid gamete-producing cells; they are just mixed up. Clearly, any zygote that is produced will have a mixture of genes and (possibly) some chromosomes which have both paternal and maternal genes.

There are other mechanisms of change which alter the genes themselves or change the number of genes present in a genome. We shall describe a mutation as any change in the sequence of genomic DNA. Gene mutations are of two types: point mutation, in which a single base is changed, and frameshift mutation, in which one or more bases (but not a multiple of three) are inserted or deleted. This changes the frame in which triplets are transcribed into RNA and ultimately translated into protein. In addition some genes are able to become transposed elsewhere in a genome. They 'jump' about and are called transposons. Chromosome changes can be caused by deletion (loss of a section), duplication (the section is repeated), inversion (the section is in the reverse order), and translocation (the section has been relocated elsewhere). There are also changes at the genome level. Ploidy is the term used to describe multiples of a chromosome complement such as haploid (n), diploid ($2n$), and tetraploid ($4n$). A good example of the influence of ploidy on evolution is among such crops as wheat and cotton. Somy describes changes to the frequency of particular chromosomes: for example, trisomy is three copies of a chromosome.

5.5 Population thinking

So far we have focused on how genes are inherited and how they or their combinations can change. In order to understand evolutionary processes (Chapter 4) we must shift our attention to looking at populations (we shall not emphasize too much whether of genes, chromosomes, genomes, or organisms). Population thinking is central to our understanding of models of evolution.

The Hardy–Weinberg theorem applies to frequencies of genes and genotypes in a population of individuals, and states that the relative frequency of each gene remains in equilibrium from one generation to the next. For a single allele, if the frequency of one form is p then that of the other (say q) is $1 - p$. The three genotypes that exist with this allele have the population proportions of

$$p^2 + 2pq + q^2 = 1.$$

This equation does not apply when a mixture of four factors changes the relative frequencies of genes in a population: mutation, selection, gene flow, and random genetic drift (drift). Drift can be described as the effect of the sampling of a population on its parents. Each generation can be thought of as a sample of its parents' population. In that the current population is a sample of its parents, we acknowledge that a statistical sampling error should be associated with gene frequencies. The effect will be small in large populations because the relative proportion of random changes will be a very small component of the large numbers. However, drift in a small population will have a marked effect.

One factor which can counteract the effect of drift is differential migration of individuals between populations which leads to gene flow. Several models of gene flow exist. For example, migration which occurs at random among a group of small populations is called the island model whereas in the stepping stone model each population receives migrants only from neighboring populations. Mutation, selection, and gene flow are deterministic factors so that if fitness, mutation rate, and rate of gene flow are the same for a number of populations that begin with same gene frequencies, they will attain the same equilibrium composition. Drift is a stochastic process because the sampling effect on the parent population is random.

Sewall Wright introduced the idea of an adaptive landscape to explain how a population's allele frequencies might evolve over time. The peaks on the landscape represent genetic compositions of a population for which the mean fitness is high and troughs are possible compositions where the mean fitness is low. As gene frequencies change and mean fitness increases the population moves uphill. Indeed, selection will operate to increase mean fitness so, on a multipeaked landscape, selection may operate to move populations to local maxima. On a fixed landscape drift and selection can act together so that populations may move uphill (through selection) or downhill (through drift). This means that the global maximum for the landscape could be reached. These ideas are formally encapsulated in Wright's (1968–1978) *shifting balance theory* of evolution. Further information on the relation of population genetics to evolutionary theory can be studied further in the books by Wright (1968–1978), Crow and Kimura (1970) and Maynard Smith (1989).

The change of gene frequencies coupled with changes in the genes themselves can lead to the emergence of new species although the process is far from simple and not fully understood (Futuyma, 1986, Maynard Smith 1993). The nature of the species concept or (for some) concepts which is central to Darwinism is complicated and will not be discussed here (see e.g. Futuyma 1986). Several mechanisms apply to promote speciation (Maynard

Smith 1993): geographical or spatial isolation, barriers preventing formation of hybrids, nonviable hybrids, hybrid infertility, and hybrid breakdown—in which post-F1 generations are weak or infertile.

Selectionist theories emphasize invariant properties of the system: the system is an internal generator of variations (Changeux and Dehaene 1989) and diversity among units of the population exists prior to any testing (Manderick 1994). We have seen how section operates to optimize fitness. Darden and Cain (1987) summarize a number of common elements in selectionist theories as follows:

- a set of a given entity type (i.e. the units of the population)

- a particular property (P) according to which members of this set vary

- an environment in which the entity type is found

- a factor in the environment to which members react differentially due to their possession or nonpossession of the property (P)

- differential benefits (both shorter and longer term) according to the possession or nonpossession of the property (P).

This scheme summarizes the selectionist approach. In addition, Maynard Smith (1989) discusses a number of selection systems (particular relevant to animals) including sexual, habitat, family, kin, group, and synergistic (cooperation). A very helpful overview of this area of ecology, behavior, and evolution is that by Sigmund (1993). Three selectionist systems in the biosciences are the neo-Darwinian theory of evolution in a population, clonal selection theory applied to the immune system, and the theory of neuronal group selection (for an excellent summary with plenty of references see that by Manderick (1994)).

There are many important aspects of evolutionary biology which have had to be omitted because of lack of space. The relevance of neutral molecular evolution theory (Kimura 1983) and nonselectionist approaches (see e.g. Goodwin and Saunders 1989, Lima de Faria 1988, Kauffman 1993) has not been discussed. In addition some important ideas have not been considered, such as evolutionary game theory (Maynard Smith 1989, Sigmund 1993), the role of sex (see e.g. Hamilton *et al* 1990), the evolution of cooperation (Axelrod 1984), the red queen (Van Valen 1973, Maynard Smith 1989), structured genomes, for example, incorporation of regulatory hierarchies (Kauffman 1993, Beaumont 1993, Clarke *et al* 1993), experiments with endosymbiotic systems (Margulis and Foster 1991, Hilario and Gogarten 1993), coevolving parasite populations (see e.g. Collins 1994; for a biological critique and further applications see Sumida and Hamilton 1994), inheritance of acquired characteristics (Landman 1991), and genomic imprinting and other epigenetic inheritance systems (for a review see Paton 1994). There are also considerable philosophical issues which must be addressed in this area which impinge on how biological sources are applied to evolutionary computing (see Sober 1984). Not least among these is the nature of adaptation.

References

Alberts B, Bray D, Lewis J, Raff M, Roberts K and Watson J D 1989 *Molecular Biology of the Cell* (New York: Garland)

Axelrod R 1984 *The Evolution of Co-operation* (Harmondsworth: Penguin)

Beaumont M A 1993 Evolution of optimal behaviour in networks of Boolean automata *J. Theor. Biol.* **165** 455–76

Changeux J-P and Dehaene S 1989 Neuronal models of cognitive functions *Cognition* **33** 63-109

Clarke B, Mittenthal J E and Senn M 1993 A model for the evolution of networks of genes *J. Theor. Biol.* **165** 269–89

Collins R 1994 Artificial evolution and the paradox of sex *Computing with Biological Metaphors* ed R C Paton (London: Chapman and Hall)

Crow J F and Kimura M 1970 *An Introduction to Population Genetics Theory* (New York: Harper and Row)

Darden L 1991 *Theory Change in Science* (New York: Oxford University Press)

Darden L and Cain J A 1987 Selection type theories *Phil. Sci.* **56** 106–29

Futuyma D J 1986 *Evolutionary Biology* (MA: Sinauer)

Goodwin B C and Saunders P T (eds) 1989 *Theoretical Biology: Epigenetic and Evolutionary Order from Complex Systems* (Edinburgh: Edinburgh University Press)

Hamilton W D, Axelrod A and Tanese R 1990 Sexual reproduction as an adaptation to resist parasites *Proc. Natl Acad. Sci. USA* **87** 3566–73

Hilario E and Gogarten J P 1993 Horizontal transfer of ATPase genes—the tree of life becomes a net of life *BioSystems* **31** 111–9

Kauffman S A 1993 *The Origins of Order* (New York: Oxford University Press)

Kimura, M 1983 *The Neutral Theory of Molecular Evolution* (Cambridge: Cambridge University Press)

Landman O E 1991 The inheritance of acquired characteristics *Ann. Rev. Genet.* **25** 1–20

Lewin B 1990 *Genes IV* (Oxford: Oxford University Press)

Lima de Faria A 1988 *Evolution without Selection* (Amsterdam: Elsevier)

Margulis L and Foster R (eds) 1991 *Symbiosis as a Source of Evolutionary Innovation: Speciation and Morphogenesis* (Cambridge, MA: MIT Press)

Manderick B 1994 The importance of selectionist systems for cognition *Computing with Biological Metaphors* ed R C Paton (London: Chapman and Hall)

Maynard Smith J 1989 *Evolutionary Genetics* (Oxford: Oxford University Press)

——1993 *The Theory of Evolution* Canto edn (Cambridge: Cambridge University Press)

Neidhart F C, Ingraham J L and Schaechter M 1990 *Physiology of the Bacterial Cell* (Sunderland, MA: Sinauer)

Paton R C 1994 Enhancing evolutionary computation using analogues of biological mechanisms *Evolutionary Computing (Lecture Notes in Computer Science 865)* ed T C Fogarty (Berlin: Springer) pp 51–64

Sigmund K 1993 *Games of Life* (Oxford: Oxford University Press)

Sinnott E W, Dunn L C and Dobzhansky T 1958 *Principles of Genetics* (New York: McGraw-Hill)

Sober E 1984 *The Nature of Selection: Evolutionary Theory in Philosophical Focus* (Chicago, IL: University of Chicago Press)

Sumida B and Hamilton W D 1994 Both Wrightian and 'parasite' peak shifts enhance genetic algorithm performance in the travelling salesman problem *Computing with Biological Metaphors* ed R C Paton (London: Chapman and Hall)

Van Valen L 1973 A new evolutionary law *Evolutionary Theory* **1** 1–30

Wright S 1968–1978 *Evolution and the Genetics of Populations* vols 1–4 (Chicago, IL: Chicago University Press)

6

A history of evolutionary computation

Kenneth De Jong, David B Fogel and Hans-Paul Schwefel

6.1 Introduction

No one will ever produce a completely accurate account of a set of past events since, as someone once pointed out, writing history is as difficult as forecasting. Thus we dare to begin our historical summary of evolutionary computation rather arbitrarily at a stage as recent as the mid-1950s.

At that time there was already evidence of the use of digital computer models to better understand the natural process of evolution. One of the first descriptions of the use of an evolutionary process for computer problem solving appeared in the articles by Friedberg (1958) and Friedberg *et al* (1959). This represented some of the early work in machine learning and described the use of an evolutionary algorithm for *automatic programming*, i.e. the task of finding a program that calculates a given input–output function. Other founders in the field remember a paper of Fraser (1957) that influenced their early work, and there may be many more such forerunners depending on whom one asks.

In the same time frame Bremermann presented some of the first attempts to apply simulated evolution to numerical optimization problems involving both linear and convex optimization as well as the solution of nonlinear simultaneous equations (Bremermann 1962). Bremermann also developed some of the early evolutionary algorithm (EA) theory, showing that the optimal mutation probability for linearly separable problems should have the value of $1/\ell$ in the case of ℓ bits encoding an individual (Bremermann *et al* 1965).

Also during this period Box developed his evolutionary operation (EVOP) ideas which involved an evolutionary technique for the design and analysis of (industrial) experiments (Box 1957, Box and Draper 1969). Box's ideas were never realized as a computer algorithm, although Spendley *et al* (1962) used them as the basis for their so-called simplex design method. It is interesting to note that the REVOP proposal (Satterthwaite 1959a, b) introducing randomness into the EVOP operations was rejected at that time.

40

As is the case with many ground-breaking efforts, these early studies were met with considerable skepticism. However, by the mid-1960s the bases for what we today identify as the three main forms of EA were clearly established. The roots of evolutionary programming (EP) (Chapter 10) were laid by Lawrence Fogel in San Diego, California (Fogel *et al* 1966) and those of genetic algorithms (GAs) (Chapter 8) were developed at the University of Michigan in Ann Arbor by Holland (1967). On the other side of the Atlantic Ocean, evolution strategies (ESs) (Chapter 9) were a joint development of a group of three students, Bienert, Rechenberg, and Schwefel, in Berlin (Rechenberg 1965).

Over the next 25 years each of these branches developed quite independently of each other, resulting in unique parallel histories which are described in more detail in the following sections. However, in 1990 there was an organized effort to provide a forum for interaction among the various EA research communities. This took the form of an international workshop entitled *Parallel Problem Solving from Nature* at Dortmund (Schwefel and Männer 1991).

Since that event the interaction and cooperation among EA researchers from around the world has continued to grow. In the subsequent years special efforts were made by the organizers of *ICGA'91* (Belew and Booker 1991), *EP'92* (Fogel and Atmar 1992), and *PPSN'92* (Männer and Manderick 1992) to provide additional opportunities for interaction.

This increased interaction led to a consensus for the name of this new field, *evolutionary computation* (EC), and the establishment in 1993 of a journal by the same name published by MIT Press. The increasing interest in EC was further indicated by the *IEEE World Congress on Computational Intelligence (WCCI)* at Orlando, Florida, in June 1994 (Michalewicz *et al* 1994), in which one of the three simultaneous conferences was dedicated to EC along with conferences on neural networks and fuzzy systems.

That brings us to the present in which the continued growth of the field is reflected by the many EC events and related activities each year, and its growing maturity reflected by the increasing number of books and articles about EC.

In order to keep this overview brief, we have deliberately suppressed many of the details of the historical developments within each of the three main EC streams. For the interested reader these details are presented in the following sections.

6.2 Evolutionary programming

Evolutionary programming (EP) was devised by Lawrence J Fogel in 1960 while serving at the National Science Foundation (NSF). Fogel was on leave from Convair, tasked as special assistant to the associate director (research), Dr Richard Bolt, to study and write a report on investing in basic research. Artificial intelligence at the time was mainly concentrated around heuristics and the simulation of primitive neural networks. It was clear to Fogel that both these approaches were limited because they model humans rather than the essential process that produces creatures of increasing intellect: evolution. Fogel considered intelligence to be based on adapting behavior to meet goals in a range of environments.   In turn, prediction was viewed as the key ingredient to intelligent behavior and this suggested a series of experiments on the use of simulated evolution of finite-state machines (Chapter 18) to forecast nonstationary time series with respect to arbitrary criteria.   These and other experiments were documented in a series of publications (Fogel 1962, 1964, Fogel *et al* 1965, 1966, and many others).

Intelligent behavior was viewed as requiring the composite ability to (i) predict one's environment, coupled with (ii) a translation of the predictions into a suitable response in light of the given goal. For the sake of generality, the environment was described as a sequence of symbols taken from a finite alphabet.   The evolutionary problem was defined as evolving an algorithm (essentially a program) that would operate on the sequence of symbols thus far observed in such a manner so as to produce an output symbol that is likely to maximize the algorithm's performance in light of both the next symbol to appear in the environment and a well-defined payoff function.   Finite-state machines provided a useful representation for the required behavior.

The proposal was as follows.   A population of finite-state machines is exposed to the environment, that is, the sequence of symbols that have been observed up to the current time. For each parent machine, as each input symbol is offered to the machine, each output symbol is compared with the next input symbol. The worth of this prediction is then measured with respect to the payoff function (e.g. all–none, absolute error, squared error, or any other expression of the meaning of the symbols). After the last prediction is made, a function of the payoff for each symbol (e.g. average payoff per symbol) indicates the fitness of the machine.

Offspring machines are created by randomly mutating each parent machine. Each parent produces offspring (this was originally implemented as only a single offspring simply for convenience).   There are five possible modes of random mutation that naturally result from the description of the machine: change an output symbol, change a state transition, add a state, delete a state, or change the initial state.   The deletion of a state and change of the initial state are only allowed when the parent machine has more than one state. Mutations are chosen with respect to a probability distribution, which is typically uniform. The number of mutations per offspring is also chosen with respect to a probability distribution or may be fixed *a priori*. These offspring are then evaluated over the existing environment in the same manner as their parents. Other mutations, such as majority logic mating operating on three or more machines, were proposed by Fogel *et al* (1966) but not implemented.

The machines that provide the greatest payoff are retained to become parents of the next generation. (Typically, half the total machines were saved so that the parent population remained at a constant size.) This process is iterated until an actual prediction of the next symbol (as yet unexperienced) in the environment is required. The best machine generates this prediction, the new symbol is added to the experienced environment, and the process is repeated. Fogel (1964) (and Fogel *et al* (1966)) used 'nonregressive' evolution. To be retained, a machine had to rank in the best half of the population. Saving lesser-adapted machines was discussed as a possibility (Fogel *et al* 1966, p 21) but not incorporated.

This general procedure was successfully applied to problems in prediction, identification, and automatic control (Fogel *et al* 1964, 1966, Fogel 1968) and was extended to simulate coevolving populations by Fogel and Burgin (1969). Additional experiments evolving finite-state machines for sequence prediction, pattern recognition, and gaming can be found in the work of Lutter and Huntsinger (1969), Burgin (1969), Atmar (1976), Dearholt (1976), and Takeuchi (1980).

In the mid-1980s the general EP procedure was extended to alternative representations including ordered lists for the traveling salesman problem (Fogel and Fogel 1986), and real-valued vectors for continuous function optimization (Fogel and Fogel 1986). This led to other applications in route planning (Fogel 1988, Fogel and Fogel 1988), optimal subset selection (Fogel 1989), and training neural networks (Fogel *et al* 1990), as well as comparisons to other methods of simulated evolution (Fogel and Atmar 1990). Methods for extending evolutionary search to a two-step process including evolution of the mutation variance were offered by Fogel *et al* (1991, 1992). Just as the proper choice of step sizes is a crucial part of every numerical process, including optimization, the internal adaptation of the mutation variance(s) is of utmost importance for the algorithm's efficiency. This process is called self-adaptation or autoadaptation in the case of no explicit control mechanism, e.g. if the variances are part of the individuals' characteristics and underlie probabilistic variation in a similar way as do the ordinary decision variables.

In the early 1990s efforts were made to organize annual conferences on EP, these leading to the first conference in 1992 (Fogel and Atmar 1992). This conference offered a variety of optimization applications of EP in robotics (McDonnell *et al* 1992, Andersen *et al* 1992), path planning (Larsen and Herman 1992, Page *et al* 1992), neural network design and training (Sebald and Fogel 1992, Porto 1992, McDonnell 1992), automatic control (Sebald *et al* 1992), and other fields.

First contacts were made between the EP and ES communities just before this conference, and the similar but independent paths that these two approaches had taken to simulating the process of evolution were clearly apparent. Members of the ES community have participated in all successive EP conferences (Bäck *et al* 1993, Sprave 1994, Bäck and Schütz 1995, Fogel *et al* 1996). There is less similarity between EP and GAs, as the latter emphasize simulating specific mechanisms that apply to natural genetic systems whereas EP emphasizes the behavioral, rather than genetic, relationships between parents and their offspring. Members of the GA and GP communities have, however, also been invited to participate in the annual conferences, making for truly interdisciplinary interaction (see e.g. Altenberg 1994, Land and Belew 1995, Koza and Andre 1996).

Since the early 1990s, efforts in EP have diversified in many directions. Applications in training neural networks have received considerable attention (see e.g. English 1994, Angeline *et al* 1994, McDonnell and Waagen 1994, Porto *et al* 1995), while relatively less attention has been devoted to evolving fuzzy systems (Haffner and Sebald 1993, Kim and Jeon 1996). Image processing applications can be found in the articles by Bhattacharjya and Roysam (1994), Brotherton *et al* (1994), Rizki *et al* (1995), and others. Recent efforts to use EP in medicine have been offered by Fogel *et al* (1995) and Gehlhaar *et al* (1995). Efforts studying and comparing methods of self-adaptation can be found in the articles by Saravanan *et al* (1995), Angeline *et al* (1996), and others. Mathematical analyses of EP have been summarized by Fogel (1995).

To offer a summary, the initial efforts of L J Fogel indicate some of the early attempts to (i) use simulated evolution to perform prediction, (ii) include variable-length encodings, (iii) use representations that take the form of a sequence of instructions, (iv) incorporate a population of candidate solutions, and (v) coevolve evolutionary programs. Moreover, Fogel (1963, 1964) and Fogel *et al* (1966) offered the early recognition that natural evolution and the human endeavor of the scientific method are essentially similar processes, a notion recently echoed by Gell-Mann (1994). The initial prescriptions for operating on finite-state machines have been extended to arbitrary representations, mutation operators, and selection methods, and techniques for self-adapting the evolutionary search have been proposed and implemented. The population size need not be kept constant and there can be a variable number of offspring per parent, much like the $(\mu + \lambda)$ methods (Section 25.4) offered in ESs. In contrast to these methods, selection is often made probabilistic in EP, giving lesser-scoring solutions some probability of surviving as parents into the next generation. In contrast to GAs, no effort is made in EP to support (some say maximize) schema processing, nor is the use of random variation constrained to emphasize specific mechanisms of genetic transfer, perhaps providing greater versatility to tackle specific problem domains that are unsuitable for genetic operators such as crossover.

6.3   Genetic algorithms

The first glimpses of the ideas underlying genetic algorithms (GAs) are found in Holland's papers in the early 1960s (see e.g. Holland 1962). In them Holland set out a broad and ambitious agenda for understanding the underlying principles of adaptive systems—systems that are capable of self-modification in response to their interactions with the environments in which they must function. Such a theory of adaptive systems should facilitate both the understanding of complex forms of adaptation as they appear in natural systems and our ability to design robust adaptive artifacts.

In Holland's view the key feature of robust natural adaptive systems was the successful use of competition and innovation to provide the ability to dynamically respond to unanticipated events and changing environments. Simple models of biological evolution were seen to capture these ideas nicely via notions of survival of the fittest and the continuous production of new offspring.

This theme of using evolutionary models both to understand natural adaptive systems and to design robust adaptive artifacts gave Holland's work a somewhat different focus than those of other contemporary groups that were exploring the use of evolutionary models in the design of efficient experimental optimization techniques (Rechenberg 1965) or for the evolution of intelligent agents (Fogel *et al* 1966), as reported in the previous section.

By the mid-1960s Holland's ideas began to take on various computational forms as reflected by the PhD students working with Holland. From the outset these systems had a distinct 'genetic' flavor to them in the sense that the objects to be evolved over time were represented internally as 'genomes' and the mechanisms of reproduction and inheritance were simple abstractions of familiar population genetics operators such as mutation, crossover, and inversion.

Bagley's thesis (Bagley 1967) involved tuning sets of weights used in the evaluation functions of game-playing programs, and represents some of the earliest experimental work in the use of diploid representations, the role of inversion, and selection mechanisms. By contrast Rosenberg's thesis (Rosenberg 1967) has a very distinct flavor of simulating the evolution of a simple biochemical system in which single-celled organisms capable of producing enzymes were represented in diploid fashion and were evolved over time to produce appropriate chemical concentrations. Of interest here is some of the earliest experimentation with adaptive crossover operators.

Cavicchio's thesis (Cavicchio 1970) focused on viewing these ideas as a form of adaptive search, and tested them experimentally on difficult search problems involving subroutine selection and pattern recognition. In his work we see some of the early studies on elitist (section 28.4) forms of selection and ideas for adapting the rates of crossover and mutation. Hollstien's thesis (Hollstien 1971) took the first detailed look at alternate selection and mating schemes. Using a test suite of two-dimensional fitness landscapes, he experimented with a variety of breeding strategies drawn from techniques used by animal breeders. Also of interest here is Hollstien's use of binary string encodings of the genome and early observations about the virtues of Gray codings.

In parallel with these experimental studies, Holland continued to work on a general theory of adaptive systems (Holland 1967). During this period he developed his now famous schema analysis of adaptive systems, relating it to the optimal allocation of trials using k-armed bandit models (Holland 1969). He used these ideas to develop a more theoretical analysis of his reproductive plans (simple GAs) (Holland 1971, 1973). Holland then pulled all of these ideas together in his pivotal book *Adaptation in Natural and Artificial Systems* (Holland 1975).

Of interest was the fact that many of the desirable properties of these algorithms being identified by Holland theoretically were frequently not observed experimentally. It was not difficult to identify the reasons for this. Hampered by a lack of computational resources and analysis tools, most of the early experimental studies involved a relatively small number of runs using small population sizes (generally less than 20). It became increasingly clear that many of the observed deviations from expected behavior could be traced to the well-known phenomenon in population genetics of *genetic drift*, the loss of genetic diversity due to the stochastic aspects of selection, reproduction, and the like in small populations.

By the early 1970s there was considerable interest in understanding better the behavior of implementable GAs. In particular, it was clear that choices of population size, representation issues, the choice of operators and operator rates all had significant effects of the observed behavior of GAs. Frantz's thesis (Frantz 1972) reflected this new focus by studying in detail the roles of crossover and inversion in populations of size 100. Of interest here is some of the earliest experimental work on multipoint crossover operators.

De Jong's thesis (De Jong 1975) broaded this line of study by analyzing both theoretically and experimentally the interacting effects of population size, crossover, and mutation on the behavior of a family of GAs being used to optimize a fixed test suite of functions. Out of this study came a strong sense that even these simple GAs had significant potential for solving difficult optimization problems.

The mid-1970s also represented a branching out of the family tree of GAs as other universities and research laboratories established research activities in this area. This happened slowly at first since initial attempts to spread the word about the progress being made in GAs were met with fairly negative perceptions from the artificial intelligence (AI) community as a result of early overhyped work in areas such as self-organizing systems and perceptrons.

Undaunted, groups from several universities including the University of Michigan, the University of Pittsburgh, and the University of Alberta organized an *Adaptive Systems Workshop* in the summer of 1976 in Ann Arbor, Michigan. About 20 people attended and agreed to meet again the following summer. This pattern repeated itself for several years, but by 1979 the organizers felt the need to broaden the scope and make things a little more formal. Holland, De Jong, and Sampson obtained NSF funding for *An Interdisciplinary Workshop in Adaptive Systems,* which was held at the University of Michigan in the summer of 1981 (Sampson 1981).

By this time there were several established research groups working on GAs. At the University of Michigan, Bethke, Goldberg, and Booker were continuing to develop GAs and explore Holland's classifier systems (Chapter 12) as part of their PhD research (Bethke 1981, Booker 1982, Goldberg 1983). At the

University of Pittsburgh, Smith and Wetzel were working with De Jong on various GA enhancements including the Pitt approach to rule learning (Smith 1980, Wetzel 1983). At the University of Alberta, Brindle continued to look at optimization applications of GAs under the direction of Sampson (Brindle 1981).

The continued growth of interest in GAs led to a series of discussions and plans to hold the first *International Conference on Genetic Algorithms (ICGA)* in Pittsburgh, Pennsylvania, in 1985. There were about 75 participants presenting and discussing a wide range of new developments in both the theory and application of GAs (Grefenstette 1985). The overwhelming success of this meeting resulted in agreement to continue *ICGA* as a biannual conference. Also agreed upon at *ICGA'85* was the initiation of a moderated electronic discussion group called *GA List*.

The field continued to grow and mature as reflected by the *ICGA* conference activities (Grefenstette 1987, Schaffer 1989) and the appearance of several books on the subject (Davis 1987, Goldberg 1989). Goldberg's book, in particular, served as a significant catalyst by presenting current GA theory and applications in a clear and precise form easily understood by a broad audience of scientists and engineers.

By 1989 the *ICGA* conference and other GA-related activities had grown to a point that some more formal mechanisms were needed. The result was the formation of the International Society for Genetic Algorithms (ISGA), an incorporated body whose purpose is to serve as a vehicle for conference funding and to help coordinate and facilitate GA-related activities. One of its first acts of business was to support a proposal to hold a theory workshop on the *Foundations of Genetic Algorithms (FOGA)* in Bloomington, Indiana (Rawlins 1991).

By this time nonstandard GAs were being developed to evolve complex, nonlinear variable-length structures such as rule sets, LISP code, and neural networks. One of the motivations for *FOGA* was the sense that the growth of GA-based applications had driven the field well beyond the capacity of existing theory to provide effective analyses and predictions.

Also in 1990, Schwefel hosted the first *PPSN* conference in Dortmund, which resulted in the first organized interaction between the ES and GA communities. This led to additional interaction at *ICGA'91* in San Diego which resulted in an informal agreement to hold *ICGA* and *PPSN* in alternating years, and a commitment to jointly initiate a journal for the field.

It was felt that in order for the journal to be successful, it must have broad scope and include other species of EA. Efforts were made to include the EP community as well (which began to organize its own conferences in 1992), and the new journal *Evolutionary Computation* was born with the inaugural issue in the spring of 1993.

The period from 1990 to the present has been characterized by tremendous growth and diversity of the GA community as reflected by the many conference activities (e.g. *ICGA* and *FOGA*), the emergence of new books on GAs, and a growing list of journal papers. New paradigms such as messy GAs (Goldberg *et al* 1991) and genetic programming (Chapter 11) (Koza 1992) were being developed. The interactions with other EC communities resulted in considerable crossbreeding of ideas and many new hybrid EAs. New GA applications continue to be developed, spanning a wide range of problem areas from engineering design problems to operations research problems to automatic programming.

6.4   Evolution strategies

In 1964, three students of the Technical University of Berlin, Bienert, Rechenberg, and Schwefel, did not at all aim at devising a new kind of optimization procedure. During their studies of aerotechnology and space technology they met at an Institute of Fluid Mechanics and wanted to construct a kind of research robot that should perform series of experiments on a flexible slender three-dimensional body in a wind tunnel so as to minimize its drag. The method of minimization was planned to be either a one variable at a time or a discrete gradient technique, gleaned from classical numerics. Both strategies, performed manually, failed, however. They became stuck prematurely when used for a two-dimensional demonstration facility, a joint plate—its optimal shape being a flat plate—with which the students tried to demonstrate that it was possible to find the optimum automatically.

Only then did Rechenberg (1965) hit upon the idea to use dice for random decisions. This was the breakthrough—on 12 June 1964. The first version of an evolutionary strategy (ES), later called the $(1 + 1)$ ES, was born, with discrete, binomially distributed mutations centered at the ancestor's position, and just one parent and one descendant per generation. This ES was first tested on a mechanical calculating machine by Schwefel before it was used for the *experimentum crucis*, the joint plate. Even then, it took a while to overcome a merely locally optimal S shape and to converge towards the expected global optimum, the flat plate. Bienert (1967), the third of the three students, later actually constructed a kind of robot that could perform the actions and decisions automatically.

Using this simple two-membered ES, another student, Lichtfuß (1965), optimized the shape of a bent pipe, also experimentally. The result was rather unexpected, but nevertheless obviously better than all shapes proposed so far.

First computer experiments, on a Zuse Z23, as well as analytical investigations using binomially distributed integer mutations, had already been performed by Schwefel (1965). The main result was that such a strategy can become stuck prematurely, i.e. at 'solutions' that are not even locally optimal. Based on this experience the use of normally instead of binomially distributed mutations became standard in most of the later computer experiments with real-valued variables and in theoretical investigations into the method's efficiency, but not however in experimental optimization using ESs. In 1966 the little ES community was destroyed by dismissal from the Institute of Fluid Mechanics ('Cybernetics as such is no longer pursued at the institute!'). Not before 1970 was it found together again at the Institute of Measurement and Control of the Technical University of Berlin, sponsored by grants from the German Research Foundation (DFG). Due to the circumstances, the group missed publishing its ideas and results properly, especially in English.

In the meantime the often-cited two-phase nozzle optimization was performed at the Institute of Nuclear Technology of the Technical University of Berlin, then in an industrial surrounding, the AEG research laboratory (Schwefel 1968, Klockgether and Schwefel 1970), also at Berlin. For a hot-water flashing flow the shape of a three-dimensional convergent–divergent (thus supersonic) nozzle with maximum energy efficiency was sought. Though in this experimental optimization an exogenously controlled binomial-like distribution was used again, it was the first time that gene duplication and deletion were incorporated into an EA, especially in a $(1 + 1)$ ES, because the optimal length of the nozzle was not known in advance. As in case of the bent pipe this experimental strategy led to highly unexpected results, not easy to understand even afterwards, but definitely much better than available before.

First Rechenberg and later Schwefel analyzed and improved their ES. For the $(1+1)$ ES, Rechenberg, in his Dr.-Ing. thesis of 1971, developed, on the basis of two convex n-dimensional model functions, a convergence rate theory for $n \gg 1$ variables. Based on these results he formulated a $\frac{1}{5}$ success rule for adapting the standard deviation of mutation (Rechenberg 1973). The hope of arriving at an even better strategy by imitating organic evolution more closely led to the incorporation of the population principle and the introduction of recombination, which of course could not be embedded in the $(1+1)$ ES. A first multimembered ES, the $(\mu + 1)$ ES—the notation was introduced later by Schwefel—was also designed by Rechenberg in his seminal work of 1973. Because of its inability to self-adapt the mutation step sizes (more accurately, standard deviations of the mutations), this strategy was never widely used.

Much more widespread became the $(\mu + \lambda)$ ES and (μ, λ) ES, both formulated by Schwefel in his Dr.-Ing. thesis of 1974–1975. It contains theoretical results such as a convergence rate theory for the $(1 + \lambda)$ ES and the $(1, \lambda)$ ES $(\lambda > 1)$, analogous to the theory introduced by Rechenberg for the $(1 + 1)$ ES (Schwefel 1977). The *multimembered* $(\mu > 1)$ ESs arose from the otherwise ineffective incorporation of mutatable mutation parameters (variances and covariances of the Gaussian distributions used). Self-adaptation was achieved with the (μ, λ) ES first, not only with respect to the step sizes, but also with respect to correlation coefficients. The enhanced ES version with correlated mutations, described already in an internal report (Schwefel 1974), was published much later (Schwefel 1981) due to the fact that the author left Berlin in 1976. A more detailed empirical analysis of the on-line self-adaptation of the internal or strategy parameters was first published by Schwefel in 1987 (the tests themselves were secretly performed on one of the first small instruction

multiple data (SIMD) parallel machines (CRAY1) at the Nuclear Research
Centre (KFA) Jülich during the early 1980s with a first parallel version of
the multimembered ES with correlated mutations). It was in this work that the
notion of *self-adaptation by collective learning* first came up. The importance of
recombination (for object as well as strategy parameters) and soft selection (or
$\mu > 1$) was clearly demonstrated. Only recently has Beyer (1995a, b) delivered
the theoretical background to that particularly important issue.

It may be worth mentioning that in the beginning there were strong objections
against increasing λ as well as μ beyond one. The argument against $\lambda > 1$
was that the exploitation of the current knowledge was unnecessarily delayed,
and the argument against $\mu > 1$ was that the survival of inferior members of
the population would unnecessarily slow down the evolutionary progress. The
hint that λ successors could be evaluated in parallel did not convince anybody
since parallel computers were neither available nor expected in the near future.
The two-membered ES and the very similar creeping random search method of
Rastrigin (1965) were investigated thoroughly with respect to their convergence
and convergence rates also by Matyas (1965) in Czechoslovakia, Born (1978)
on the Eastern side of the Berlin wall (!), and Rappl (1984) in Munich.

Since this early work many new results have been produced by the ES
community consisting of the group at Berlin (Rechenberg, since 1972) and that
at Dortmund (Schwefel, since 1985). In particular, strategy variants concerning
other than only real-valued parameter optimization, i.e. real-world problems,
were invented. The first use of an ES for binary optimization using multicellular
individuals was presented by Schwefel (1975). The idea of using several
subpopulations and niching mechanisms for global optimization was propagated
by Schwefel in 1977; due to a lack of computing resources, however, it could
not be tested thoroughly at that time. Rechenberg (1978) invented a notational
scheme for such nested ESs.

Beside these nonstandard approaches there now exists a wide range of
other ESs, e.g. several parallel concepts (Hoffmeister and Schwefel 1990,
Lohmann 1991, Rudolph 1991, 1992, Sprave 1994, Rudolph and Sprave 1995),
ESs for multicriterion problems (Kursawe 1991, 1992), for mixed-integer tasks
(Lohmann 1992, Rudolph 1994, Bäck and Schütz 1995), and even for problems
with a variable-dimensional parameter space (Schütz and Sprave 1996), and
variants concerning nonstandard step size and direction adaptation schemes (see
e.g. Matyas 1967, Stewart *et al* 1967, Fürst *et al* 1968, Heydt 1970, Rappl 1984,
Ostermeier *et al* 1994). Comparisons between ESs, GAs, and EP may be found
in the articles by Bäck *et al* (1991, 1993). It was Bäck (1996) who introduced
a common algorithmic scheme for all brands of current EAs.

Omitting all these other useful nonstandard ESs—a commented collection of
literature concerning ES applications was made at the University of Dortmund
(Bäck *et al* 1992)—the history of ESs is closed with a mention of three recent
books by Rechenberg (1994), Schwefel (1995), and Bäck (1996) as well as
three recent contributions that may be seen as written tutorials (Schwefel and

Rudolph 1995, Bäck and Schwefel 1995, Schwefel and Bäck 1995), which on the one hand define the actual standard ES algorithms and on the other hand present some recent theoretical results.

References

Altenberg L 1994 Emergent phenomena in genetic programming *Proc. 3rd Annu. Conf. on Evolutionary Programming (San Diego, CA, 1994)* ed A V Sebald and L J Fogel (Singapore: World Scientific) pp 233–41

Andersen B, McDonnell J and Page W 1992 Configuration optimization of mobile manipulators with equality constraints using evolutionary programming *Proc. 1st Ann. Conf. on Evolutionary Programming (La Jolla, CA, 1992)* ed D B Fogel and W Atmar (La Jolla, CA: Evolutionary Programming Society) pp 71–9

Angeline P J, Fogel D B and Fogel L J 1996 A comparison of self-adaptation methods for finite state machines in a dynamic environment *Evolutionary Programming V—Proc. 5th Ann. Conf. on Evolutionary Programming (1996)* ed L J Fogel, P J Angeline and T Bäck (Cambridge, MA: MIT Press)

Angeline P J, Saunders G M and Pollack J B 1994 An evolutionary algorithm that constructs recurrent neural networks *IEEE Trans. Neural Networks* **NN-5** 54–65

Atmar J W 1976 *Speculation of the Evolution of Intelligence and Its Possible Realization in Machine Form* ScD Thesis, New Mexico State University

Bäck T 1996 *Evolutionary Algorithms in Theory and Practice* (New York: Oxford University Press)

Bäck T, Hoffmeister F and Schwefel H-P 1991 A survey of evolution strategies *Proc. 4th Int. Conf. on Genetic Algorithms (San Diego, CA, 1991)* ed R K Belew and L B Booker (San Mateo, CA: Morgan Kaufmann) pp 2–9

——1992 *Applications of Evolutionary Algorithms* Technical Report of the University of Dortmund Department of Computer Science Systems Analysis Research Group SYS-2/92

Bäck T, Rudolph G and Schwefel H-P 1993 Evolutionary programming and evolution strategies: similarities and differences *Proc. 2nd Ann. Conf. on Evolutionary Programming (San Diego, CA, 1993)* ed D B Fogel and W Atmar (La Jolla, CA: Evolutionary Programming Society) pp 11–22

Bäck T and Schütz M 1995 Evolution strategies for mixed-integer optimization of optical multilayer systems *Evolutionary Programming IV—Proc. 4th Ann. Conf on Evolutionary Programming (San Diego, CA, 1995)* ed J R McDonnell, R G Reynolds and D B Fogel (Cambridge, MA: MIT Press) pp 33–51

Bäck T and Schwefel H-P 1995 Evolution strategies I: variants and their computational implementation *Genetic Algorithms in Engineering and Computer Science, Proc. 1st Short Course EUROGEN-95* ed G Winter, J Périaux, M Galán and P Cuesta (New York: Wiley) pp 111–26

Bagley J D 1967 *The Behavior of Adaptive Systems which Employ Genetic and Correlation Algorithms* PhD Thesis, University of Michigan

Belew R K and Booker L B (eds) 1991 *Proc. 4th Int. Conf. on Genetic Algorithms (San Diego, CA, 1991)* (San Mateo, CA: Morgan Kaufmann)

Bethke A D 1981 *Genetic Algorithms as Function Optimizers* PhD Thesis, University of Michigan

Beyer H-G 1995a *How GAs do Not Work—Understanding GAs Without Schemata and Building Blocks* Technical Report of the University of Dortmund Department of Computer Science Systems Analysis Research Group SYS-2/95

——1995b Toward a theory of evolution strategies: on the benefit of sex—the $(\mu/\mu, \lambda)$-theory *Evolutionary Comput.* **3** 81–111

Bhattacharjya A K and Roysam B 1994 Joint solution of low-, intermediate- and high-level vision tasks by evolutionary optimization: application to computer vision at low SNR *IEEE Trans. Neural Networks* **NN-5** 83–95

Bienert P 1967 *Aufbau einer Optimierungsautomatik für drei Parameter* Dipl.-Ing. Thesis, Technical University of Berlin, Institute of Measurement and Control Technology

Booker L 1982 *Intelligent Behavior as an Adaptation to the Task Environment* PhD Thesis, University of Michigan

Born J 1978 *Evolutionsstrategien zur numerischen Lösung von Adaptationsaufgaben* PhD Thesis, Humboldt University at Berlin

Box G E P 1957 Evolutionary operation: a method for increasing industrial productivity *Appl. Stat.* **6** 81–101

Box G E P and Draper N P 1969 *Evolutionary Operation. A Method for Increasing Industrial Productivity* (New York: Wiley)

Bremermann H J 1962 Optimization through evolution and recombination *Self-Organizing Systems* ed M C Yovits *et al* (Washington, DC: Spartan)

Bremermann H J, Rogson M and Salaff S 1965 Search by evolution *Biophysics and Cybernetic Systems—Proc. 2nd Cybernetic Sciences Symp.* ed M Maxfield, A Callahan and L J Fogel (Washington, DC: Spartan) pp 157–67

Brindle A 1981 *Genetic Algorithms for Function Optimization* PhD Thesis, University of Alberta

Brotherton T W, Simpson P K, Fogel D B and Pollard T 1994 Classifier design using evolutionary programming *Proc. 3rd Ann. Conf. on Evolutionary Programming (San Diego, CA, 1994)* ed A V Sebald and L J Fogel (Singapore: World Scientific) pp 68–75

Burgin G H 1969 On playing two-person zero-sum games against nonminimax players *IEEE Trans. Syst. Sci. Cybernet.* **SSC-5** 369–70

Cavicchio D J 1970 *Adaptive Search Using Simulated Evolution* PhD Thesis, University of Michigan

Davis L 1987 *Genetic Algorithms and Simulated Annealing* (London: Pitman)

Dearholt D W 1976 Some experiments on generalization using evolving automata *Proc. 9th Int. Conf. on System Sciences (Honolulu, HI)* pp 131–3

De Jong K A 1975 *Analysis of Behavior of a Class of Genetic Adaptive Systems* PhD Thesis, University of Michigan

English T M 1994 Generalization in populations of recurrent neural networks *Proc. 3rd Ann. Conf. on Evolutionary Programming (San Diego, CA, 1994)* ed A V Sebald and L J Fogel (Singapore: World Scientific) pp 26–33

Fogel D B 1988 An evolutionary approach to the traveling salesman problem *Biol. Cybernet.* **60** 139–44

——1989 Evolutionary programming for voice feature analysis *Proc. 23rd Asilomar Conf. on Signals, Systems and Computers (Pacific Grove, CA)* pp 381–3

——1995 *Evolutionary Computation: Toward a New Philosophy of Machine Intelligence* (New York: IEEE)

Fogel D B and Atmar J W 1990 Comparing genetic operators with Gaussian mutations in simulated evolutionary processing using linear systems *Biol. Cybernet.* **63** 111–4

——(eds) 1992 *Proc. 1st Ann. Conf. on Evolutionary Programming (La Jolla, CA, 1992)* (La Jolla, CA: Evolutionary Programming Society)

Fogel D B and Fogel L J 1988 Route optimization through evolutionary programming *Proc. 22nd Asilomar Conf. on Signals, Systems and Computers (Pacific Grove, CA)* pp 679–80

Fogel D B, Fogel L J and Atmar J W 1991 Meta-evolutionary programming *Proc. 25th Asilomar Conf. on Signals, Systems and Computers (Pacific Grove, CA)* ed R R Chen pp 540–5

Fogel D B, Fogel L J, Atmar J W and Fogel G B 1992 Hierarchic methods of evolutionary programming *Proc. 1st Ann. Conf. on Evolutionary Programming (La Jolla, CA, 1992)* ed D B Fogel and W Atmar (La Jolla, CA: Evolutionary Programming Society) pp 175–82

Fogel D B, Fogel L J and Porto V W 1990 Evolving neural networks *Biol. Cybernet.* **63** 487–93

Fogel D B, Wasson E C and Boughton E M 1995 Evolving neural networks for detecting breast cancer *Cancer Lett.* **96** 49–53

Fogel L J 1962 Autonomous automata *Industrial Res.* **4** 14–9

——1963 *Biotechnology: Concepts and Applications* (Englewood Cliffs, NJ: Prentice-Hall)

——1964 *On the Organization of Intellect* PhD Thesis, University of California at Los Angeles

——1968 Extending communication and control through simulated evolution *Bioengineering—an Engineering View Proc. Symp. on Engineering Significance of the Biological Sciences* ed G Bugliarello (San Francisco, CA: San Francisco Press) pp 286–304

Fogel L J, Angeline P J and Bäck T (eds) 1996 *Evolutionary Programming V—Proc. 5th Ann. Conf. on Evolutionary Programming (1996)* (Cambridge, MA: MIT Press)

Fogel L J and Burgin G H 1969 *Competitive Goal-seeking through Evolutionary Programming* Air Force Cambridge Research Laboratories Final Report Contract AF 19(628)-5927

Fogel L J and Fogel D B 1986 *Artificial Intelligence through Evolutionary Programming* US Army Research Institute Final Report Contract PO-9-X56-1102C-1

Fogel L J, Owens A J and Walsh M J 1964 On the evolution of artificial intelligence *Proc. 5th Natl Symp. on Human Factors in Electronics* (San Diego, CA: IEEE)

——1965 Artificial intelligence through a simulation of evolution *Biophysics and Cybernetic Systems* ed A Callahan, M Maxfield and L J Fogel (Washington, DC: Spartan) pp 131–56

——1966 *Artificial Intelligence through Simulated Evolution* (New York: Wiley)

Frantz D R 1972 *Non-linearities in Genetic Adaptive Search* PhD Thesis, University of Michigan

Fraser A S 1957 Simulation of genetic systems by automatic digital computers *Aust. J. Biol. Sci.* **10** 484–99

Friedberg R M 1958 A learning machine: part I *IBM J.* **2** 2–13

Friedberg R M, Dunham B and North J H 1959 A learning machine: part II *IBM J.* **3** 282–7

Fürst H, Müller P H and Nollau V 1968 Eine stochastische Methode zur Ermittlung der Maximalstelle einer Funktion von mehreren Veränderlichen mit experimentell ermittelbaren Funktionswerten und ihre Anwendung bei chemischen Prozessen *Chem.–Tech.* **20** 400–5

Gehlhaar *et al* 1995Gehlhaar D K *et al* 1995 Molecular recognition of the inhibitor AG-1343 by HIV-1 protease: conformationally flexible docking by evolutionary programming *Chem. Biol.* **2** 317–24

Gell-Mann M 1994 *The Quark and the Jaguar* (New York: Freeman)

Goldberg D E 1983 *Computer-Aided Gas Pipeline Operation using Genetic Algorithms and Rule Learning* PhD Thesis, University of Michigan

——1989 *Genetic Algorithms in Search, Optimization and Machine Learning* (Reading, MA: Addison-Wesley)

Goldberg D E, Deb K and Korb B 1991 Don't worry, be messy *Proc. 4th Int. Conf. on Genetic Algorithms (San Diego, CA, 1991)* ed R K Belew and L B Booker (San Mateo, CA: Morgan Kaufmann) pp 24–30

Grefenstette J J (ed) 1985 *Proc. 1st Int. Conf. on Genetic Algorithms and Their Applications (Pittsburgh, PA, 1985)* (Hillsdale, NJ: Erlbaum)

——1987 *Proc. 2nd Int. Conf. on Genetic Algorithms and Their Applications (Cambridge, MA, 1987)* (Hillsdale, NJ: Erlbaum)

Haffner S B and Sebald A V 1993 Computer-aided design of fuzzy HVAC controllers using evolutionary programming *Proc. 2nd Ann. Conf. on Evolutionary Programming (San Diego, CA, 1993)* ed D B Fogel and W Atmar (La Jolla, CA: Evolutionary Programming Society) pp 98–107

Heydt G T 1970 *Directed Random Search* PhD Thesis, Purdue University

Hoffmeister F and Schwefel H-P 1990 A taxonomy of parallel evolutionary algorithms *Parcella '90, Proc. 5th Int. Workshop on Parallel Processing by Cellular Automata and Arrays* vol 2, ed G Wolf, T Legendi and U Schendel (Berlin: Academic) pp 97–107

Holland J H 1962 Outline for a logical theory of adaptive systems *J. ACM* **9** 297–314

——1967 Nonlinear environments permitting efficient adaptation *Computer and Information Sciences II* (New York: Academic)

——1969 Adaptive plans optimal for payoff-only environments *Proc. 2nd Hawaii Int. Conf. on System Sciences* pp 917–20

——1971 Processing and processors for schemata *Associative information processing* ed E L Jacks (New York: Elsevier) pp 127–46

——1973 Genetic algorithms and the optimal allocation of trials *SIAM J. Comput.* **2** 88–105

——1975 *Adaptation in Natural and Artificial Systems* (Ann Arbor, MI: University of Michigan Press)

Hollstien R B 1971 *Artificial Genetic Adaptation in Computer Control Systems* PhD Thesis, University of Michigan

Kim J-H and Jeon J-Y 1996 Evolutionary programming-based high-precision controller design *Evolutionary Programming V—Proc. 5th Ann. Conf. on Evolutionary Programming (1996)* ed L J Fogel, P J Angeline and T Bäck (Cambridge, MA: MIT Press)

Klockgether J and Schwefel H-P 1970 Two-phase nozzle and hollow core jet experiments *Proc. 11th Symp. on Engineering Aspects of Magnetohydrodynamics* ed D G Elliott (Pasadena, CA: California Institute of Technology) pp 141–8

Koza J R 1992 *Genetic Programming* (Cambridge, MA: MIT Press)

Koza J R and Andre D 1996 Evolution of iteration in genetic programming *Evolutionary Programming V—Proc. 5th Ann. Conf. on Evolutionary Programming (1996)* ed L J Fogel, P J Angeline and T Bäck (Cambridge, MA: MIT Press)

Kursawe F 1991 A variant of evolution strategies for vector optimization *Parallel Problem Solving from Nature—Proc. 1st Workshop PPSN I (Lecture Notes in Computer Science 496) (Dortmund, 1991)* ed H-P Schwefel and R Männer (Berlin: Springer) pp 193–7

——1992 Naturanaloge Optimierverfahren—Neuere Entwicklungen in der Informatik *Studien zur Evolutorischen Ökonomik II (Schriften des Vereins für Socialpolitik 195 II)* ed U Witt (Berlin: Duncker and Humblot) pp 11–38

Land M and Belew R K 1995 Towards a self-replicating language for computation *Evolutionary Programming IV—Proc. 4th Ann. Conf on Evolutionary Programming (San Diego, CA, 1995)* ed J R McDonnell, R G Reynolds and D B Fogel (Cambridge, MA: MIT Press) pp 403–13

Larsen R W and Herman J S 1992 A comparison of evolutionary programming to neural networks and an application of evolutionary programming to a navy mission planning problem *Proc. 1st Ann. Conf. on Evolutionary Programming (La Jolla, CA, 1992)* ed D B Fogel and W Atmar (La Jolla, CA: Evolutionary Programming Society) pp 127–33

Lichtfuß H J 1965 *Evolution eines Rohrkrümmers* Dipl.-Ing. Thesis, Technical University of Berlin, Hermann Föttinger Institute for Hydrodynamics

Lohmann R 1991 Application of evolution strategy in parallel populations *Parallel Problem Solving from Nature—Proc. 1st Workshop PPSN I (Dortmund, 1991) (Lecture Notes in Computer Science 496)* ed H-P Schwefel and R Männer (Berlin: Springer) pp 198–208

——1992 Structure evolution and incomplete induction *Parallel Problem Solving from Nature 2 (Brussels, 1992)* ed R Männer and B Manderick (Amsterdam: Elsevier–North-Holland) pp 175–85

Lutter B E and Huntsinger R C 1969 Engineering applications of finite automata *Simulation* **13** 5–11

Männer R and Manderick B (eds) 1992 *Parallel Problem Solving from Nature 2 (Brussels, 1992)* (Amsterdam: Elsevier–North-Holland)

Matyas J 1965 Random optimization *Automation Remote Control* **26** 244–51

——1967 Das zufällige Optimierungsverfahren und seine Konvergenz *Proc. 5th Int. Analogue Computation Meeting (Lausanne, 1967)* **1** 540–4

McDonnell J R 1992 Training neural networks with weight constraints *Proc. 1st Ann. Conf. on Evolutionary Programming (La Jolla, CA, 1992)* ed D B Fogel and W Atmar (La Jolla, CA: Evolutionary Programming Society) pp 111–9

McDonnell J R, Andersen B D, Page W C and Pin F 1992 Mobile manipulator configuration optimization using evolutionary programming *Proc. 1st Ann. Conf. on Evolutionary Programming (La Jolla, CA, 1992)* ed D B Fogel and W Atmar (La Jolla, CA: Evolutionary Programming Society) pp 52–62

McDonnell J R and Waagen D 1994 Evolving recurrent perceptrons for time-series prediction *IEEE Trans. Neural Networks* **NN-5** 24–38

Michalewicz Z *et al* (eds) 1994 *Proc. 1st IEEE Conf. on Evolutionary Computation (Orlando, FL, 1994)* (Piscataway, NJ: IEEE)

Ostermeier A, Gawelczyk A and Hansen N 1994 Step-size adaptation based on non-local use of selection information *Parallel Problem Solving from Nature—PPSN III Int. Conf. on Evolutionary Computation (Jerusalem, 1994) (Lecture notes in Computer Science 866)* ed Y Davidor, H-P Schwefel and R Männer (Berlin: Springer) pp 189–98

Page W C, Andersen B D and McDonnell J R 1992 An evolutionary programming approach to multi-dimensional path planning *Proc. 1st Ann. Conf. on Evolutionary Programming (La Jolla, CA, 1992)* ed D B Fogel and W Atmar (La Jolla, CA: Evolutionary Programming Society) pp 63–70

Porto V W 1992 Alternative methods for training neural networks *Proc. 1st Ann. Conf. on Evolutionary Programming (La Jolla, CA, 1992)* ed D B Fogel and W Atmar (La Jolla, CA: Evolutionary Programming Society) pp 100–10

Porto V W, Fogel D B and Fogel L J 1995 Alternative neural network training methods *IEEE Expert* **10** 16–22

Rappl G 1984 *Konvergenzraten von Random-Search-Verfahren zur globalen Optimierung* PhD Thesis, Bundeswehr University

Rastrigin L A 1965 *Random Search in Optimization Problems for Multiparameter Systems (translated from the Russian original: Sluchainyi poisk v zadachakh optimisatsii mnogoarametricheskikh sistem, Zinatne, Riga)* Air Force System Command Foreign Technology Division FTD-HT-67-363

Rawlins G J E (ed) 1991 *Foundations of Genetic Algorithms* (San Mateo, CA: Morgan Kaufmann)

Rechenberg I 1965 *Cybernetic Solution Path of an Experimental Problem* Royal Aircraft Establishment Library Translation 1122

——1973 *Evolutionsstrategie: Optimierung technischer Systeme nach Prinzipien der biologischen Evolution* (Stuttgart: Frommann–Holzboog)

——1978 Evolutionsstrategien *Simulationsmethoden in der Medizin und Biologie* ed B Schneider and U Ranft (Berlin: Springer) pp 83–114

——1994 *Evolutionsstrategie '94* (Stuttgart: Frommann–Holzboog)

Rizki M M, Tamburino L A and Zmuda M A 1995 Evolution of morphological recognition systems *Evolutionary Programming IV—Proc. 4th Ann. Conf on Evolutionary Programming (San Diego, CA, 1995)* ed J R McDonnell, R G Reynolds and D B Fogel (Cambridge, MA: MIT Press) pp 95–106

Rosenberg R 1967 *Simulation of Genetic Populations with Biochemical Properties* PhD Thesis, University of Michigan

Rudolph G 1991 Global optimization by means of distributed evolution strategies *Parallel Problem Solving from Nature—Proc. 1st Workshop PPSN I (Dortmund, 1991) (Lecture Notes in Computer Science 496)* ed H-P Schwefel and R Männer (Berlin: Springer) pp 209–13

——1992 Parallel approaches to stochastic global optimization *Parallel Computing: from Theory to Sound Practice, Proc. Eur. Workshop on Parallel Computing* ed W Joosen and E Milgrom (Amsterdam: IOS) pp 256–67

——1994 An evolutionary algorithm for integer programming *Parallel Problem Solving from Nature—PPSN III Int. Conf. on Evolutionary Computation (Jerusalem, 1994) (Lecture notes in Computer Science 866)* ed Y Davidor, H-P Schwefel and R Männer (Berlin: Springer) pp 139–48

Rudolph G and Sprave J 1995 A cellular genetic algorithm with self-adjusting acceptance threshold *Proc. 1st IEE/IEEE Int. Conf. on Genetic Algorithms in Engineering*

Systems: Innovations and Applications (GALESIA '95) (Sheffield, 1995) (London: IEE) pp 365–72

Sampson J R 1981 *A Synopsis of the Fifth Annual Ann Arbor Adaptive Systems Workshop* Department of Computing and Communication Science, Logic of Computers Group Technical Report University of Michigan

Saravanan N, Fogel D B and Nelson K M 1995 A comparison of methods for self-adaptation in evolutionary algorithms *BioSystems* **36** 157–66

Satterthwaite F E 1959a Random balance experimentation *Technometrics* **1** 111–37

——1959b *REVOP or Random Evolutionary Operation* Merrimack College Technical Report 10-10-59

Schaffer J D (ed) 1989 *Proc. 3rd Int. Conf. on Genetic Algorithms (Fairfax, WA, 1989)* (San Mateo, CA: Morgan Kaufmann)

Schütz M and Sprave J 1996 Application of parallel mixed-integer evolution strategies with mutation rate pooling *Evolutionary Programming V—Proc. 5th Ann. Conf. on Evolutionary Programming (1996)* ed L J Fogel, P J Angeline and T Bäck (Cambridge, MA: MIT Press)

Schwefel H-P 1965 *Kybernetische Evolution als Strategie der experimentellen Forschung in der Strömungstechnik* Dipl.-Ing. Thesis, Technical University of Berlin, Hermann Föttinger Institute for Hydrodynamics

——1968 *Experimentelle Optimierung einer Zweiphasendüse Teil I* AEG Research Institute Project MHD-Staustrahlrohr 11034/68 Technical Report 35

——1974 *Adaptive Mechanismen in der biologischen Evolution und ihr Einfluß auf die Evolutionsgeschwindigkeit* Technical University of Berlin Working Group of Bionics and Evolution Techniques at the Institute for Measurement and Control Technology Technical Report Re 215/3

——1975 *Binäre Optimierung durch somatische Mutation* Working Group of Bionics and Evolution Techniques at the Institute of Measurement and Control Technology of the Technical University of Berlin and the Central Animal Laboratory of the Medical Highschool of Hannover Technical Report

——1977 *Numerische Optimierung von Computer-Modellen mittels der Evolutionsstrategie (Interdisciplinary Systems Research 26)* (Basle: Birkhäuser)

——1981 *Numerical Optimization of Computer Models* (Chichester: Wiley)

——1987 Collective phenomena in evolutionary systems *Problems of Constancy and Change—the Complementarity of Systems Approaches to Complexity, Papers Presented at the 31st Ann. Meeting Int. Society Gen. Syst. Res.* vol 2, ed P Checkland and I Kiss (Budapest: International Society for General System Research) pp 1025–33

——1995 *Evolution and Optimum Seeking* (New York: Wiley)

Schwefel H-P and Bäck T 1995 Evolution strategies II: theoretical aspects *Genetic Algorithms in Engineering and Computer Science Proc. 1st Short Course EUROGEN-95* ed G Winter, J Périaux, M Galán and P Cuesta (New York: Wiley) pp 127–40

Schwefel H-P and Männer R (eds) 1991 *Parallel Problem Solving from Nature—Proc. 1st Workshop PPSN I (Dortmund, 1991) (Lecture Notes in Computer Science 496)* (Berlin: Springer)

Schwefel H-P and Rudolph G 1995 Contemporary evolution strategies *Advances in Artificial Life—Proc. 3rd Eur. Conf. on Artificial Life (ECAL'95) (Lecture Notes in*

Computer Science 929) ed F Morán, A Moreno, J J Merelo and P Chacón (Berlin: Springer) pp 893–907

Sebald A V and Fogel D B 1992 Design of fault-tolerant neural networks for pattern classification *Proc. 1st Ann. Conf. on Evolutionary Programming (La Jolla, CA, 1992)* ed D B Fogel and W Atmar (La Jolla, CA: Evolutionary Programming Society) pp 90–9

Sebald A V, Schlenzig J and Fogel D B 1992 Minimax design of CMAC encoded neural controllers for systems with variable time delay *Proc. 1st Ann. Conf. on Evolutionary Programming (La Jolla, CA, 1992)* ed D B Fogel and W Atmar (La Jolla, CA: Evolutionary Programming Society) pp 120–6

Smith S F 1980 *A Learning System Based on Genetic Adaptive Algorithms* PhD Thesis, University of Pittsburgh

Spendley W, Hext G R and Himsworth F R 1962 Sequential application of simplex designs in optimisation and evolutionary operation *Technometrics* **4** 441–61

Sprave J 1994 Linear neighborhood evolution strategy *Proc. 3rd Ann. Conf. on Evolutionary Programming (San Diego, CA, 1994)* ed A V Sebald and L J Fogel (Singapore: World Scientific) pp 42–51

Stewart E C, Kavanaugh W P and Brocker D H 1967 Study of a global search algorithm for optimal control *Proc. 5th Int. Analogue Computation Meeting (Lausanne, 1967)* vol 1, pp 207–30

Takeuchi A 1980 Evolutionary automata—comparison of automaton behavior and Restle's learning model *Information Sci.* **20** 91–9

Wetzel A 1983 *Evaluation of the Effectiveness of Genetic Algorithms in Combinatorial Optimization* unpublished manuscript, University of Pittsburgh

7

Introduction to evolutionary algorithms

Thomas Bäck

7.1 General outline of evolutionary algorithms

Since they are gleaned from the model of organic evolution, all basic instances of evolutionary algorithms share a number of common properties, which are mentioned here to characterize the prototype of a general evolutionary algorithm:

(i) Evolutionary algorithms utilize the collective learning process of a population of individuals. Usually, each individual represents (or encodes) a search point in the space of potential solutions to a given problem. Additionally, individuals may also incorporate further information; for example, strategy parameters (Sections 16.2 and 32.2) of the evolutionary algorithm.

(ii) Descendants of individuals are generated by randomized processes intended to model mutation (Chapter 32) and recombination (Chapter 33). Mutation corresponds to an erroneous self-replication of individuals (typically, small modifications are more likely than large ones), while recombination exchanges information between two or more existing individuals.

(iii) By means of evaluating individuals in their environment, a measure of quality or fitness value can be assigned to individuals. As a minimum requirement, a comparison of individual fitness is possible, yielding a binary decision (better or worse). According to the fitness measure, the selection process favors better individuals to reproduce more often than those that are relatively worse.

These are just the most general properties of evolutionary algorithms, and the instances of evolutionary algorithms as described in the following chapters use the components in various different ways and combinations. Some basic differences in the utilization of these principles characterize the mainstream instances of evolutionary algorithms; that is, genetic algorithms (Chapter 8), evolution strategies (Chapter 9), and evolutionary programming (Chapter 10). See D B Fogel (1995) and Bäck (1996) for a detailed overview of similarities and differences of these instances and Bäck and Schwefel (1993) for a brief comparison.

- Genetic algorithms (originally described by Holland (1962, 1975) at Ann Arbor, Michigan, as so-called adaptive or reproductive plans) emphasize recombination (crossover) (Chapter 33) as the most important search operator and apply mutation (Chapter 32) with very small probability solely as a 'background operator.' They also use a probabilistic selection operator (proportional selection) (Chapter 23) and often rely on a binary representation (Chapter 15) of individuals.
- Evolution strategies (developed by Rechenberg (1965, 1973) and Schwefel (1965, 1977) at the Technical University of Berlin) use normally distributed mutations to modify real-valued vectors (Chapter 16) and emphasize mutation (Section 32.2) and recombination (Section 33.2) as essential operators for searching in the search space and in the strategy parameter space at the same time. The selection operator (Chapter 25) is deterministic, and parent and offspring population sizes usually differ from each other.
- Evolutionary programming (originally developed by Lawrence J Fogel (1962) at the University of California in San Diego, as described in Fogel *et al* (1966) and refined by David B Fogel (1992) and others) emphasizes mutation and does not incorporate the recombination of individuals. Similarly to evolution strategies, when approaching real-valued optimization problems, evolutionary programming also works with normally distributed mutations and extends the evolutionary process to the strategy parameters. The selection operator (Section 27.1) is probabilistic, and presently most applications are reported for search spaces involving real-valued vectors, but the algorithm was originally developed to evolve finite-state machines (Chapter 18) .

In addition to these three mainstream methods, which are described in detail in the next three chapters, genetic programming, classifier systems, and hybridizations of evolutionary algorithms with other techniques are considered in chapters 11–13, respectively. As an introductory remark, we only mention that genetic programming applies the evolutionary search principle to automatically develop computer programs in suitable languages (Chapter 10) (often LISP, but others are possible as well), while classifier systems search the space of production rules (or sets of rules) of the form 'IF <condition> THEN <action>'.

A variety of different representations of individuals and corresponding operators are presently known in evolutionary algorithm research, and it is the aim of Chapters 14–34 to present all these in detail. Here, we will use these chapters as a construction kit to assemble the basic instances of evolutionary algorithms.

As a general framework for these basic instances, we define I to denote an arbitrary space of individuals $a \in I$, and $F : I \to \mathbb{R}$ to denote a real-valued fitness function of individuals. Using μ and λ to denote parent and offspring population sizes, $P(t) = (a_1(t), \ldots, a_\mu(t)) \in I^\mu$ characterizes a population at generation t. Selection, mutation, and recombination are described as operators

$s : I^\lambda \rightarrow I^\mu$, $m : I^\kappa \rightarrow I^\lambda$, and $r : I^\mu \rightarrow I^\kappa$ that transform complete populations. By describing all operators on the population level (though this is counterintuitive for mutation), a high-level perspective is adopted, which is sufficently general to cover different instances of evolutionary algorithms. For mutation, the operator can of course be reduced to the level of single individuals by defining m through a multiple application of a suitable operator $m' : I \rightarrow I$ on individuals.

These operators typically depend on additional sets of parameters Θ_s, Θ_m, and Θ_r which are characteristic for the operator and the representation of individuals. Additionally, an initialization procedure generates a population of individuals (typically at random, but an initialization with known starting points should of course also be possible), an evaluation routine determines the fitness values of the individuals of a population, and a termination criterion is applied to determine whether or not the algorithm should stop.

Putting all this together, a basic evolutionary algorithm reduces to the simple recombination–mutation–selection loop as outlined below:

Input: $\mu, \lambda, \Theta_\iota, \Theta_r, \Theta_m, \Theta_s$
Output: a^*, the best individual found during the run, or
             P^*, the best population found during the run.
1     $t \leftarrow 0$;
2     $P(t) \leftarrow$ initialize(μ);
3     $F(t) \leftarrow$ evaluate($P(t), \mu$);
4   **while** ($\iota(P(t), \Theta_\iota) \neq$ true) **do**
5             $P'(t) \leftarrow$ recombine($P(t), \Theta_r$);
6             $P''(t) \leftarrow$ mutate($P'(t), \Theta_m$);
7             $F(t) \leftarrow$ evaluate($P''(t), \lambda$);
8             $P(t+1) \leftarrow$ select($P''(t), F(t), \mu, \Theta_s$);
9             $t \leftarrow t + 1$;
      od

After initialization of t (line 1) and the population $P(t)$ of size μ (line 2) as well as its fitness evaluation (line 3), the while-loop is entered. The termination criterion ι might depend on a variety of parameters, which are summarized here by the argument Θ_ι. Similarly, recombination (line 5), mutation (line 6), and selection (line 8) depend on a number of algorithm-specific additional parameters. While $P(t)$ consists of μ individuals, $P'(t)$ and $P''(t)$ are assumed to be of size κ and λ, respectively. Of course, $\lambda = \kappa = \mu$ is allowed and is the default case in genetic algorithms. The setting $\kappa = \mu$ is also often used in evolutionary programming (without recombination), but it depends on the application and the situation is quickly changing. Either recombination or mutation might be absent from the main loop, such that $\kappa = \mu$ (absence of recombination) or $\kappa = \lambda$ (absence of mutation) is required in these cases. The selection operator selects μ individuals from $P''(t)$ according to the fitness values $F(t)$, t is incremented (line 9), and the body of the main loop is repeated.

The input parameters of this general evolutionary algorithm include the population sizes μ and λ as well as the parameter sets Θ_ι, Θ_r, Θ_m, and Θ_s of the basic operators. Notice that we allow recombination to equal the identity mapping; that is, $P''(t) = P'(t)$ is possible.

The following sections of this chapter present the common evolutionary algorithms as particular instances of the general scheme.

References

Bäck T 1996 *Evolutionary Algorithms in Theory and Practice* (New York: Oxford University Press)

Bäck T and Schwefel H-P 1993 An overview of evolutionary algorithms for parameter optimization *Evolutionary Computation* **1**(1) 1–23

Fogel D B 1992 *Evolving Artificial Intelligence* PhD Thesis, University of California, San Diego

——1995 *Evolutionary Computation: Toward a New Philosophy of Machine Intelligence* (Piscataway, NJ: IEEE)

Fogel L J 1962 Autonomous automata *Industr. Res.* **4** 14–9

Fogel L J, Owens A J and Walsh M J 1966 *Artificial Intelligence through Simulated Evolution* (New York: Wiley)

Holland J H 1962 Outline for a logical theory of adaptive systems *J. ACM* **3** 297–314

——1975 *Adaptation in Natural and Artificial Systems* (Ann Arbor, MI: University of Michigan Press)

Rechenberg I 1965 Cybernetic solution path of an experimental problem *Library Translation No 1122* Royal Aircraft Establishment, Farnborough, UK

——1973 *Evolutionsstrategie: Optimierung technischer Systeme nach Prinzipien der biologischen Evolution* (Stuttgart: Frommann-Holzboog)

Schwefel H-P 1965 *Kybernetische Evolution als Strategie der experimentellen Forschung in der Strömungstechnik* Diplomarbeit, Technische Universität, Berlin

——1977 *Numerische Optimierung von Computer-Modellen mittels der Evolutionsstrategie* Interdisciplinary Systems Research, vol 26 (Basel: Birkhäuser)

Further reading

The introductory section to evolutionary algorithms certainly provides the right place to mention the most important books on evolutionary computation and its subdisciplines. The following list is not intended to be complete, but only to guide the reader to the literature.

1. Bäck T 1996 *Evolutionary Algorithms in Theory and Practice* (New York: Oxford University Press)

    A presentation and comparison of evolution strategies, evolutionary programming, and genetic algorithms with respect to their behavior as parameter optimization methods. Furthermore, the role of mutation and selection in genetic algorithms is discussed in detail, arguing that mutation is much more useful than usually claimed in connection with genetic algorithms.

2. Goldberg D E 1989 *Genetic Algorithms in Search, Optimization, and Machine Learning* (Reading, MA: Addison-Wesley)

    An overview of genetic algorithms and classifier systems, discussing all important techniques and operators used in these subfields of evolutionary computation.

3. Rechenberg I 1994 *Evolutionsstrategie '94* Werkstatt Bionik und Evolutionstechnik, vol 1 (Stuttgart: Frommann-Holzboog)

    A description of evolution strategies in the form used by Rechenberg's group in Berlin, including a reprint of (Rechenberg 1973).

4. Schwefel H-P 1995 *Evolution and Optimum Seeking* Sixth-Generation Computer Technology Series (New York: Wiley)

    The most recent book on evolution strategies, covering the (μ, λ)-strategy and all aspects of self-adaptation of strategy parameters as well as a comparison of evolution strategies with classical optimization methods.

5. Fogel D B 1995 *Evolutionary Computation: Toward a New Philosophy of Machine Intelligence* (Piscataway, NJ: IEEE)

    The book covers all three main areas of evolutionary computation (i.e. genetic algorithms, evolution strategies, and evolutionary programming) and discusses the potential for using simulated evolution to achieve machine intelligence.

6. Michalewicz Z 1994 *Genetic Algorithms + Data Structures = Evolution Programs* (Berlin: Springer)

    Michalewicz also takes a more general view at evolutionary computation, thinking of evolutionary heuristics as a principal method for search and optimization, which can be applied to any kind of data structure.

7. Kinnear K E 1994 *Advances in Genetic Programming* (Cambridge, MA: MIT Press)

    This collection of articles summarizes the state of the art in genetic programming, emphasizing other than LISP-based approaches to genetic programming.

8. Koza J R 1992 *Genetic Programming: On the Programming of Computers by Means of Natural Selection* (Cambridge, MA: MIT Press)

9. Koza J R 1994 *Genetic Programming II* (Cambridge, MA: MIT Press)

    The basic books for genetic programming using LISP programs, demonstrating the feasibility of the method by presenting a variety of application examples from diverse fields.

8

Genetic algorithms

Larry J Eshelman

8.1 Introduction

Genetic algorithms (GAs) are a class of evolutionary algorithms first proposed
and analyzed by John Holland (1975). There are three features which distinguish
GAs, as first proposed by Holland, from other evolutionary algorithms: (i)
the representation used—bitstrings (Chapter 15); (ii) the method of selection—
proportional selection (Chapter 23) ; and (iii) the primary method of producing
variations—crossover (Chapter 33). Of these three features, however, it
is the emphasis placed on crossover which makes GAs distinctive. Many
subsequent GA implementations have adopted alternative methods of selection,
and many have abandoned bitstring representations for other representations
more amenable to the problems being tackled. Although many alternative
methods of crossover have been proposed, in almost every case these variants
are inspired by the spirit which underlies Holland's original analysis of GA
behavior in terms of the processing of schemata or building blocks. It should be
pointed out, however, that the evolution strategy paradigm (Chapter 9) has added
crossover to its repertoire, so that the distinction between classes of evolutionary
algorithms has become blurred (Bäck *et al* 1991).

We shall begin by outlining what might be called the canonical GA, similar
to that described and analyzed by Holland (1975) and Goldberg (1987). We
shall introduce a framework for describing GAs which is richer than needed
but which is convenient for describing some variations with regard to the
method of selection. First we shall introduce some terminology. The individual
structures are often referred to as chromosomes. They are the genotypes that
are manipulated by the GA. The evaluation routine decodes these structures
into some phenotypical structure and assigns a fitness value. Typically, but not
necessarily, the chromosomes are bitstrings. The value at each locus on the
bitstring is referred to as an allele. Sometimes the individuals loci are also
called genes. At other times genes are combinations of alleles that have some
phenotypical meaning, such as parameters.

8.2 Genetic algorithm basics and some variations

An initial population of individual structures $P(0)$ is generated (usually randomly) and each individual is evaluated for fitness. Then some of these individuals are selected for mating and copied (select_repro) to the mating buffer $C(t)$. In Holland's original GA, individuals are chosen for mating probabilistically, assigning each individual a probability proportional to its observed performance. Thus, better individuals are given more opportunities to produce offspring (reproduction with emphasis). Next the genetic operators (usually mutation and crossover) are applied to the individuals in the mating buffer, producing offspring $C'(t)$. The rates at which mutation and crossover are applied are an implementation decision. If the rates are low enough, it is likely that some of the offspring produced will be identical to their parents. Other implementation details are how many offspring are produced by crossover (one or two), and how many individuals are selected and paired in the mating buffer. In Holland's original description, only one pair is selected for mating per cycle. The pseudocode for the genetic algorithm is as follows:

```
begin
    t = 0;
    initialize P(t);
    evaluate structures in P(t);
    while termination condition not satisfied do
    begin
        t = t + 1;
        select_repro C(t) from P(t-1);
        recombine and mutate structures in C(t)
        forming C'(t);
        evaluate structures in C'(t);
        select_replace P(t) from C'(t) and P(t-1);
    end
end
```

After the new offspring have been created via the genetic operators the two populations of parents and children must be merged to create a new population. Since most GAs maintain a fixed-sized population M, this means that a total of M individuals need to be selected from the parent and child populations to create a new population. One possibility is to use all the children generated (assuming that the number is not greater than M) and randomly select (without any bias) individuals from the old population to bring the new population up to size M. If only one or two new offspring are produced, this in effect means randomly replacing one or two individuals in the old population with the new offspring. (This is what Holland's original proposal did.) On the other hand, if the number of offspring created is equal to M, then the old parent population is completely replaced by the new population.

There are several opportunities for biasing selection: selection for reproduction (or mating) and selection from the parent and child populations to produce the new population. The GAs most closely associated with Holland do all their biasing at the reproduction selection stage. Even among these GAs, however, there are a number of variations. If reproduction with emphasis is used, then the probability of an individual being chosen is a function of its observed fitness. A straightforward way of doing this would be to total the fitness values assigned to all the individuals in the parent population and calculate the probability of any individual being selected by dividing its fitness by the total fitness. One of the properties of this way of assigning probabilities is that the GA will behave differently on functions that seem to be equivalent from an optimization point of view such as $y = ax^2$ and $y = ax^2 + b$. If the b value is large in comparison to the differences in the value produced by the ax^2 term, then the differences in the probabilities for selecting the various individuals in the population will be small, and selection pressure will be very weak. This often happens as the population converges upon a narrow range of values. One way of avoiding this behavior is to scale the fitness function, typically to the worst individual in the population (De Jong 1975). Hence the measure of fitness used in calculating the probability for selecting an individual is not the individual's absolute fitness, but its fitness relative to the worst individual in the population.

Although scaling can eliminate the problem of not enough selection pressure, often GAs using fitness proportional selection suffer from the opposite problem—too much selection pressure. If an individual is found which is much better than any other, the probability of selecting this individual may become quite high (especially if scaling to the worst is used). There is the danger that many copies of this individual will be placed in the mating buffer, and this individual (and its similar offspring) will rapidly take over the population (premature convergence). One way around this is to replace fitness proportional selection with ranked selection (Whitley 1989). The individuals in the parent population are ranked, and the probability of selection is a linear function of rank rather than fitness, where the 'steepness' of this function is an adjustable parameter.

Another popular method of performing selection is tournament selection (Goldberg and Deb 1991). A small subset of individuals is chosen at random, and then the best individual (or two) in this set is (are) selected for the mating buffer. Tournament selection, like rank selection, is less subject to rapid takeover by good individuals, and the selection pressure can be adjusted by controlling the size of the subset used.

Another common variation of those GAs that rely upon reproduction selection for their main source of selection bias is to maintain one copy of the best individual found so far (De Jong 1975). This is referred to as the elitist strategy (Section 28.4). It is actually a method of biased parent selection, where the best member of the parent population is chosen and all but one of the M members of the child population are chosen. Depending upon the implementation, the selection of the child to be replaced by the best individual from the parent population may or may not be biased.

A number of GA variations make use of biased replacement selection. Whitley's GENITOR, for example, creates one child each cycle, selecting the parents using ranked selection, and then replacing the worst member of the population with the new child (Whitley 1989). Syswerda's steady-state GA creates two children each cycle, selecting parents using ranked selection, and then stochastically choosing two individuals to be replaced, with a bias towards the worst individuals in the parent population (Syswerda 1989). Eshelman's CHC uses unbiased reproductive selection by randomly pairing all the members of the parent population, and then replacing the worst individuals of the parent population with the better individuals of the child population. (In effect, the offspring and parent populations are merged and the best M (population size) individuals are chosen.) Since the new offspring are only chosen by CHC if they are better than the members of the parent population, the selection of both the offspring and parent populations is biased (Eshelman 1991).

These methods of replacement selection, and especially that of CHC, resemble the $(\mu + \lambda)$ ES method of selection (Section 25.4) sometimes originally used by evolution strategies (ESs) (Bäck *et al* 1991). From μ parents λ offspring are produced; the μ parents and λ offspring are merged; and the best μ individuals are chosen to form the new parent population. The other ES selection method, (μ, λ) ES (Section 25.4), places all the bias in the child selection stage. In this case, μ parents produce λ offspring $(\lambda > \mu)$, and the best μ offspring are chosen to replace the parent population. Mühlenbein's breeder GA also uses this selection mechanism (Mühlenbein and Schlierkamp-Voosen 1993).

Often a distinction is made between generational and *steady-state* GAs (Section 28.3). Unfortunately, this distinction tends to merge two properties that are quite independent: whether the replacement strategy of the GA is biased or not and whether the GA produces one (or two) versus many (usually M) offspring each cycle. Syswerda's steady-state GA, like Whitley's GENITOR, allows only one mating per cycle and uses a biased replacement selection, but there are also GAs that combine multiple matings per cycle with biased replacement selection (CHC) as well as a whole class of ESs $((\mu + \lambda)$ ES). Furthermore, the GA described by Holland (1975) combined a single mating per cycle and unbiased replacement selection. Of these two features, it would seem that the most significant is the replacement strategy. De Jong and Sarma (1993) found that the main difference between GAs allowing many matings versus few matings per cycle is that the latter have a higher variance in performance.

The choice between a biased and an unbiased replacement strategy, on the other hand, is a major determinant of GA behavior. First, if biased replacement is used in combination with biased reproduction, then the problem of premature convergence is likely to be compounded. (Of course this will depend upon other factors, such as the size of the population, whether ranked selection is used, and, if so, the setting of the selection bias parameter.) Second, the obvious shortcoming of unbiased replacement selection can turn out to be a strength. On the negative side, replacing the parents by the children, with no mechanism for keeping those parents that are better than any of the children, risks losing, perhaps forever, very good individuals. On the other hand, replacing the parents by the children can allow the algorithm to wander, and it may be able to wander out of a local minimum that would trap a GA relying upon biased replacement selection. Which is the better strategy cannot be answered except in the context of the other mechanisms of the algorithm (as well as the nature of the problem being solved). Both Syswerda's steady-state GA and Whitley's GENITOR combine a biased replacement strategy with a mechanism for eliminating children which are duplicates of any member in the parent population. CHC uses unbiased reproductive selection, relying solely upon biased replacement selection as its only source of selection pressure, and uses several mechanisms for maintaining diversity (not mating similar individuals and seeded restarts), which allow it to take advantage of the preserving properties of a deterministic replacement strategy without suffering too severely from its shortcomings.

8.3   Mutation and crossover

All evolutionary algorithms work by combining selection with a mechanism for producing variations. The best known mechanism for producing variations is mutation, where one allele of a gene is randomly replaced by another. In other words, new trial solutions are created by making small, random changes in the representation of prior trial solutions. If a binary representation is used, then mutation is achieved by 'flipping' bits at random. A commonly used rate of mutation is one over the string length. For example, if the chromosome is one hundred bits long, then the mutation rate is set so that each bit has a probability of 0.01 of being flipped.

Although most GAs use mutation along with crossover, mutation is sometimes treated as if it were a background operator for assuring that the population will consist of a diverse pool of alleles that can be exploited by crossover. For many optimization problems, however, an evolutionary algorithm using mutation without crossover can be very effective (Mathias and Whitley 1994). This is not to suggest that crossover never provides an added benefit, but only that one should not disparage mutation.

The intuitive idea behind crossover is easy to state: given two individuals who are highly fit, but for different reasons, ideally what we would like to do is create a new individual that combines the best features from each. Of course, since we presumably do not know which features account for the good performance (if we did we would not need a search algorithm), the best we can do is to recombine features at random. This is how crossover operates. It treats these features as building blocks scattered throughout the population and tries to recombine them into better individuals via crossover. Sometimes crossover will combine the worst features from the two parents, in which case these children will not survive for long. But sometimes it will recombine the best features from two good individuals, creating even better individuals, provided these features are compatible.

Suppose that the representation is the classical bitstring representation: individual solutions in our population are represented by binary strings of zeros and ones of length L. A GA creates new individuals via crossover by choosing two strings from the parent population, lining them up, and then creating two new individuals by swapping the bits at random between the strings. (In some GAs only one individual is created and evaluated, but the procedure is essentially the same.) Holland originally proposed that the swapping be done in segments, not bit by bit. In particular, he proposed that a single locus be chosen at random and all bits after that point be swapped. This is known as one-point crossover. Another common form of crossover is two-point crossover which involves choosing two points at random and swapping the corresponding segments from the two parents defined by the two points. There are of course many possible variants. The best known alternative to one- and two-point crossover is *uniform crossover*. Uniform crossover randomly swaps individual bits between the two parents (i.e. exchanges between the parents the values at loci chosen at random).

Following Holland, GA behavior is typically analyzed in terms of schemata. Given a space of structures represented by bitstrings of length L, schemata represent partitions of the search space. If the bitstrings of length L are interpreted as vectors in a L-dimensional hypercube, then schemata are hyperplanes of the space. A schema can be represented by a string of L symbols from the set $0, 1, \#$ where $\#$ is a 'wildcard' matching either 0 or 1. Each string of length L may be considered a sample from the partition defined by a schema if it matches the schema at each of the defined positions (i.e. the non-# loci). For example, the string 011001 instantiates the schema 01##0#. Each string, in fact, instantiates 2^L schemata.

Two important schema properties are order and defining length. The order of a schema is the number of defined loci (i.e. the number of non-# symbols). For example the schema #01##1### is an order 3 schema. The defining length is the distance between the loci of the first and last defined positions. The defining length of the above schema is four since the loci of the first and last defined positions are 2 and 6.

From the hyperplane analysis point of view, a GA can be interpreted as focusing its search via crossover upon those hyperplane partition elements that have on average produced the best-performing individuals. Over time the search becomes more and more focused as the population converges since the degree of variation used to produce new offspring is constrained by the remaining variation in the population. This is because crossover has the property that Radcliffe refers to as respect—if two parents are instances of the same schema, the child will also be an instance (Radcliffe 1991). If a particular schema conveys high fitness values to its instances, then the population is likely to converge on the defining bits of this schema. Once it so converges, all offspring will be instances of this schema. This means that as the population converges, the search becomes more and more focused on smaller and smaller partitions of the search space.

It is useful to contrast crossover with mutation in this regard. Whereas mutation creates variations by flipping bits randomly, crossover is restricted to producing variations at those loci on which the population has not yet converged. Thus crossover, and especially bitwise versions of crossover, can be viewed as a form of adaptive mutation, or convergence-controlled variation (CCV).

The standard explanation of how GAs operate is often referred to as the building block hypothesis. According to this hypothesis, GAs operate by combining small building blocks into larger building blocks. The intuitive idea behind recombination is that by combining features (or building blocks) from two good parents crossover will often produce even better children; for example, a mother with genes for sharp teeth and a father with genes for sharp claws will have the potential of producing some children who have both features. More formally, the building blocks are the schemata discussed above.

Loosely interpreted, the building block hypothesis is another way of asserting that GAs operate through a process of CCV. The building block hypothesis, however, is often given a stronger interpretation. In particular, crossover is seen as having the added value of being able to recombine middle-level building blocks that themselves cannot be built from lower-level building blocks (where level refers to either the defining length or order, depending on the crossover operator). We shall refer to this explanation as to how GAs work as the strict building block hypothesis (SBBH), and contrast it with the weaker convergence-controlled variation hypothesis (CCVH).

To differentiate these explanations, it is useful to compare crossover with an alternative mechanism for achieving CCV. Instead of pairing individuals and swapping segments or bits, a more direct method of generating CCVs is to use the distribution of the allele values in the population to generate new offspring. This is what Syswerda's bitwise simulated crossover (BSC) algorithm does (Syswerda 1993). In effect, the distribution of allele values is used to generate a vector of allele probabilities, which in turn is used to generate a string of ones and zeros. Baluja's PBIL goes one step further and eliminates the population, and simply keeps a probability vector of allele values, using an update rule to modify it based on the fitness of the samples generated (Baluja 1995).

The question is, if one wants to take advantage of CCV with its ability to adapt, why use crossover, understood as involving pairwise mating, rather than one of these poolwise schemes? One possible answer is that the advantage is only one of implementation. The pairwise implementation does not require any centralized bookkeeping mechanism. In other words, crossover (using pairwise mating) is simply nature's way of implementing a decentralized version of CCV.

A more theoretically satisfying answer is that pairwise mating is better able to preserve essential linkages among the alleles. One manifestation of this is that there is no obvious way to implement a segment-based version of poolwise mating, but this point also applies if we compare poolwise mating with only crossover operators that operate at the bit level, such as uniform crossover. If two allele values are associated in some individual, the probability of these values being associated in the children is much higher for pairwise mating than poolwise. To see this consider an example. Suppose the population size is 100, and that an individual of average fitness has some unique combination of allele values, say all ones in the first three positions. This individual will have a 0.01 probability (one out of 100) of being selected for mating, assuming it is of average fitness. If uniform crossover is being used, with a 0.5 probability of swapping the values at each locus, and one offspring is being produced per mating, then the probability of the three allele values being propagated without disruption has a lower bound of 0.125 (0.5^3). This is assuming the worst-case scenario that every other member in the population has all zeros in the first three positions (and ignoring the possibility of mating this individual with a copy of itself). Thus, the probability of propagating this schema is 0.00125 ($0.01 * 0.125$). On the other hand, if BSC is being used, then the probability of propagating this schema is much lower. Since there is only one instance of this individual in the population, there is only one chance in 100 of propagating each allele and only 0.000 001 (0.01^3) of propagating all three.

Ultimately, one is faced with a tradeoff: the enhanced capability of pairwise mating to propagate difficult-to-find schemata is purchased at the risk of increased hitchhiking; that is, the population may prematurely converge on bits that do not convey additional fitness but happen to be present in the individuals that are instances of good schemata. According to both the CCVH and the SBBH, crossover must not simply preserve and propagate good schemata, but must also recombine them with other good schemata. Recombination, however, requires that these good schemata be tried in the context of other schemata. In order to determine which schemata are the ones contributing to fitness, we must test them in many different contexts, and this involves prying apart the defining positions that contribute to fitness from those that are spurious, but the price for this reduced hitchhiking is higher disruption (the breaking up of the good schemata). This price will be too high if the algorithm cannot propagate critical, highly valued, building blocks or, worse yet, destroys them in the next crossover cycle.

This tradeoff applies not only to the choice between poolwise and pairwise methods of producing variation, but also to the choice between various methods of crossover. Uniform crossover, for example, is less prone to hitchhiking than two-point crossover, but is also more disruptive, and poolwise mating schemes are even more disruptive than uniform crossover. In Holland's original analysis this tradeoff between preserving the good schemata while performing vigorous recombination is downplayed by using a segment-based crossover operator such as one- or two-point crossover and assuming that the important building blocks are of short defining length. Unfortunately, for the types of problem to which GAs are supposedly ideally suited—those that are highly complex with no tractable analytical solution—there is no *a priori* reason to assume that the problem will, or even can, be represented so that important building blocks will be those with short defining length. To handle this problem Holland proposed an inversion operator that could reorder the loci on the string, and thus be capable of finding a representation that had building blocks with short defining lengths. The inversion operator, however, has not proven sufficiently effective in practice at recoding strings on the fly. To overcome this linkage problem, Goldberg has proposed what he calls messy GAs, but, before discussing messy GAs, it will be helpful to describe a class of problems that illustrate these linkage issues: deceptive problems.

Deception is a notion introduced by Goldberg (1987). Consider two incompatible schemata, A and B. A problem is deceptive if the average fitness of A is greater than B even though B includes a string that has a greater fitness than any member of A. In practice this means that the lower-order building blocks lead the GA away from the global optimum. For example, consider a problem consisting of five-bit segments for which the fitness of each is determined as follows (Liepins and Vose 1991). For each *one* the segment receives a point, and thus five points for all *ones*, but for all *zeros* it receives a value greater than five. For problems where the value of the optimum is between five and eight the problem is fully deceptive (i.e. all relevant lower-order hyperplanes lead toward the deceptive attractor). The total fitness is the sum of the fitness of the segments.

It should be noted that it is probably a mistake to place too much emphasis on the formal definition of deception (Grefenstette 1993). What is really important is the concept of being misled by the lower-order building blocks. Whereas the formal definition of deception stresses the average fitness of the hyperplanes taken over the entire search space, selection only takes into account the observed average fitness of hyperplanes (those in the actual population). The interesting set of problems is those that are misleading in that manipulation of the lower-order building blocks is likely to lead the search away from the middle-level building blocks that constitute the optimum solution, whether these middle-level building blocks are deceptive in the formal sense or not. In the above class of functions, even when the value of the optimum is greater than eight (and so not fully deceptive), but still not very large, e.g. ten, the problem is solvable by a GA using segment-based crossover, very difficult for a GA using bitwise uniform crossover, and all but impossible for a poolwise-based algorithm like BSC.

As long as the deceptive problem is represented so that the loci of the positions defining the building blocks are close together on the string, it meets Holland's original assumption that the important building blocks are of short defining length. The GA will be able to exploit this information using one- or two-point crossover—the building blocks will have a low probability of being disrupted, but will be vigorously recombined with other building blocks along the string. If, on the other hand, the bits constituting the deceptive building blocks are maximally spread out on the chromosome, then a crossover operator such as one- or two-point crossover will tend to break up the good building blocks. Of course, maximally spreading the deceptive bits along the string is the extreme case, but bunching them together is the opposite extreme.

Since one is not likely to know enough about the problem to be able to guarantee that the building blocks are of short defining length, segmented crossover loses its advantage over bitwise crossover. It is true that bitwise crossover operators are more disruptive, but there are several solutions to this problem. First, there are bitwise crossover operators that are much less disruptive than the standard uniform crossover operator (Spears and De Jong 1991, Eshelman and Schaffer 1995). Second, the problem of preservation can often be ameliorated by using some form of replacement selection so that good individuals survive until they are replaced by better individuals (Eshelman and Schaffer 1995). Thus a disruptive form of crossover such as uniform crossover can be used and good schemata can still be preserved. Uniform crossover will still make it difficult to propagate these high-order, good schemata once they are found, but, provided the individuals representing these schemata are not replaced by better individuals that represent incompatible schemata, they will be preserved and may eventually be able to propagate their schemata on to their offspring. Unfortunately, this proviso is not likely to be met by any but low-order deceptive problems. Even for deceptive problems of order five, the difficulty of propagating optimal schemata is such that the suboptimal schemata tend to crowd out the optimum ones.

Perhaps the ultimate GA for tackling deceptive problems is Goldberg's messy GA (mGA) (Goldberg *et al* 1991). Whereas in more traditional GAs the manipulation of building blocks is implicit, mGAs explicitly manipulate the building blocks. This is accomplished by using variable-length strings that may be underspecified or overspecified; that is, some bit positions may not be defined, and some positions may have conflicting specifications. This is what makes mGAs messy.

These strings constitute the building blocks. They consist of a set of position–value pairs. Overspecified strings are evaluated by a simple conflict resolution strategy such as first-come-first-served rules. Thus, ((1 0) (2 1) (1 1) (3 0)) would be interpreted as 010, ignoring the third pair, since the first position has already been defined. Underspecified strings are interpreted by filling in the missing values using a competitive template, a locally optimal structure. For example, if the locally optimal structure, found by testing one bit at a time, is 111, then the string ((1 0) (3 0)) would be interpreted by filling in the value for the (missing) second position with the value of the second position in the template. The resulting 010 string would then be evaluated.

mGAs have an outer and an inner loop. The inner loop consists of three phases: the initialization, primordial, and juxtaposition phases. In the initialization phase all substrings of length k are created and evaluated, i.e. all combinations of strings with k defining positions (where k is an estimate of the highest order of deception in the problem). As was explained above the missing values are filled in using the competitive template. (As will be explained below, the template for the k level of the outer loop is the solution found at the $k - 1$ level.)

In the primordial phase, selection is applied to the population of individuals produced during the initialization phase without any operators. Thus the substrings that have poor evaluations are eliminated and those with good evaluations have multiple copies in the resulting population.

In the juxtapositional phase selection in conjunction with cut and splice operators is used to evolve improved variations. Again, the competitive template is used for filling in missing values, and the first-come-first-served rule is used for handling overspecified strings created by the splice operator. The cut and splice operators act much like one-point crossover in a traditional GA, keeping in mind that the strings are of variable length and may be underspecified or overspecified.

The outer loop is over levels. It starts at the level of $k = 1$, and continues through each level until it reaches a user-specified stopping criterion. At each level, the solution found at the previous level is used as the competitive template.

One of the limitations of mGAs as originally conceived is that the initialization phase becomes extremely expensive as the mGA progresses up the levels. A new variant of the mGA speeds up the process by eliminating the need to process all the variants in the initialization stage (Goldberg *et al* 1993). The initialization and primordial phases of the original mGA are replaced by a 'probabilistically complete initialization' procedure. This procedure is divided into several steps. During the first step strings of nearly length L are evaluated (using the template to fill in the missing values). Then selection is applied to these strings without any operators (much as was done in the primordial phase of the original mGA, but for only a few generations). Then the algorithm enters a filtering step where some of the genes in the strings are deleted, and the shortened strings are evaluated using the competitive template. Then selection is applied again. This process is repeated until the resulting strings are of length k. Then the mGA goes into the juxtaposition stage like the original mGA. By replacing the original initialization and primordial stages with stepwise filtering and selection, the number of evaluations required is drastically reduced for problems of significant size. (Goldberg *et al* (1993) provide analytical methods for determining the population and filtering reduction constants.) This new version of the mGA is very effective at solving loosely linked deceptive problems, i.e. those problems where the defining positions of the deceptive segments are spread out along the bitstring.

mGAs were designed to operate according to the SBBH, and deceptive problems illustrate that there are problems where being able to manipulate building blocks can provide an added value over CCV. It still is an open question, however, as to how representative deceptive problems are of the types of real-world problem that GAs might encounter. No doubt, many difficult real-world problems have deceptive or misleading elements in them. If they did not, they could be easily solved by local search methods. However it does not necessarily follow that such problems can be solved by a GA that is good at solving deceptive problems. The SBBH assumes that the misleading building blocks will exist in the initial population, that they can be identified early in the search before they are lost, and that the problem can be solved incrementally by combining these building blocks, but perhaps the building blocks that have misleading alternatives have little meaning until late in the search and so cannot be expected to survive in the population.

Even if the SBBH turns out not to be as useful an hypothesis as originally supposed, the increased propagation capabilities of pairwise mating may give a GA (using pairwise mating) an advantage over a poolwise CCV algorithm. To see why this is the case it is useful to define the prototypical individual for a given population: for each locus we assign a one or a zero depending upon which value is most frequent in the population (randomly assigning a value if they are equally frequent). Suppose the population contains some maverick individual that is quite far from the prototypical individual although it is near the optimum (as measured by Hamming distance) but is of only average fitness. Since an algorithm using a poolwise method of producing offspring will tend to produce individuals that are near the prototypical individual, such an algorithm is unlikely to explore the region around the maverick individual. On the other hand, a GA using pairwise mating is more likely to explore the region around the maverick individual, and so more likely to discover the optimum. Ironically, pairwise mating is, in this respect, more mutation-like than poolwise mating. While pairwise mating retains the benefits of CCV, it less subject to the majoritarian tendencies of poolwise mating.

8.4 Representation

Although GAs typically use a bitstring representation, GAs are not restricted to bitstrings. A number of early proponents of GAs developed GAs that use other representations, such as real-valued parameters (Davis 1991, Janikow and Michalewicz 1991, Wright 1991; see Chapter 16), permutations (Davis 1985, Goldberg and Lingle 1985, Grefenstette et al 1985; see Chapter 17), and treelike hierarchies (Antonisse and Keller 1987; see Chapter 19). Koza's genetic programming (GP) paradigm (Koza 1992; see Chapter 11) is a GA-based method for evolving programs, where the data structures are LISP S-expressions, and crossover creates new LISP S-expressions (offspring) by exchanging subtrees from the two parents.

In the case of combinatorial problems such as the traveling salesman problem (TSP), a number of order-based or sequencing crossover operators have been proposed. The choice of operator will depend upon one's goal. If the goal is to solve a TSP, then preserving adjacency information will be the priority, which suggests a crossover operator that operates on common edges (links between cities shared by the two parents) (Whitley *et al* 1989). On the other hand, if the goal is to solve a scheduling problem, then preserving relative order is likely to be the priority, which suggests an order preserving crossover operator. Syswerda's order crossover operator (Syswerda 1991), for example, chooses several positions at random in the first parent, and then produces a child so that the relative order of the chosen elements in the first parent is imposed upon the second parent.

Even if binary strings are used, there is still a choice to be made as to which binary coding scheme to use for numerical parameters. Empirical studies have usually found that Gray code is superior to the standard power-of-two binary coding (Caruana and Schaffer 1988), at least for the commonly used test problems. One reason is that the latter introduces Hamming cliffs—two numerically adjacent values may have bit representations that are many bits apart (up to $L-1$). This will be a problem if there is some degree of gradualness in the function, i.e. small changes in the variables usually correspond to small changes in the function. This is often the case for functions with numeric parameters.

As an example, consider a five-bit parameter, with a range from 0 to 31. If it is encoded using the standard binary coding, then 15 is encoded as 01111, whereas 16 is encoded as 10000. In order to move from 15 to 16, all five bits need to be changed. On the other hand, using Gray coding, 15 would be represented as 01000 and 16 as 11000, differing only by 1 bit.

When choosing an alternative representation, it is critical that a crossover operator be chosen that is appropriate for the representation. For example, if real-valued parameters are used, then a possible crossover operator is one that for each parameter uses the parameter values of the two parents to define an interval from which a new parameter is chosen (Eshelman and Schaffer 1993). As the GA makes progress it will narrow the range over which it searches for new parameter values.

If, for the chosen representation and crossover operator, the building blocks are unlikely to be instantiated independently of each other in the population, then a GA may not be appropriate. This problem has plagued finding crossover operators that are good for solving TSPs. The natural building blocks, it would seem, are subtours. However, what counts as a good subtour will almost always depend upon what the other subtours are. In other words, two good, but suboptimal solutions to a TSP may not have many subtours (other than very short ones) that are compatible with each other so that they can be spliced together to form a better solution. This hurdle is not unique to combinatorial problems.

Given the importance of the representation, a number of researches have suggested methods for allowing the GA to adapt its own coding. We noted earlier that Holland proposed the inversion operator for rearranging the loci in the string. Another approach to adapting the representation is Shaefer's ARGOT system (Shaefer 1987). ARGOT contains an explicit parameterized representation of the mappings from bitstrings to real numbers and heuristics for triggering increases and decreases in resolution and for shifts in the ranges of these mappings. A similar idea is employed by Schraudolph and Belew (1992) who provide a heuristic for increasing the resolution triggered when the population begins to converge. Mathias and Whitley (1994) have proposed what they call delta coding. When the population converges, the numeric representation is remapped so that the parameter ranges are centered around the best value found so far, and the algorithm is restarted. There are also heuristics for narrowing or extending the range.

There are also GAs with mechanisms for dynamically adapting the rate at which GA operators are used or which operator is used. Davis, who has developed a number of nontraditional operators, proposed a mechanism for adapting the rate at which these operators are applied based on the past success of these operators during a run of the algorithm (Davis 1987).

8.5   Parallel genetic algorithms

All evolutionary algorithms, because they maintain a population of solutions, are naturally parallelizable. However, because GAs use crossover, which is a way of sharing information, there are two other variations that are unique to GAs (Gordon and Whitley 1993). The first, most straightforward, method is to simply have one global population with multiple processors for evaluating individual solutions. The second method, often referred to as the island model (alternatively, the migration or coarse-grain model), maintains separate subpopulations. Selection and crossover take place in each subpopulation in isolation from the other subpopulations. Every so often an individual from one of the subpopulations is allowed to migrate to another subpopulation. This way information is shared among subpopulations.

The third method, often referred to as the neighborhood model (alternatively, the diffusion or fine-grain model), maintains overlapping neighborhoods. The neighborhood for which selection (for reproduction and replacement) applies is restricted to a region local to each individual. What counts as a neighborhood will depend upon the neighborhood topology used. For example, if the population is arranged upon some type of spherical structure, individuals might be allowed to mate with (and forced to compete with) neighbors within a certain radius.

8.6 Conclusion

Although the above discussion has been in the context of GAs as potential function optimizers, it should be pointed out that Holland's initial GA work was in the broader context of exploring GAs as adaptive systems (De Jong 1993). GAs were designed to be a simulation of evolution, not to solve problems. Of course, evolution has come up with some wonderful designs, but one must not lose sight of the fact that evolution is an opportunistic process operating in an environment that is continuously changing. Simon has described evolution as a process of searching where there is no goal (Simon 1983). This is not to question the usefulness of GAs as function optimizers, but only to emphasize that the perspective of function optimization is somewhat different from that of adaptation, and that the requirements of the corresponding algorithms will be somewhat different.

References

Antonisse H J and Keller K S 1987 Genetic operators for high-level knowledge representations *Proc. 2nd Int. Conf. on Genetic Algorithms (Cambridge, MA, 1987)* ed J J Grefenstette (Hillsdale, NJ: Erbaum) pp 69–76

Bäck T, Hoffmeister F and Schwefel H 1991 A survey of evolution strategies *Proc. 4th Int. Conf. on Genetic Algorithms (San Diego, CA, 1991)* ed R K Belew and L B Booker (San Mateo, CA: Morgan Kaufmann) pp 2–9

Baluja S 1995 *An Empirical Comparison of Seven Iterative and Evolutionary Function Optimization Heuristics* Carnegie Mellon University School of Computer Science Technical Report CMU-CS-95-193

Caruana R A and Schaffer J D 1988 Representation and hidden bias: Gray vs. binary coding for genetic algorithms *Proc. 5th Int. Conf. on Machine Learning* (San Mateo, CA: Morgan Kaufmann) pp 153–61

Davis L 1985 Applying adaptive algorithms to epistatic domains *Proc. Int. Joint Conference on Artificial Intelligence* pp 162–4

——1987 Adaptive operator probabilities in genetic algorithms *Proc. 3rd Int. Conf. on Genetic Algorithms (Fairfax, VA, 1989)* ed J D Schaffer (San Mateo, CA: Morgan Kaufmann) pp 61–9

——1991 Hybridization and numerical representation *The Handbook of Genetic Algorithms* ed L Davis (New York: Van Nostrand Reinhold) pp 61–71

De Jong K 1975 *An Analysis of the Behavior of a Class of Genetic Adaptive Systems* Doctoral Thesis, Department of Computer and Communication Sciences, University of Michigan

——1993 Genetic algorithms are not function optimizers *Foundations of Genetic Algorithms 2* ed D Whitley (San Mateo, CA: Morgan Kaufmann) pp 5–17

De Jong K and Sarma J 1993 Generation gaps revisited *Foundations of Genetic Algorithms 2* ed D Whitley (San Mateo, CA: Morgan Kaufmann) pp 19–28

Eshelman L J 1991 The CHC adaptive search algorithm: how to have safe search when engaging in nontraditional genetic recombination *Foundations of Genetic Algorithms* ed G J E Rawlins (San Mateo, CA: Morgan Kaufmann) pp 265–83

Eshelman L J and Schaffer J D 1993 Real-coded genetic algorithms and interval schemata *Foundations of Genetic Algorithms 2* ed D Whitley (San Mateo, CA: Morgan Kaufmann) pp 187–202

——1995 Productive recombination and propagating and preserving schemata *Foundations of Genetic Algorithms 3* ed D Whitley (San Mateo, CA: Morgan Kaufmann) pp 299–313

Goldberg D E 1987 Simple genetic algorithms and the minimal, deceptive problem *Genetic Algorithms and Simulated Annealing* ed L Davis (San Mateo, CA: Morgan Kaufmann) pp 74–88

——1989 *Genetic Algorithms in Search, Optimization, and Machine Learning* (Reading, MA: Addison-Wesley)

Goldberg D E and Deb K 1991 A comparative analysis of selection schemes used in genetic algorithms *Foundations of Genetic Algorithms* ed G J E Rawlins (San Mateo, CA: Morgan Kaufmann) pp 69–93

Goldberg D E, Deb K, Kargupta H and Harik G 1993 Rapid, accurate optimization of difficult problems using fast messy genetic algorithms *Proc. 5th Int. Conf. on Genetic Algorithms (Urbana-Champaign, IL, 1993)* ed S Forrest (San Mateo, CA: Morgan Kaufmann) pp 56–64

Goldberg D E, Deb K and Korb B 1991 Don't worry, be messy *Proc. 4th Int. Conf. on Genetic Algorithms (San Diego, CA, 1991)* ed R K Belew and L B Booker (San Mateo, CA: Morgan Kaufmann) pp 24–30

Goldberg D E and Lingle R L 1985 Alleles, loci, and the traveling salesman problem *Proc. 1st Int. Conf. on Genetic Algorithms (Pittsburgh, PA, 1985)* ed J J Grefenstette (Hillsdale, NJ: Erbaum) pp 154–9

Gordon V S and Whitley 1993 Serial and parallel genetic algorithms and function optimizers *Proc. 5th Int. Conf. on Genetic Algorithms (Urbana-Champaign, IL, 1993)* ed S Forrest (San Mateo, CA: Morgan Kaufmann) pp 177–83

Grefenstette J J 1993 Deception considered harmful *Foundations of Genetic Algorithms 2* ed D Whitley (San Mateo, CA: Morgan Kaufmann) pp 75–91

Grefenstette J J, Gopal R, Rosmaita B J and Van Gucht D 1985 Genetic algorithms for the traveling salesman problem *Proc. 1st Int. Conf. on Genetic Algorithms (Pittsburgh, PA, 1985)* ed J J Grefenstette (Hillsdale, NJ: Erbaum) pp 160–8

Holland J H 1975 *Adaptation in Natural and Artificial Systems* (Ann Arbor, MI: University of Michigan Press)

Janikow C Z and Michalewicz Z 1991 An experimental comparison of binary and floating point representations in genetic algorithms *Proc. 4th Int. Conf. on Genetic Algorithms (San Diego, CA, 1991)* ed R K Belew and L B Booker (San Mateo, CA: Morgan Kaufmann) pp 31–6

Koza J 1992 *Genetic Programming: on the Programming of Computers by Means of Natural Selection and Genetics* (Cambridge, MA: MIT Press)

Liepins G E and Vose M D 1991 Representational issues in genetic optimization *J. Exp. Theor. AI* **2** 101–15

Mathias K E and Whitley L D 1994 Changing representations during search: a comparative study of delta coding *Evolutionary Comput.* **2**

Mühlenbein H and Schlierkamp-Voosen 1993 The science of breeding and its application to the breeder genetic algorithm *Evolutionary Comput.* **1**

Radcliffe N J 1991 Forma analysis and random respectful recombination *Proc. 4th Int. Conf. on Genetic Algorithms (San Diego, CA, 1991)* ed R K Belew and L B Booker (San Mateo, CA: Morgan Kaufmann) pp 222–9

Schaffer J D, Eshelman L J and Offutt D 1991 Spurious correlations and premature convergence in genetic algorithms *Foundations of Genetic Algorithms* ed G J E Rawlins (San Mateo, CA: Morgan Kaufmann) pp 102–12

Schraudolph N N and Belew R K 1992 Dynamic parameter encoding for genetic algorithms *Machine Learning* **9** 9–21

Shaefer C G 1987 The ARGOT strategy: adaptive representation genetic optimizer technique *Genetic Algorithms and Their Applications: Proc. 2nd Int. Conf. on Genetic Algorithms (Cambridge, MA, 1987)* ed J J Grefenstette (Hillsdale, NJ: Erlbaum) pp 50–8

Simon H A 1983 *Reason in Human Affairs* (Stanford, CA: Stanford University Press)

Spears W M and De Jong K A 1991 On the virtues of parameterized uniform crossover *Proc. 4th Int. Conf. on Genetic Algorithms (San Diego, CA, 1991)* ed R K Belew and L B Booker (San Mateo, CA: Morgan Kaufmann) pp 230–6

Syswerda G 1989 Uniform crossover in genetic algorithms *Proc. 3rd Int. Conf. on Genetic Algorithms (Fairfax, VA, 1989)* ed J D Schaffer (San Mateo, CA: Morgan Kaufmann) pp 2–9

——1991 Schedule optimization using genetic algorithms *Handbook of Genetic Algorithms* ed L Davis (New York: Van Nostrand Reinhold) pp 332–49

——1993 Simulated crossover in genetic algorithms *Foundations of Genetic Algorithms 2* ed D Whitley (San Mateo, CA: Morgan Kaufmann) pp 239–55

Whitley D 1989 The GENITOR algorithm and selection pressure: why rank-based allocation of reproductive trials is best *Proc. 3rd Int. Conf. on Genetic Algorithms (Fairfax, VA, 1989)* ed J D Schaffer (San Mateo, CA: Morgan Kaufmann) pp 116–21

Whitley D, Starkweather T and Fuquay D 1989 Scheduling problems and traveling salesmen: the genetic edge recombination operator *Proc. 3rd Int. Conf. on Genetic Algorithms (Fairfax, VA, 1989)* ed J D Schaffer (San Mateo, CA: Morgan Kaufmann) pp 116–21

Wright A 1991 Genetic algorithms for real parameter optimization *Foundations of Genetic Algorithms* ed G J E Rawlins (San Mateo, CA: Morgan Kaufmann) pp 205–18

9

Evolution strategies

Günter Rudolph

9.1 The archetype of evolution strategies

Minimizing the total drag of three-dimensional slender bodies in a turbulent flow was, and still is, a general goal of research in institutes of hydrodynamics. Three students (Peter Bienert, Ingo Rechenberg, and Hans-Paul Schwefel) met each other at such an institute, the Hermann Föttinger Institute of the Technical University of Berlin, in 1964. Since they were fascinated not only by aerodynamics, but also by cybernetics, they hit upon the idea to solve the analytically (and at that time also numerically) intractable form design problem with the help of some kind of robot. The robot should perform the necessary experiments by iteratively manipulating a flexible model positioned at the outlet of a wind tunnel. An *experimentum crucis* was set up with a two-dimensional foldable plate. The iterative search strategy—first performed by hand, a robot was developed later on by Peter Bienert—was expected to end up with a flat plate: the form with minimal drag. But it did not, since a one-variable-at-a-time as well as a discrete gradient-type strategy always got stuck in a local minimum: an S-shaped folding of the plate. Switching to small random changes that were only accepted in the case of improvements—an idea of Ingo Rechenberg— brought the breakthrough, which was reported at the joint annual meeting of WGLR and DGRR in Berlin, 1964 (Rechenberg 1965). The interpretation of binomially distributed changes as mutations and of the decision to step back or not as selection (on 12 June 1964) was the seed for all further developments leading to evolution strategies (ESs) as they are known today. So much about the birth of the ES.

It should be mentioned that the domain of the decision variables was not fixed or even restricted to real variables at that time. For example, the experimental optimization of the shape of a supersonic two-phase nozzle by means of mutation and selection required discrete variables and mutations (Klockgether and Schwefel 1970) whereas first numerical experiments with the early ES on a Zuse Z 23 computer (Schwefel 1965) employed discrete mutations of real variables. The apparent fixation of ESs to Euclidean search spaces nowadays is probably due to the fact that Rechenberg (1973) succeeded in analyzing the simple version in Euclidean space with continuous mutation for several test problems.

Within this setting the archetype of ESs takes the following form. An individual a consisting of an element $X \in \mathbb{R}^n$ is mutated by adding a normally distributed random vector $Z \sim N(0, I_n)$ that is multiplied by a scalar $\sigma > 0$ (I_n denotes the unit matrix with rank n). The new point is accepted if it is better than or equal to the old one, otherwise the old point passes to the next iteration. The selection decision is based on a simple comparison of the objective function values of the old and the new point. Assuming that the objective function $f : \mathbb{R}^n \to \mathbb{R}$ is to be minimized, the simple ES, starting at some point $X_0 \in \mathbb{R}^n$, is determined by the following iterative scheme:

$$X_{t+1} = \begin{cases} X_t + \sigma_t Z_t & \text{if } f(X_t + \sigma_t Z_t) \leq f(X_t) \\ X_t & \text{otherwise} \end{cases} \tag{9.1}$$

where $t \in \mathbb{N}_0$ denotes the iteration counter and where $(Z_t : t \geq 0)$ is a sequence of independent and identically distributed standard normal random vectors.

The general algorithmic scheme (9.1) was not a novelty: Schwefel (1995, pp 94–5), presents a survey of forerunners and related versions of (9.1) since the late 1950s. Most methods differed in the mechanism of adjusting the parameter σ_t, that is used to control the strength of the mutations (i.e. the length of the mutation steps in n-dimensional space). Rechenberg's solution to control parameter σ_t is known as the $1/5$ success rule: Increase σ_t if the relative frequency of successful mutations over some period in the past is larger than $1/5$, otherwise decrease σ_t. Schwefel (1995, p 112), proposed the following implementation. Let $t \in \mathbb{N}$ be the generation (or mutation) counter and assume that $t \geq 10n$.

(i)   If $t \bmod n = 0$ then determine the number s of successful mutations that have occurred during the steps $t - 10n$ to $t - 1$.

(ii) If $s < 2n$ then multiply the step lengths by the factor 0.85.

(iii) If $s > 2n$ then divide the step lengths by the factor 0.85.

First ideas to extend the simple ES (9.1) can be found in the book by Rechenberg (1973, pp 78–86). The population consists of $\mu > 1$ parents. Two parents are selected at random and recombined by multipoint crossover and the resulting individual is finally mutated. The offspring is added to the population. The selection operation chooses the μ best individuals out of the $\mu + 1$ in total to serve as parents of the next iteration. Since the search space was binary, this ES was exactly the same evolutionary algorithm as became known later under the term steady-state genetic algorithm (Section 28.1). The usage of this algorithmic scheme for Euclidean search spaces poses the problem of how to control the step length control parameter σ_t. Therefore, the 'steady-state' ES is no longer in use.

9.2 Contemporary evolution strategies

The general algorithmic frame of contemporary ESs is easily presented by the symbolic notation introduced by Schwefel (1977). The abbreviation $(\mu + \lambda)$ ES denotes an ES that generates λ offspring from μ parents and selects the μ best individuals from the $\mu + \lambda$ individuals (parents and offspring) in total. This notation can be used to express the simple ES by $(1 + 1)$ ES and the 'steady-state' ES by $(\mu + 1)$ ES. Since the latter is not in use it is convention that the abbreviation $(\mu + \lambda)$ ES always refers to an ES parametrized according to the relation $1 \leq \mu \leq \lambda < \infty$.

The abbreviation (μ, λ) ES denotes an ES that generates λ offspring from μ parents but selects the μ best individuals only from the λ offspring. As a consequence, λ must be necessarily at least as large as μ. However, since the parameter setting $\mu = \lambda$ represents nothing more than a random walk, it is convention that the abbreviation (μ, λ) ES always refers to an ES parametrized according to the relation $1 \leq \mu < \lambda < \infty$.

Apart from the population concept contemporary ESs differ from early ESs in that an individual consists of an element $x \in \mathbb{R}^n$ of the search space plus several individual parameters controlling the individual mutation distribution. Usually, mutations are distributed according to a multivariate normal distribution with zero mean and some covariance matrix $\mathbf{C}$ that is symmetric and positive definite. Unless matrix $\mathbf{C}$ is a diagonal matrix, the mutations in each coordinate direction are correlated (Schwefel 1995, p 240). It was shown in Rudolph (1992) that a matrix is symmetric and positive definite if and only if it is decomposable via $\mathbf{C} = (\mathbf{ST})'\mathbf{ST}$ where $\mathbf{S}$ is a diagonal matrix with positive diagonal entries and

$$\mathbf{T} = \prod_{i=1}^{n-1} \prod_{j=i+1}^{n} \mathbf{R}_{ij}(\omega_k) \tag{9.2}$$

is an orthogonal matrix built by a product of $n(n-1)/2$ elementary rotation matrices $\mathbf{R}_{ij}$ with angles $\omega_k \in (0, 2\pi]$. An elementary rotation matrix $\mathbf{R}_{ij}(\omega)$ is a unit matrix where four specific entries are replaced by $r_{ii} = r_{jj} = \cos \omega$ and $r_{ij} = -r_{ji} = -\sin \omega$.

As a consequence, $n(n-1)/2$ angles and n scaling parameters are sufficient to generate arbitrary correlated normal random vectors with zero mean and covariance matrix $\mathbf{C} = (\mathbf{ST})'\mathbf{ST}$ via $Z = \mathbf{T'S'}N$, where N is a standard normal random vector (since matrix multiplication is associative, random vector Z can be created in $\mathrm{O}(n^2)$ time by multiplication from right to left).

There remains, however, the question of how to choose and adjust these individual strategy parameters. The idea that a population-based ES could be able to adapt σ_t individually by including these parameters in the mutation–selection process came up early (Rechenberg 1973, pp 132–7). Although first experiments with the $(\mu + 1)$ ES provided evidence that this approach works in principle, the first really successful implementation of the idea of self-adaptation was presented by Schwefel (1977) and it is based on the observation that a surplus of offspring (i.e. $\lambda > \mu$) is a good advice to establish self-adaptation of individual parameters.

To start with a simple case let $\mathbf{C} = \sigma^2 \mathbf{I}_n$. Thus, the only parameter to be self-adapted for each individual is the step length control parameter σ. For this purpose let the the genome of each individual be represented by the tuple $(\mathbf{X}, \sigma) \in \mathbb{R}^n \times \mathbb{R}_+$, that undergoes the genetic operators. Now mutation is a two-stage operation:

$$\sigma_{t+1} = \sigma_t \exp(\tau Z_\tau)$$
$$\mathbf{X}_{t+1} = \mathbf{X}_t + \sigma_{t+1} \mathbf{Z}$$

where $\tau = n^{-1/2}$ and Z_τ is a standard normal random variable whereas $\mathbf{Z}$ is a standard normal random vector. This scheme can be extended to the general case with $n(n+1)/2$ parameters.

(i) Let $\omega \in (0, 2\pi]^{n(n-1)/2}$ denote the angles that are necessary to build the orthogonal rotation matrix $\mathbf{T}(\omega)$ via (9.2). The mutated angles $\omega_{t+1}^{(i)}$ are obtained by

$$\omega_{t+1}^{(i)} = (\omega_t^{(i)} + \varphi Z_\omega^{(i)}) \bmod 2\pi$$

where $\varphi > 0$ and the independent random numbers $Z_\omega^{(i)}$ with $i = 1, \ldots, n(n-1)/2$ are standard normally distributed.

(ii) Let $\sigma \in \mathbb{R}_+^n$ denote the standard deviations that are represented by the diagonal matrix $\mathbf{S}(\sigma) = \mathrm{diag}(\sigma^{(1)}, \ldots, \sigma^{(n)})$. The mutated standard deviations are obtained as follows. Draw a standard normally distributed random number Z_τ. For each $i = 1, \ldots, n$ set

$$\sigma_{t+1}^{(i)} = \sigma_t^{(i)} \exp(\tau Z_\tau + \eta Z_\sigma^{(i)})$$

where $(\tau, \eta) \in \mathbb{R}_+^2$ and the independent random numbers $Z_\sigma^{(i)}$ are standard normally distributed. Note that Z_τ is drawn only once.

(iii) Let $\mathbf{X} \in \mathbb{R}^n$ be the object variables and $\mathbf{Z}$ be a standard normal random vector. The mutated object variable vector is given by

$$\mathbf{X}_{t+1} = \mathbf{X}_t + \mathbf{T}(\omega_{t+1})\mathbf{S}(\sigma_{t+1})\mathbf{Z}.$$

According to Schwefel (1995) a good heuristic for the choice of the constants appearing in the above mutation operation is

$$(\varphi, \tau, \eta) = (5\pi/180, (2n)^{-1/2}, (4n)^{-1/4})$$

but recent extensive simulation studies (Kursawe 1996) revealed that the above recommendation is not the best choice—especially in the case of multimodal objective functions it seems to be better to use weak selection pressure (μ/λ not too small) and a parametrization obeying the relation $\tau > \eta$. As a consequence, a final recommendation cannot be given here, yet.

As soon as $\mu > 1$, the decision variables as well as the internal strategy parameters can be recombined with usual recombination operators. Notice that there is no reason to employ the same recombination operator for the angles, standard deviations, and object variables. For example, one could apply intermediate recombination (Chapter 33) to the angles as well as standard deviations and uniform crossover to the object variables. With this choice recombination of two parents works as follows. Choose two parents (X, σ, ω) and (X', σ', ω') at random. Then the preliminary offspring resulting from the recombination process is

$$\left(UX + (I - U)X', \; \frac{\sigma + \sigma'}{2}, \; \frac{(\omega + \omega') \bmod 4\pi}{2} \right)$$

where I is the unit matrix and U is a random diagonal matrix whose diagonal entries are either zero or one with the same probability. Note that the angles must be adjusted to the interval $(0, 2\pi]$.

After these preparations a sketch of a contemporary ES can be presented:

Generate μ initial parents of the type (X, σ, ω) and determine their objective function values $f(X)$.

repeat

  do λ times:

    Choose $\rho \geq 2$ parents at random.

    Recombine their angles, standard deviations, and object variables.

    Mutate the angles, standard deviations, and object variables of the preliminary offspring obtained via recombination.

    Determine the offspring's objective function value.

    Put the offspring into the offspring population.

  end do

  Select the μ best individuals either from the offspring population or from the union of the parent and offspring population.

  The selected individuals represent the new parents.

until some stopping criterion is satisfied.

It should be noted that there are other proposals to adapt σ_t. In the case of a $(1, \lambda)$ ES with $\lambda = 3k$ and $k \in \mathbb{N}$, Rechenberg (1994, p 47) devised the following rule: Generate k offspring with σ_t, k offspring with $c\sigma_t$ and k offspring with σ_t/c for some $c > 0$ ($c = 1.3$ is recommended for $n \leq 100$, for larger n the constant c should decrease).

Further proposals, that are however still in an experimental state, try to derandomize the adaptation process by exploiting information gathered in preceding iterations (Ostermeier *et al* 1995). This approach is related to (deterministic) variable metric (or quasi-Newton) methods, where the Hessian matrix is approximated iteratively by certain update rules. The inverse of the Hessian matrix is in fact the optimal choice for the covariance matrix C. A large variety of update rules is given by the *Oren–Luenberger class* (Oren and

Luenberger 1974) and it might be useful to construct probabilistic versions of these update rules, but it should be kept in mind that ESs are designed to tackle difficult nonconvex problems and not convex ones: the usage of such techniques increases the risk that ESs will be attracted by local optima.

Other ideas that have not yet achieved a standard include the introduction of an additional age parameter κ for individuals in order to have intermediate forms of selection between the $(\mu + \lambda)$ ES with $\kappa = \infty$ and the (μ, λ) ES with $\kappa = 1$ (Schwefel and Rudolph 1995), as well as the huge variety of ESs whose population possesses a spatial structure. Since the latter is important for parallel implementations and applies to other evolutionary algorithms as well the description is omitted here.

9.3   Nested evolution strategies

The shorthand notation $(\mu \overset{+}{,} \lambda)$ ES was extended by Rechenberg (1978) to the expression

$$[\, \mu' \overset{+}{,} \lambda' \, (\mu \overset{+}{,} \lambda)^\gamma \,]^{\gamma'} \text{ ES}$$

with the following meaning. There are μ' populations of μ parents. These are used to generate (e.g. by merging) λ' initial populations of μ individuals each. For each of these λ' populations a $(\mu \overset{+}{,} \lambda)$ ES is run for γ generations. The criterion to rank the λ' populations after termination might be the average fitness of the individuals in each population. This scheme is repeated γ' times. The obvious generalization to higher levels of nesting is described by Rechenberg (1994), where it is also attempted to develop a shorthand notation to specify the parametrization completely.

This nesting technique is of course not limited to ESs: other evolutionary algorithms and even mixtures of them can be used instead. In fact, the somewhat artificial distinction between ESs, genetic algorithms, and evolutionary programs becomes more and more blurred when higher concepts enter the scene. Finally, some fields of application of nested evolutionary algorithms will be described briefly.

9.3.1   Alternative method to control internal parameters

Herdy (1992) used λ' subpopulations, each of them possessing its own different and fixed step size σ. Thus, there is no step size control at the level of individuals. After γ generations the improvements (in terms of fitness) achieved by each subpopulation is compared to each other and the best μ' subpopulations are selected. Then the process repeats with slightly modified values of σ. Since subpopulations with a near-optimal step size will achieve larger improvements, they will be selected (i.e. better step sizes will survive), resulting in an alternative method to control the step size.

9.3.2 Mixed-integer optimization

Lohmann (1992) considered optimization problems in which the decision variables are partially discrete and partially continuous. The nested approach worked as follows. The ESs in the inner loop were optimizing over the continuous variables while the discrete variables were held fixed. After termination of the inner loop, the evolutionary algorithm in the outer loop compared the fitness values achieved in the subpopulations, selected the best ones, mutated the discrete variables and passed them as fixed parameters to the subpopulations in the inner loop.

It should be noted that this approach to mixed-integer optimization may cause some problems. In essence, a Gauß–Seidel-like optimization strategy is realized, because the search alternates between the subspace of discrete variables and the subspace of continuous variables. Such a strategy must fail whenever simultaneous changes in discrete *and* continuous variables are necessary to achieve further improvements.

9.3.3 Minimax optimization

Sebald and Schlenzig (1994) used nested optimization to tackle minimax problems of the type

$$\min_{x \in X}\{\max_{y \in Y}\{f(x, y)\}\}$$

where $X \subseteq \mathbb{R}^n$ and $Y \subseteq \mathbb{R}^m$. Equivalently, one may state the problem as follows:

$$\min\{g(x) : x \in X\} \qquad \text{where} \quad g(x) = \max\{f(x, y) : y \in Y\}.$$

The evolutionary algorithm in the inner loop maximizes $f(x, y)$ with fixed parameters x, while the outer loop is responsible for minimize $g(x)$ing over the set X.

Other applications of this technique are imaginable. An additional aspect touches the evident degree of independence of executing the evolutionary algorithms in the inner loop. As a consequence, nested evolutionary algorithms are well suited for parallel computers.

References

Herdy M 1992 Reproductive isolation as strategy parameter in hierachically organized evolution strategies *Parallel Problem Solving from Nature, 2 (Proc. 2nd Int. Conf. on Parallel Problem Solving from Nature, Brussels, 1992)* ed R Männer and B Manderick (Amsterdam: Elsevier) pp 207–17

Klockgether J and Schwefel H-P 1970 Two-phase nozzle and hollow core jet experiments *Proc. 11th Symp. on Engineering Aspects of Magnetohydrodynamics* ed D Elliott (Pasadena, CA: California Institute of Technology) pp 141–8

Kursawe F 1996 Breeding evolution strategies—first results, talk presented at Dagstuhl lectures *Applications of Evolutionary Algorithms (March 1996)*

Lohmann R 1992 Structure evolution and incomplete induction *Parallel Problem Solving from Nature, 2 (Proc. 2nd Int. Conf. on Parallel Problem Solving from Nature, Brussels, 1992)* ed R Männer and B Manderick (Amsterdam: Elsevier) pp 175–85

Oren S and Luenberger D 1974 Self scaling variable metric (SSVM) algorithms, Part II: criteria and sufficient conditions for scaling a class of algorithms *Management Sci.* **20** 845–62

Ostermeier A, Gawelczyk A and Hansen N 1995 A derandomized approach to self-adaptation of evolution strategies *Evolut. Comput.* **2** 369–80

Rechenberg I 1965 *Cybernetic solution path of an experimental problem* Library Translation 1122, Royal Aircraft Establishment, Farnborough, UK

——1973 *Evolutionsstrategie: Optimierung technischer Systeme nach Prinzipien der biologischen Evolution* (Stuttgart: Frommann-Holzboog)

——1978 Evolutionsstrategien *Simulationsmethoden in der Medizin und Biologie* ed B Schneider and U Ranft (Berlin: Springer) pp 83–114

——1994 *Evolutionsstrategie '94* (Stuttgart: Frommann-Holzboog)

Rudolph G 1992 On correlated mutations in evolution strategies *Parallel Problem Solving from Nature, 2 (Proc. 2nd Int. Conf. on Parallel Problem Solving from Nature, Brussels, 1992)* ed R Männer and B Manderick (Amsterdam: Elsevier) pp 105–14

Schwefel H-P 1965 *Kybernetische Evolution als Strategie der experimentellen Forschung in der Strömungstechnik* Diplomarbeit, Technical University of Berlin

——1977 *Numerische Optimierung von Computer-Modellen mittels der Evolutionsstrategie* (Basel: Birkhäuser)

——1995 *Evolution and Optimum Seeking* (New York: Wiley)

Schwefel H-P and Rudolph G 1995 Contemporary evolution strategies *Advances in Artificial Life* ed F Morana *et al* (Berlin: Springer) pp 893–907

Sebald A V and Schlenzig J 1994 Minimax design of neural net controllers for highly uncertain plants *IEEE Trans. Neural Networks* **NN-5** 73–82

10

Evolutionary programming

V William Porto

10.1 Introduction

Evolutionary programming (EP) is one of a class of paradigms for simulating evolution which utilizes the concepts of Darwinian evolution to iteratively generate increasingly appropriate solutions (organisms) in light of a static or dynamically changing environment. This is in sharp contrast to earlier research into artificial intelligence research which largely centered on the search for simple heuristics. Instead of developing a (potentially) complex set of rules which were derived from human experts, EP evolves a set of solutions which exhibit optimal behavior with regard to an environment and desired payoff function. In a most general framework, EP may be considered an optimization technique wherein the algorithm iteratively optimizes behaviors, parameters, or other constructs. As in all optimization algorithms, it is important to note that the point of optimality is completely independent of the search algorithm, and is solely determined by the adaptive topography (i.e. response surface) (Atmar 1992).

In its standard form, the basic evolutionary program utilizes the four main components of all evolutionary computation (EC) algorithms: initialization, variation, evaluation (scoring), and selection. At the basis of this, as well as other EC algorithms, is the presumption that, at least in a statistical sense, learning is encoded phylogenically versus ontologically in each member of the population. 'Learning' is a byproduct of the evolutionary process as successful individuals are retained through stochastic trial and error. Variation (e.g. mutation) provides the means for moving solutions around on the search space, preventing entrapment in local minima. The evaluation function directly measures fitness, or equivalently the behavioral error, of each member in the population with regard to the environment. Finally, the selection process probabilistically culls suboptimal solutions from the population, providing an efficient method for searching the topography.

The basic EP algorithm starts with a population of trial solutions which are initialized by random, heuristic, or other appropriate means. The size of the population, μ, may range over a broadly distributed set, but is in general larger than one. Each of these trial solutions is evaluated with regard to the specified fitness function. After the creation of a population of initial solutions, each of the parent members is altered through application of a mutation process; in strict EP, recombination is not utilized. Each parent member i generates λ_i progeny which are replicated with a stochastic error mechanism (mutation). The fitness or behavioral error is assessed for all offspring solutions with the selection process performed by one of several general techniques including: (i) the best μ solutions are retained to become the parents for the next generation (elitist, see Section 28.4), or (ii) μ of the best solutions are statistically retained (tournament, see Chapter 24), or (iii) proportional-based selection (Chapter 23). In most applications, the size of the population remains constant, but there is no restriction in the general case. The process is halted when the solution reaches a predetermined quality, a specified number of iterations has been achieved, or some other criterion (e.g. sufficient convergence) stops the algorithm.

EP differs philosophically from other evolutionary computational techniques such as genetic algorithms (GAs) (Chapter 8) in a crucial manner. EP is a top-down versus bottom-up approach to optimization. It is important to note that (according to neo-Darwinism) selection operates only on the phenotypic expressions of a genotype; the underlying coding of the phenotype is only affected indirectly. The realization that a sum of optimal parts rarely leads to an optimal overall solution is key to this philosophical difference. GAs rely on the identification, combination, and survival of 'good' building blocks (schemata) iteratively combining to form larger 'better' building blocks. In a GA, the coding structure (genotype) is of primary importance as it contains the set of optimal building blocks discovered through successive iterations. The building block hypothesis is an implicit assumption that the fitness is a separable function of the parts of the genome. This successively iterated local optimization process is different from EP, which is an entirely global approach to optimization. Solutions (or organisms) in an EP algorithm are judged solely on their fitness with respect to the given environment. No attempt is made to partition credit to individual components of the solutions. In EP (and in evolution strategies (ESs), see Chapter 9), the variation operator allows for simultaneous modification of all variables at the same time. Fitness, described in terms of the behavior of each population member, is evaluated directly, and is the sole basis for survival of an individual in the population. Thus, a crossover operation designed to recombine building blocks is not utilized in the general forms of EP.

10.2 History

The genesis of EP (Section 6.2) was motivated by the desire to generate an alternative approach to artificial intelligence. Fogel (1962) conceived of using the simulation of evolution to develop artificial intelligence which did not rely on heuristics, but instead generated organisms of increasing intellect over time. Fogel (1964, Fogel *et al* 1966) made the observation that intelligent behavior requires the ability of an organism to make correct predictions within its environment, while being able to translate these predictions into a suitable response for a given goal. This early work focused on evolving finite-state machines (Chapter 18; see the articles by Mealy (1955), and Moore (1957) for a discussion of these automata) which provided a most generic testbed for this approach. A finite-state machine (figure 10.1) is a mechanism which operates on a finite set (i.e. alphabet) of input symbols, possesses a finite number of internal states, and produces output symbols from a finite alphabet. As in all finite-state machines, the corresponding input–output symbol pairs and state transitions from every state define the specific behavior of the machine.

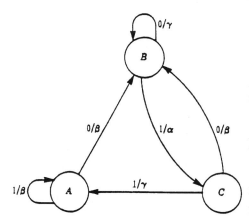

Figure 10.1. A simple finite-state machine diagram. Input symbols are shown to the left of the slash. Output symbols are to the right of the slash. The finite-state machine is presumed to start in state A.

In a series of experiments (Fogel *et al* 1966), an environment was simulated by a sequence of symbols from a finite-length alphabet. The problem was defined as follows: evolve an algorithm which would operate on the sequence of symbols previously observed in a manner that would produce an output symbol which maximizes the benefit to the algorithm in light of the next symbol to appear in the environment, relative to a well-defined payoff function.

EP was originally defined by Fogel (1964) in the following manner. A population of parent finite-state machines, after initialization, is exposed to the sequence of symbols (i.e. environment) which have been observed up to the current time. As each input symbol is presented to each parent machine, the output symbol is observed (predicted) and compared to the next input symbol. A predefined payoff function provides a means for measuring the worth of each prediction. After the last prediction is made, some function of the sequence of payoff values is used to indicate the overall fitness of each machine. Offspring machines are created by randomly mutating each parent machine. As defined above, there are five possible resulting modes of random mutation for a finite-state machine. These are: (i) change an output symbol; (ii) change a state transition; (iii) add a state; (iv) delete an existing state; and (v) change the initial state. Other mutations were proposed but results of experiments with these mutations were not described by Fogel *et al* (1966). To prevent the possibility of creating null machines, the deletion of a state and the changing of the initial state were allowed only when a parent machine had more than one state.

Mutation operators are chosen with respect to a specified probability distribution which may be uniform, or another desired distribution. The number of mutation operations applied to each offspring is also determined with respect to a specified probability distribution function (e.g. Poisson) or may be fixed *a priori*. Each of the mutated offspring machines is evaluated over the existing environment (set of input–output symbol pairs) in the same manner as the parent machines.

After offspring have been created through application of the mutation operator(s) on the members of the parent population, the machines providing the greatest payoff with respect to the payoff function are retained to become parent members for the next generation. Typically, one offspring is created for each parent, and half of the total machines are retained in order to maintain a constant population size. The process is iterated until it is required to make an actual prediction of the next output symbol in the environment, which has yet to be encountered. This is analogous to the presentation of a naive exemplar to a previously trained neural network. Out of the entire population of machines, only the best machine, in terms of its overall worth, is chosen to generate the new output symbol. Fogel originally proposed selection of machines which score in the top half of the entire population, i.e. a nonregressive selection mechanism. Although discussed as a possibility to increase variance, the retention of lesser-quality machines was not incorporated in these early experiments.

Since the topography (response surface) is changed after each presentation of a symbol, the fitness of the evolved machines must change to reflect the payoff from the previous prediction. This prevents evolutionary stagnation as the adaptive topography is experiencing continuous change. As is evidenced in nature, the complexity of the representation is increased as the finite-state machines learn to recognize more subtle features in the experienced sequence of symbols.

10.2.1 Early foundations

Fogel (see Fogel 1964, Fogel *et al* 1966) used EP on a series of successively more difficult prediction tasks. These experiments ranged from simple two-symbol cyclic sequences, eight-symbol cyclic sequences degraded by addition

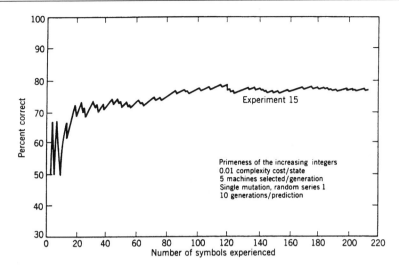

Figure 10.2. A plot showing the convergence of EP on finite-state machines evolved to predict primeness of numbers.

of noise, and sequences of symbols generated by other finite-state machines to nonstationary sequences and sequences taken from the article by Flood (1962).

In one example, the capability for predicting the 'primeness', i.e. whether or not a number is prime, was tested. A nonstationary sequence of symbols was generated by classifying each of the monotonically increasing set of integers as prime (with symbol 1) or nonprime (with symbol 0). The payoff function consisted of an all-or-none function where one point was provided for each correct prediction. No points or penalty were assessed for incorrect predictions. A small penalty term was added to maximize parsimony, through the subtraction of 0.01 multiplied by the number of states in the machine. This complexity penalty was added due to the limited memory available on the computers at that time. After presentation of 719 symbols, the iterative process was halted with the best machine possessing one state, with both output symbols being zero. Figure 10.2 indicates the prediction score achieved in this nonstationary environment. Because prime numbers become increasingly infrequent (Burton 1976), the asymptotic worth of this machine, given the defined payoff function, approaches 100%.

After initial, albeit qualified, success with this experiment, the goal was altered to provide a greater payoff for correct prediction of a rare event. Correct prediction of a prime was worth one plus the number of nonprimes preceding it. For the first 150 symbols, 30 correct predictions were made (primes predicted as primes), 37 false positives (nonprimes predicted as primes), and five primes were missed. On predicting the 151st through 547th symbols there were 65 correct predictions of primes, and 67 false positives. Of the first 35 prime numbers, five were missed, but of the next 65 primes, none were missed. Fogel *et al* (1966) indicated that the machines demonstrated the capability to quickly recognize numbers which are divisible by two and three as being nonprime, and that some capability to recognize divisibility by five as being indicative of nonprimes was also evidenced. Thus, the machines generated evidence of learning a definition of primeness without prior knowledge of the explicit nature of a prime number, or any ability to explicitly divide.

Fogel and Burgin (1969) researched the use of EP in game theory. In a number of experiments, EP was consistently able to discover the globally optimal strategy in simple two-player, zero-sum games involving a small number of possible plays. This research also showed the ability of the technique to outperform human subjects in more complicated games. Several extensions were made to the simulations to address nonzero-sum games (e.g. pursuit evasion.) A three-dimensional model was constructed where EP was used to guide an interceptor towards a moving target. Since the target was, in most circumstances, allowed a greater degree of maneuverability, the success or failure of the interceptor was highly dependent upon the learned ability to predict the position of the target without *a priori* knowledge of the target's dynamics.

A different aspect of EP was researched by Walsh *et al* (1970) where EP was used for prediction as a precursor to automatic control. This research concentrated on decomposing a finite-state machine into submachines which could be executed in parallel to obtain the overall output of the evolved system. A primary goal of this research was to maximize parsimony in the evolving machines. In these experiments, finite-state machines containing seven and eight states were used as the generating function for three output symbols. The performance of three human subjects was compared to the evolved models when predicting the next symbol in the respective environments. In these experiments, EP was consistently able to outperform the human subjects.

10.2.2 Extensions

The basic EP paradigm may be described by the following EP algorithm:

$$t := 0;$$
$$\text{initialize } P(0) := \left\{ a_1'(0), a_2'(0), \dots, a_\mu'(0) \right\}$$
$$\text{evaluate } P(0) : \left\{ \Phi(a_1'(0)), \Phi(a_2'(0)), \dots, \Phi(a_\mu'(0)) \right\}$$
$$\text{iterate}$$
$$\quad \{$$
$$\qquad \text{mutate: } P'(t) := m_{\Theta_m}(P(t))$$
$$\qquad \text{evaluate: } P'(t) : \left\{ \Phi(a_1'(t)), \Phi(a_2'(t)), \dots, \Phi(a_\lambda'(t)) \right\}$$
$$\qquad \text{select: } P(t+1) := s_{\Theta_s}(P'(t) \cup Q)$$
$$\qquad t := t + 1;$$
$$\quad \}$$

where:

a' is an individual member in the population

$\mu \geq 1$ is the size of the parent population

$\lambda \geq 1$ is the size of the offspring population

$P(t) := \{a_1'(t), a_2'(t), \ldots, a_\mu'(t)\}$ is the population at time t

$\Phi : I \rightarrow \Re$ is the fitness mapping

m_{Θ_m} is the mutation operator with controlling parameters Θ_m

s_{Θ_s} is the selection operator $\ni s_{\Theta_s} : (I^\lambda \cup I^{\mu+\lambda}) \rightarrow I^\mu$

$Q \in \{\emptyset, P(t)\}$ is a set of individuals additionally accounted for in the selection step, i.e. parent solutions.

Other than on initialization, the search space is generally unconstrained; constraints are utilized for generation and initialization of starting parent solutions. Constrained optimization may be addressed through imposition of the requirement that (i) the mutation operator (Section 32.4) is formulated to only generate legitimate solutions (often impossible) or (ii) a penalty function is applied to offspring mutations lying outside the constraint bounds in such a manner that they do not become part of the next generation. The objective function explicitly defines the fitness values which may be scaled to positive values (although this is not a requirement, it is sometimes performed to alter the range for ease of implementation).

In early versions of EP applied to continuous parameter optimization (Fogel 1992) the mutation operator is Gaussian with a zero mean and variance obtained for each component of the object variable vector as the square root of a linear transform of the fitness value $\varphi(x)$.

$$x_i(k+1) := x_i(k) + \sqrt{\beta_i(k)\varphi(x_i(k) + \gamma_i)} + N_i(0, 1)$$

where $x(k)$ is the object variable vector, β is the proportionality constant, and γ is an offset parameter. Both β and γ must be set externally for each problem. $N_i(0, 1)$ is the ith independent sample from a Gaussian distribution with zero mean and unit variance.

Several extensions to the finite-state machine formulation of Fogel *et al* (1966) have been offered to address continuous optimization problems as well as to allow for various degrees of parametric self-adaptation (Fogel 1991a, 1992, 1995). There are three main variants of the basic paradigm, identified as follows:

(i) original EP, where continuous function optimization is performed without any self-adaptation mechanism;

(ii) continuous EP where new individuals in the population are inserted directly without iterative generational segmentation (i.e. an individual becomes part of the existing (surviving) population without waiting for the conclusion of a discrete generation; this also known as steady-state selection (Section 28.3) in GAs and $(\mu + 1)$ selection (Chapter 9) in ES);

(iii) self-adaptive EP, which augments the solution vectors with one or more parameters governing the mutation process (e.g. variances, covariances) to permit self-adaptation of these parameters through the same iterative mutation, scoring, and selection process. In addition, self-adaptive EP may also be continuous in the sense of (ii) above.

The original EP is an extension of the formulation of Fogel *et al* (1966) wherein continuous-valued functions replace the discrete alphabets of finite-state machines. The continuous form of EP was investigated by Fogel and Fogel (1993). To properly simulate this algorithmic variant, it is necessary to insert new population members by asynchronous methods (e.g. event-driven interrupts in a true multitasking, real-time operating system). Iterative algorithms running on a single central processing unit (CPU) are much more prevalent, since they are easier to program on today's computers, hence most implementations of EP are performed on a generational (epoch-to-epoch) basis.

Self-adaptive EP is an important extension of the algorithm in that it successfully overcomes the need for explicit user-tuning of the parameters associated with mutation. Global convergence may be obtained even in the presence of suboptimal parameterization, but available processing time is most often a precious resource and any mechanism for optimizing the convergence rate is helpful. As proposed by Fogel (1992, 1995), EP can self-adapt the variances for each individual in the following way:

$$x_i(k+1) := x_i(k) + \upsilon_i(k) * N_i(0, 1)$$
$$\upsilon_i(k+1) := \upsilon_i(k) + [\sigma \upsilon_i(k)]^{1/2} * N_i(0, 1).$$

The variable σ ensures that the variance υ_i remains nonnegative. Fogel (1992) suggests a simple rule wherein $\forall \upsilon_i(k) \leq 0$, $\upsilon_i(k)$ is set to ξ, a value close to but not identically equal to zero (to allow some degree of mutation). The sequence of updating the object variable x_i and variance υ_i was proposed to occur in opposite order from that of ESs (Bäck and Schwefel 1993, Rechenberg 1965, Schwefel 1981). Gehlhaar and Fogel (1996) provide evidence favoring the ordering commonly found in ES.

Further development of this theme led Fogel (1991a, 1992) to extend the procedure to alter the correlation coefficients between components of the object vector. A symmetric correlation coefficient matrix **P** is incorporated into the evolutionary paradigm in addition to the self-adaptation of the standard deviations. The components of **P** are initialized over the interval $[-1, 1]$ and mutated by perturbing each component, again, through the addition of independent realizations from a Gaussian random distribution. Bounding limits are placed upon the resultant mutated variables wherein any mutated coefficient which exceeds the bounds $[-1, 1]$ is reset to the upper or lower limit, respectively. Again, this methodology is similar to that of Schwefel (1981), as perturbations of both the standard deviations and rotation angles (determined by the covariance matrix **P**) allow adaptation to arbitrary contours on the error surface. This self-adaptation through the incorporation of correlated mutations across components of each parent object vector provides a mechanism for expediting the convergence rate of EP.

Fogel (1988) developed different selection operators which utilized tournament competition (Chapter 24) between solution organisms. These operators assigned a number of wins for each solution organism based on a set of individual competitions (using fitness scores as the determining factor) among each solution and each of the q competitors randomly selected from the total population.

10.3   Current directions

Since the explosion of research into evolutionary algorithms in the late 1980s and early 1990s, EP has been applied to a wide range of problem domains with considerable success. Application areas in the current literature include training, construction, and optimization of neural networks, optimal routing (in two, three, and higher dimensions), drug design, bin packing, automatic control, game theory, and optimization of intelligently interactive behaviors of autonomous entities, among many others. Beginning in 1992, annual conferences on EP have brought much of this research into the open where these and other applications as well as basic research have expanded into numerous interdisciplinary realms.

Notable within a small sampling of the current research is the work in neural network design. Early efforts (Porto 1989, Fogel *et al* 1990, McDonnell 1992, and others) focused on utilizing EP for training neural networks to prevent entrapment in local minima. This research showed not only that EP was well suited to training a range of network topologies, but also that it was often more efficient than conventional (e.g. gradient-based) methods and was capable of finding optimal weight sets while escaping local minima points. Later research (Fogel 1992, Angeline *et al* 1994, McDonnell and Waagen 1993) involved simultaneous evolution of both the weights and structure of feedforward and feedback networks. Additional research into the areas of using EP for robustness training (Sebald and Fogel 1992), and for designing fuzzy neural networks for feature selection, pattern clustering, and classification (Brotherton and Simpson 1995) have been very successful as well as instructive.

EP has been also used to solve optimal routing problems. The traveling salesman problem (TSP), one of many in the class of nondeterministic-polynomial-time- (NP-) complete (see Aho *et al* 1974) problems, has been studied extensively. Fogel (1988, 1993) demonstrated the capability of EP to address such problems. A representation was used wherein each of the cities to be visited was listed in order, with candidate solutions being permutations of this listing. A population of random tours is scored with respect to their Euclidean length. Each of the tours is mutated using one of many possible inversion operations (e.g. select two cities in the tour at random and reverse the order of the segment defined by the two cities) to generate an offspring. All of the offspring are then scored, with either elitist or stochastic competition used to cull lower-scoring members from the population. Optimization of the tours was quite rapid. In one such experiment with 1000 cities uniformly distributed, the best tour (after only 4×10^7 function evaluations) was estimated to be within 5–7% of the optimal tour length. Thus, excellent solutions were obtained after searching only an extremely small portion of the total potential search space.

EP has also been utilized in a number of medical applications. For example, the issue of optimizing drug design was researched by Gehlhaar *et al* (1995). EP was utilized to perform a conformational and position search within the binding site of a protein. The search space of small molecules which could potentially dock with the crystallographically determined binding site was explored iteratively guided by a database of crystallographic protein–ligand complexes. Geometries were constrained by known physical (in three dimensions) and chemical bounds. Results demonstrated the efficacy of this technique as it was orders of magnitude faster in finding suitable ligands than previous hands-on methodologies. The probability of successfully predicting the proper binding modes for these ligands was estimated at over 95% using nominal values for the crystallographic binding mode and number of docks attempted. These studies have permitted the rapid development of several candidate drugs which are currently in clinical trials.

The issue of utilizing EP to control systems has been addressed widely (Fogel and Fogel 1990, Fogel 1991a, Page *et al* 1992, and many others). Automatic control of fuzzy heating, ventilation, and air conditioning (HVAC) controllers was addressed by Haffner and Sebald (1993). In this study, a nonlinear, multiple-input, multiple-output (MIMO) model of a HVAC system was used and controlled by a fuzzy controller designed using EP. Typical fuzzy controllers often use trial and error methods to determine parameters and transfer functions, hence they can be quite time consuming with a complex MIMO HVAC system. These experiments used EP to design the membership functions and values (later studies were extended to find rules as well as responsibilities of the primary controller) to automate the tuning procedure. EP worked in an overall search space containing 76 parameters, 10 controller inputs, seven controller outputs, and 80 rules. Simulation results demonstrated that EP was quite effective at choosing the membership functions of the control laboratory and corridor pressures in the model. The synergy of combining EP with fuzzy set constructs proved quite fruitful in reducing the time required to design a stable, functioning HVAC system.

Game theory has always been at the forefront of artificial intelligence research. One interesting game, the iterated prisoner's dilemma, has been studied by numerous investigators (Axelrod 1987, Fogel 1991b, Harrald and Fogel 1996, and others). In this two-person game, each player can choose one of two possible behavioral policies: defection or cooperation. Defection implies increasing one's own reward at the expense of the opposing player, while cooperation implies increasing the reward for both players. If the game is played over a single iteration, the dominant move is defection. If the players' strategies depend on the results of previous iterations, mutual cooperation may possibly become a rational policy, whereas if the sequence of strategies is not correlated, the game degenerates into a series of single plays with defection being the end result. Each player must choose to defect or cooperate on each trial. Table 10.1 describes a general form of the payoff function in the prisoner's dilemma.

Table 10.1. A general form of the payoff matrix for the prisoner's dilemma problem. γ_1 is the payoff to each player for mutual cooperation. γ_2 is the payoff for cooperating when the other player defects. γ_3 is the payoff for defecting when the other player cooperates. γ_4 is the payoff to each player for mutual defection. Entries (α, β) indicate payoffs to players A and B, respectively.

| | | Player B | |
		C	D
	C	(γ_1, γ_1)	(γ_2, γ_3)
Player A			
	D	(γ_3, γ_2)	(γ_4, γ_4)

In addition, the payoff matrix defining the game is subject to the following constraints (Rapoport 1966):

$$2\gamma_1 > \gamma_2 + \gamma_3$$
$$\gamma_3 > \gamma_1 > \gamma_4 > \gamma_2.$$

Both neural network approaches (Harrald and Fogel 1996) and finite-state machine approaches (Fogel 1991b) have been applied to this problem. Finite-state machines are typically used where there are discrete choices between cooperation and defection. Neural networks allow for a continuous range of choices between these two opposite strategies. Results of these preliminary experiments using EP, in general, indicated that mutual cooperation is more likely to occur when the behaviors are limited to the extremes (the finite-state machine representation of the problem), whereas in the neural network continuum behavioral representation of the problem, it is easier to slip into a state of mutual defection.

Development of interactively intelligent behaviors was investigated by Fogel *et al* (1996). EP was used to optimize computer-generated force (CGF) behaviors such that they learned new courses of action adaptively as changes in the environment (i.e. presence or absence of opposing side forces) were encountered. The actions of the CGFs were created at the response of an event scheduler which recognized significant changes in the environment as perceived by the forces under evolution. New plans of action were found during these event periods by invoking an evolutionary program. The iterative EP process was stopped when time or CPU limits were met, and relinquished control of the simulated forces back to the CGF simulator after transmitting newly evolved instruction sets for each simulated unit. This process proved quite successful and offered a significant improvement over other rule-based systems.

10.4 Future research

Important research is currently being conducted into the understanding of the convergence properties of EP, as well as the basic mechanisms of different mutation operators and selection mechanisms. Certainly of great interest is the potential for self-adaptation of exogeneous parameters of the mutation operation (meta and Rmeta-EP), as this not only frees the user from the often difficult task of parameterization, but also provides a built-in, automated mechanism for providing optimal settings throughout a range of problem domains. The number of application areas of this optimization technique is constantly growing. EP, along with the other EC techniques, is being used on previously untenable, often NP-complete, problems which occur quite often in commercial and military problems.

References

Aho A V, Hopcroft J E and Ullman J D 1974 *The Design and Analysis of Computer Algorithms* (Reading, MA: Addison-Wesley) pp 143–5, 318–26

Angeline P, Saunders G and Pollack J 1994 Complete induction of recurrent neural networks *Proc. 3rd Ann. Conf. on Evolutionary Programming (San Diego, CA, 1994)* ed A V Sebald and L J Fogel (Singapore: World Scientific) pp 1–8

Atmar W 1992 On the rules and nature of simulated evolutionary programming *Proc. 1st Ann. Conf. on Evolutionary Programming (La Jolla, CA, 1992)* ed D B Fogel and W Atmar (San Diego, CA: Evolutionary Programming Society) pp 17–26

Axelrod R 1987 The evolution of strategies in the iterated prisoner's dilemma *Genetic Algorithms and Simulated Annealing* ed L Davis (London) pp 32–42

Bäck T and Schwefel H-P 1993 An overview of evolutionary algorithms for parameter optimization *Evolutionary Comput.* **1** 1–23

Brotherton T W and Simpson P K 1995 Dynamic feature set training of neural networks for classification *Evolutionary Programming IV: Proc. 4th Ann. Conf. on Evolutionary Programming (San Diego, CA, 1995)* ed J R McDonnell, R G Reynolds and D B Fogel (Cambridge, MA: MIT Press) pp 83–94

Burton D M 1976 *Elementary Number Theory* (Boston, MA: Allyn and Bacon) p 136–52

Flood M M 1962 Stochastic learning theory applied to choice experiments with cats, dogs and men *Behavioral Sci.* **7** 289–314

Fogel D B 1988 An evolutionary approach to the traveling salesman problem *Biol. Cybernet.* **60** 139–44

——1991a *System Identification through Simulated Evolution* (Needham, MA: Ginn)

——1991b The evolution of intelligent decision making in gaming *Cybernet. Syst.* **22** 223–36

——1992 *Evolving Artificial Intelligence* PhD Dissertation, University of California

——1993 Applying evolutionary programming to selected traveling salesman problems *Cybernet. Syst.* **24** 27–36

——1995 *Evolutionary Computation, Toward a New Philosophy of Machine Intelligence* (Piscataway, NJ: IEEE)

Fogel D B and Fogel L J 1990 Optimal routing of multiple autonomous underwater vehicles through evolutionary programming *Proc. Symp. on Autonomous Underwater Vehicle Technology* (Washington, DC: IEEE Oceanic Engineering Society) pp 44–7

Fogel D B, Fogel L J and Porto V W 1990 Evolving neural networks *Biol. Cybernet.* **63** 487–93

Fogel G B and Fogel D B 1993 Continuous evolutionary programming: analysis and experiments *Cybernet. Syst.* **26** 79–90

Fogel L J 1962 Autonomous automata *Industrial Res.* **4** 14–9

Fogel L J 1964 *On The Organization of Intellect* PhD Dissertation, University of California

Fogel L J and Burgin G H 1969 *Competitive Goal-Seeking through Evolutionary Programming* Air Force Cambridge Research Labratories Final Report Contract AF19(628)-5927

Fogel L J, Owens A J and Walsh M J 1966 *Artificial Intelligence through Simulated Evolution* (New York: Wiley)

Fogel L J, Porto V W and Owen M 1996 An intelligently interactive non-rule-based computer generated force *Proc. 6th Conf. on Computer Generated Forces and Behavioral Representation* (Orlando, FL: Institute for Simulation and Training STRICOM-DMSO) pp 265–70

Gehlhaar D K and Fogel D B 1996 Tuning evolutionary programming for conformationally flexible molecular docking *Proc. 5th Ann. Conf. on Evolutionary Programming (1996)* ed L J Fogel, P J Angeline and T Bäck (Cambridge, MA: MIT Press) pp 419–29

Gehlhaar D K, Verkhivker G, Rejto P A, Fogel D B, Fogel L J and Freer S T 1995 Docking conformationally flexible small molecules into a protein binding site through evolutionary programming *Evolutionary Programming IV: Proc. 4th Ann. Conf. on Evolutionary Programming (San Diego, CA, 1995)* (Cambridge, MA: MIT Press) pp 615–27

Harrald P G and Fogel D B 1996 Evolving continuous behaviors in the iterated prisoner's dilemma *BioSystems* **37** 135–45

Haffner S B and Sebald A V 1993 Computer-aided design of fuzzy HVAC controllers using evolutionary programming *Proc. 2nd Ann. Conf. on Evolutionary Programming (San Diego, CA, 1993)* ed D B Fogel and W Atmar (San Diego, CA: Evolutionary Programming Society) pp 98–107

McDonnell J M 1992 Training neural networks with weight constraints *Proc. 1st Ann. Conf. on Evolutionary Programming (La Jolla, CA, 1992)* ed D B Fogel and W Atmar (San Diego, CA: Evolutionary Programming Society) pp 111–9

McDonnell J M and Waagen D 1993 Neural network structure design by evolutionary programming *Proc. 2nd Ann. Conf. on Evolutionary Programming (San Diego, CA, 1993)* ed D B Fogel and W Atmar (San Diego, CA: Evolutionary Programming Society) pp 79–89

Mealy G H 1955 A method of synthesizing sequential circuits *Bell Syst. Tech. J.* **34** 1054–79

Moore E F 1957 Gedanken-experiments on sequential machines: automata studies *Annals of Mathematical Studies* vol 34 (Princeton, NJ: Princeton University Press) pp 129–53

Page W C, McDonnell J M and Anderson B 1992 An evolutionary programming approach to multi-dimensional path planning *Proc. 1st Ann. Conf. on Evolutionary Programming (La Jolla, CA, 1992)* ed D B Fogel and W Atmar (San Diego, CA: Evolutionary Programming Society) pp 63–70

Porto V W 1989 Evolutionary methods for training neural networks for underwater pattern classification *24th Ann. Asilomar Conf. on Signals, Systems and Computers* vol 2 pp 1015–19

Rapoport A 1966 *Optimal Policies for the Prisoner's Dilemma* University of North Carolina Psychometric Laboratory Technical Report 50

Rechenberg I 1965 *Cybernetic Solution Path of an Experimental Problem* Royal Aircraft Establishment Translation 1122

Schwefel H-P 1981 *Numerical Optimization of Computer Models* (Chichester: Wiley)

Sebald A V and Fogel D B 1992 Design of fault tolerant neural networks for pattern classification *Proc. 1st Ann. Conf. on Evolutionary Programming (La Jolla, CA, 1992)* ed D B Fogel and W Atmar (San Diego, CA: Evolutionary Programming Society) pp 90–9

Walsh M J, Burgin G H and Fogel L J 1970 *Prediction and Control through the Use of Automata and their Evolution* US Navy Final Report Contract N00014-66-C-0284

Further reading

There are several excellent general references available to the reader interested in furthering his or her knowledge in this exciting area of EC. The following books are a few well-written examples providing a good theoretical background in EP as well as other evolutionary algorithms.

1. Bäck T 1996 *Evolutionary Algorithms in Theory and Practice* (New York: Oxford University Press)

2. Fogel D B 1995 *Evolutionary Computation, Toward a New Philosophy of Machine Intelligence* (Piscataway, NJ: IEEE)

3. Schwefel H-P 1981 *Numerical Optimization of Computer Models* (Chichester: Wiley)

4. Schwefel H-P 1995 *Evolution and Optimization Seeking* (New York: Wiley)

11

Derivative methods in genetic programming

Kenneth E Kinnear, Jr

11.1   Introduction

This chapter describes the fundamental concepts of genetic programming (GP) (Koza 1989, 1992). Genetic programming is a form of evolutionary algorithm which is distinguished by a particular set of choices as to representation, genetic operator design, and fitness evaluation. When examined in isolation, these choices define an approach to evolutionary computation (EC) which is considered by some to be a specialization of the genetic algorithm (GA). When considered together, however, these choices define a conceptually different approach to evolutionary computation which leads researchers to explore new and fruitful lines of research and practical applications.

11.2   Genetic programming defined and explained

Genetic programming is implemented as an evolutionary algorithm in which the data structures that undergo adaptation are executable computer programs. Fitness evaluation in genetic programming involves executing these evolved programs. Genetic programming, then, involves an evolution-directed search of the space of possible computer programs for ones which, when executed, will produce the best fitness.

In short, genetic programming breeds computer programs. To create the initial population a large number of computer programs are generated at random. Each of these programs is executed and the results of that execution are used to assign a fitness value to each program. Then a new population of programs, the next generation, is created by directly copying certain selected existing programs, where the selection is based on their fitness values. This population is filled out by creating a number of new offspring programs through genetic operations on existing parent programs which are selected based, again, on their fitness. Then, this new population of programs is again evaluated and a fitness is assigned to each program based on the results of its evaluation. Eventually this process is terminated by the creation and evaluation of a 'correct' program or the recognition of some other specific termination criteria.

More specifically, at the most basic level, genetic programming is defined as a genetic algorithm with some unusual choices made as to the representation of the problem, the genetic operators used to modify that representation, and the fitness evaluation techniques employed.

11.2.1 A specialized representation: executable programs

Any evolutionary algorithm is distinguished by the structures used to represent the problem to be solved. These are the structures which undergo transformation, and in which the potential solutions reside.

Originally, most genetic algorithms used linear strings of bits (Chapter 15) as the structures which evolved (Holland 1975), and the representation of the problem was typically the encoding of these bits as numeric or logical parameters of a variety of algorithms. The evolving structures were often used as parameters to human-coded algorithms. In addition, the bitstrings used were frequently of fixed length, which aided in the translation into parameters for the algorithms involved.

More recently, genetic algorithms have appeared with real-valued numeric sequences used as the evolvable structures, still frequently used as parameters to particular algorithms. In recent years, many genetic algorithm implementations have appeared with sequences which are of variable length, sometimes based on the order of the sequences, and which contain more complex and structured information than parameters to existing algorithms.

The representation used by genetic programming is that of an executable program. There is no single form of executable program which is used by all genetic programming implementations, although many implementations use a tree-structured representation (Chapter 19) highly reminiscent of a LISP functional expression. These representations are almost always of a variable size, though for implementation purposes a maximum size is usually specified.

Figure 11.1 shows an example of a tree-structured representation for a genetic programming implementation. The specific task for which this is a reasonable representation is the learning of a Boolean function from a set of inputs. This figure contains two different types of node (as do most genetic programming representations) which are called functions and terminals. Terminals are usually inputs to the program, although they may also be constants. They are the variables which are set to values external to the program itself prior to the fitness evaluation performed by executing the program. In this example d0 and d1 are the terminals. They can take on binary values of either zero or one.

Functions take inputs and produce outputs and possibly produce side-effects. The inputs can be either a terminal or the output of another function. In the above example, the functions are AND, OR, and NOT. Two of these functions

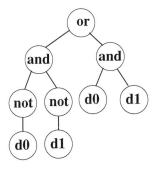

Figure 11.1. Tree-structured representation used in genetic programming.

are functions of two inputs, and one is a function of one input. Each produces a single output and no side effect.

The fitness evaluation for this particular *individual* is determined by the effectiveness with which it will produce the correct logical output for all of the test cases against which it is tested.

One way to characterize the design of a representation for an application of genetic programming to a particular problem is to view it as the design of a language, and this can be a useful point of view. Perhaps it is more useful, however, to view the design of a genetic programming representation as that of the design of a virtual machine—since usually the execution engine must be designed and constructed as well as the representation or language that is executed.

The representation for the program (i.e. the definition of the functions and terminals) must be designed along with the virtual machine that is to execute them. Rarely are the programs evolved in genetic programming given direct control of the central processor of a computer (although see the article by Nordin (1994)). Usually, these programs are interpreted under control of a virtual machine which defines the functions and terminals. This includes the functions which process the data, the terminals that provide the inputs to the programs, and any control functions whose purpose is to affect the execution flow of the program.

As part of this virtual machine design task, it is important to note that the output of any function or the value of any terminal may be used as the input to any function. Initially, this often seems to be a trivial problem, but when actually performing the design of the representation and virtual machine to execute that representation, it frequently looms rather large. Two solutions are typically used for this problem. One approach is to design the virtual machine, represented by the choice of functions and terminals, to use only a single data type. In this way, the output of any function or the value of any terminal is acceptable as input to any function. A second approach is to allow more than one data type to exist in the virtual machine. Each function must then be defined to operate on any of the existing data types. Implicit coercions are performed by each function on its input to convert the data type that it receives to one that it is more normally defined to process. Even after handling the data type problem, functions must be defined over the entire possible range of argument values. Simple arithmetic division must be defined to return some value even when division by zero is attempted.

It is important to note that the definition of functions and the virtual machine that executes them is not restricted to functions whose only action is to provide a single output value based on their inputs. Genetic programming functions are often defined whose primary purpose is the actions they take by virtue of their side-effects. These functions must return some value as well, but their real purpose is interaction with an environment external to the genetic programming system.

An additional type of side-effect producing function is one that implements a control structure within the virtual machine defined to execute the genetically evolved program. All of the common programming control constructs such as if–then–else, while–do, for, and others have been implemented as evolvable control constructs within genetic programming systems. Looping constructs must be protected in such a way that they will never loop forever, and usually have an arbitrary limit set on the number of loops which they will execute.

As part of the initialization of a genetic programming run, a large number of individual programs are generated at random. This is relatively straightforward, since the genetic programming system is supplied with information about the number of arguments required by each function, as well as all of the available terminals. Random program trees are generated using this information, typically of a relatively small size. The program trees will tend to grow quickly to be quite large in the absence of some explicit evolutionary pressure toward small size or some simple hard-coded limits to growth.

11.2.2   Genetic operators for evolving programs

The second specific design approach that distinguishes genetic programming from other types of genetic algorithm is the design of the genetic operators. Having decided to represent the problem to be solved as a population of computer programs, the essence of an evolutionary algorithm is to evaluate the fitness of the individuals in the population and then to create new members of the population based in some way on the individuals which have the highest fitness in the current population.

In genetic algorithms, *recombination* is typically the key genetic operator employed, with some utility ascribed to mutation as well. In this way, genetic programming is no different from any other genetic algorithm. Genetic algorithms usually have genetic material organized in a linear fashion and the recombination, or crossover, algorithm defined for such genetic material is quite straightforward (see Section 33.1). The usual representation of genetic programming programs as tree-structured combinations of functions and terminals requires a different form of recombination algorithm. A major step in the invention of genetic programming was the design of a recombination operator which would simply and easily allow the creation of an offspring program tree using as inputs the program trees of two individuals of generally high fitness as parents (Cramer 1985, Koza 1989, 1992).

In any evolutionary algorithm it is vitally important that the fitness of the offspring be related to that of the parents, or else the process degenerates into one of random search across whatever representation space was chosen. It is equally vital that some variation, indeed heritable variation, be introduced into the offspring's fitness, otherwise no improvement toward an optimum is possible.

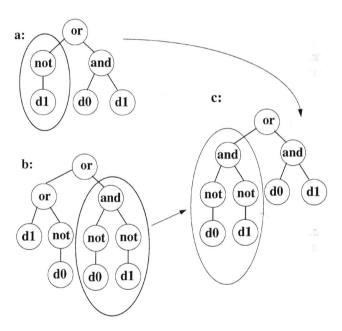

Figure 11.2. Recombination in genetic programming.

The tree-structured genetic material usually used in genetic programming has a particularly elegant recombination operator that may be defined for it. In figure 11.2, there are two parent program trees, (a) and (b). They are to be recombined through crossover to create an offspring program tree (c). A subtree is chosen in each of the parents, and the offspring is created by inserting the subtree chosen from (b) into the place where the subtree was chosen in (a). This very simply creates an offspring program tree which preserves the same constraints concerning the number of inputs to each function as each parent tree. In practice it yields a offspring tree whose fitness has enough relationship to that of its parents to support the evolutionary search process. Variations in this crossover approach are easy to imagine, and are currently the subject of considerable active research in the genetic programming community (D'haeseleer 1994, Teller 1996).

Mutation (Chapter 32) is a genetic operator which can be applied to a single parent program tree to create an offspring tree. The typical mutation operator used selects a point inside a parent tree, and generates a new random subtree to replace the selected subtree. This random subtree is usually generated by the same procedure used to generate the initial population of program trees.

11.2.3 Fitness evaluation of genetically evolved programs

Finally, then, the last detailed distinction between genetic programming and a more usual implementation of the genetic algorithm is that of the assignment of a fitness value for a individual.

In genetic programming, the representation of the individual is a program which, when executed under control of a defined virtual machine, implements some algorithm. It may do this by returning some value (as would be the case for a system to learn a specific Boolean function) or it might do this through the performance of some task through the use of functions which have side-effects that act on a simulated (or even the real) world.

The results of the program's execution are evaluated in some way, and this evaluation represents the fitness of the individual. This fitness is used to drive the selection process for copying into the next generation or for the selection of parents to undergo genetic operations yielding offspring. Any selection operator from those presented in Chapters 22–33 can be used.

There is certainly a desire to evolve programs using genetic programming that are 'general', that is to say that they will not only correctly process the fitness cases on which they are evolved, but will process correctly any fitness cases which could be presented to them. Clearly, in the cases where there are infinitely many possible cases, such as evolving a general sorting algorithm (Kinnear 1993), the evolutionary process can only be driven by a very limited number of fitness cases. Many of the lessons from machine learning on the tradeoffs between generality and performance on training cases have been helpful to genetic programming researchers, particularly those from decision tree approaches to machine learning (Iba et al 1994).

11.3 The development of genetic programming

LISP was the language in which the ideas which led to genetic programming were first developed (Cramer 1985, Koza 1989, 1992). LISP has always been one of the preeminent language choices for implementation where programs need to be treated as data. In this case, programs are data while they are being evolved, and are only considered executable when they are undergoing fitness evaluation.

As genetic programming itself evolved in LISP, the programs that were executed began to look less and less like LISP programs. They continued to be tree structured but soon few if any of the functions used in the evolved programs were standard LISP functions. Around 1992 many people implemented genetic programming systems in C and C++, along with many other programming languages. Today, other than a frequent habit of printing the representation of tree-structured genetic programs in a LISP-like syntax, there is no particular connection between genetic programming and LISP.

There are many public domain implementations of genetic programming in a wide variety of programming languages. For further details, see the reading list at the end of this section.

11.4   The value of genetic programming

Genetic programming is defined as a variation on the theme of genetic algorithms through some specific selections of representation, genetic operators appropriate to that representation, and fitness evaluation as execution of that representation in a virtual machine. Taken in isolation, these three elements do not capture the value or promise of genetic programming. What makes genetic programming interesting is the conceptual shift of the problem being solved by the genetic algorithm. A genetic algorithm searches for something, and genetic programming shifts the search from that of parameter discovery for some existing algorithm designed to solve a problem to a search for a program (or algorithm) to solve the problem directly. This shift has a number of ramifications.

- This conceptualization of evolving computer programs is powerful in part because it can change the way that we think about solving problems. Through experience, it has become natural to think about solving problems through a process of human-oriented program discovery. Genetic programming allows us to join this approach to problem solving with powerful EC-based search techniques.

  An example of this is a variation of genetic programming called stack genetic programming (Perkis 1994), where the program is a variable-length linear string of functions and terminals, and the argument passing is defined to be on a stack. The genetic operators in a linear system such as this are much closer to the traditional genetic algorithm operators, but the execution and fitness evaluation (possibly including side-effects) is equivalent to any other sort of genetic programming. The characteristics of stack genetic programming have not yet been well explored but it is clear that it has rather different strengths and weaknesses than does traditional genetic programming.

Many of the approaches to simulation of adaptive behavior involve simple programs designed to control *animats*. The conceptualization of evolving computer programs as presented by genetic programming fits well with work on evolving adaptive entities (Reynolds 1994, Sims 1994).

- There has been a realization that not only can we evolve programs that are built from human-created functions and terminals, but that the functions from which they are built can evolve as well. Koza's invention of automatically defined functions (ADFs) (Koza 1994) is one such example of this realization. ADFs allow the definitions of certain subfunctions to evolve even while the functions that call them are evolving. For certain classes of problems, ADFs result in considerable increases in performance (Koza 1994, Angeline and Pollack 1993, Kinnear 1994).

- Genetic programming is capable of integrating a wide variety of existing capabilities, and has potential to tie together several complementary subsystems into an overall hybrid system. The functions need not be simple arithmetic or logical operators, but could instead be fast Fourier transforms, GMDH systems, or other complex building blocks. They could even be the results of other evolutionary computation algorithms.

- The genetic operators that create offspring programs from parent programs are themselves programs. These programs can also be evolved either as part of a separate process, or in a coevolutionary way with the programs on which they operate. While any evolutionary computation algorithm could have parameters that affect the genetic operators be part of the evolutionary process, genetic programming provides a natural way to let the operators (defined as programs) evolve directly (Teller 1996, Angeline 1996).

- Genetic programming naturally enhances the possibility for increasingly indirect evolution. As an example of the possibilities, genetic programming has been used to evolve grammars which, when executed, produce the structure of an untrained neural network. These neural networks are then trained, and the trained networks are then evaluated on a test set. The results of this evaluation are then used as the fitnesses of the evolved grammars (Gruau 1993).

This last example is a step along the path toward modeling embryonic development in genetic programming. The opportunity exists to evolve programs whose results are themselves programs. These resulting programs are then executed and their values or side-effects are evaluated—and become the fitness for the original evolving, program creating programs. The analogy to natural embryonic development is clear, where the genetic material, the genotype, produces through development a body, the phenotype, which then either does or does not produce offspring, the fitness (Kauffman 1993).

Genetic programming is valuable in part because we find it natural to examine issues such as those mentioned above when we think about evolutionary computation from the genetic programming perspective.

References

Angeline P J 1996 Two self-adaptive crossover operators for genetic programming *Advances in Genetic Programming 2* ed P J Angeline and K E Kinnear Jr (Cambridge, MA: MIT Press)

Angeline P J and Pollack J B 1993 Competitive environments evolve better solutions for complex tasks *Proc. 5th Int. Conf. on Genetic Algorithms (Urbana-Champaign, IL, July 1993)* ed S Forrest (San Mateo, CA: Morgan Kaufmann)

Cramer N L 1985 A representation of the adaptive generation of simple sequential programs *Proc. 1st Int. Conf. on Genetic Algorithms (Pittsburgh, PA, July 1985)* ed J J Grefenstette (Hillsdale, NJ: Erlbaum)

D'haeseleer P 1994 Context preserving crossover in genetic programming *1st IEEE Conf. on Evolutionary Computation (Orlando, FL, June 1994)* (Piscataway, NJ: IEEE)

Gruau F 1993 Genetic synthesis of modular neural networks *Proc. 5th Int. Conf. on Genetic Algorithms (Urbana-Champaign, IL, July 1993)* ed S Forrest (San Mateo, CA: Morgan Kaufmann)

Holland J H 1975 *Adaptation in Natural and Artificial Systems* (Ann Arbor, MI: The University of Michigan Press)

Iba H, de Garis H and Sato T 1994 Genetic programming using a minimum description length principle *Advances in Genetic Programming* ed K E Kinnear Jr (Cambridge, MA: MIT Press)

Kauffman S A 1993 *The Origins of Order: Self-Organization and Selection in Evolution* (New York: Oxford University Press)

Kinnear K E Jr 1993 Generality and difficulty in genetic programming: evolving a sort *Proc. 5th Int. Conf. on Genetic Algorithms (Urbana-Champaign, IL, July, 1993)* ed S Forrest (San Mateo, CA: Morgan Kaufmann)

——1994 Alternatives in automatic function definition: a comparison of performance *Advances in Genetic Programming* ed K E Kinnear Jr (Cambridge, MA: MIT Press)

Koza J R 1989 Hierarchical genetic algorithms operating on populations of computer programs *Proc. 11th Int. Joint Conf. on Artificial Intelligence* (San Mateo, CA: Morgan Kaufmann)

Nordin P 1994 A compiling genetic programming system that directly manipulates the machine code *Advances in Genetic Programming* ed K E Kinnear Jr (Cambridge, MA: MIT Press)

Perkis T 1994 Stack-based genetic programming *Proc. 1st IEEE Int. Conf. on Evolutionary Computation (Orlando, FL, June 1994)* (Piscataway, NJ: IEEE)

Reynolds C R 1994 Competition, coevolution and the game of tag *Artificial Life IV: Proc. 4th Int. Workshop on the Synthesis and Simulation of Living Systems* ed R A Brooks and P Maes (Cambridge, MA: MIT Press)

Sims K 1994 Evolving 3D morphology and behavior by competition *Artificial Life IV: Proc. 4th Int. Workshop on the Synthesis and Simulation of Living Systems* ed R A Brooks and P Maes (Cambridge, MA: MIT Press)

Teller A 1996 Evolving programmers: the co-evolution of intelligent recombination operators *Advances in Genetic Programming 2* ed P J Angeline and K E Kinnear Jr (Cambridge, MA: MIT Press)

Further reading

1. Koza J R 1992 *Genetic Programming* (Cambridge, MA: MIT Press)

    The first book on the subject. Contains full instructions on the possible details of
    carrying out genetic programming, as well as a complete explanation of genetic
    algorithms (on which genetic programming is based). Also contains 11 chapters
    showing applications of genetic programming to a wide variety of typical artificial
    intelligence, machine learning, and sometimes practical problems. Gives many
    examples of how to design a representation of a problem for genetic programming.

2. Koza J R 1994 *Genetic Programming II* (Cambridge, MA: MIT Press)

    A book principally about automatically defined functions (ADFs). Shows the
    applications of ADFs to a wide variety of problems. The problems shown in this
    volume are considerably more complex than those shown in *Genetic Programming*,
    and there is much less introductory material.

3. Kinnear K E Jr (ed) 1994 *Advances in Genetic Programming* (Cambridge, MA:
    MIT Press)

    Contains a short introduction to genetic programming, and 22 research papers on
    the theory, implementation, and application of genetic programming. The papers are
    typically longer than those in a technical conference and allow a deeper exploration
    of the topics involved. Shows a wide range of applications of genetic programming,
    as well as useful theory and practice in genetic programming implementation.

4. Forrest S (ed) 1993 *Proc. 5th Int. Conf. on Genetic Algorithms (Urbana-Champaign,
    IL, July 1993)* (San Mateo, CA: Morgan Kaufmann)

    Contains several interesting papers on genetic programming.

5. 1994 *1st IEEE Conf. on Evolutionary Computation (Orlando, FL, June 1994)*
    (Piscataway, NJ: IEEE)

    Contains many papers on genetic programming as well as a wide assortment of
    other EC-based papers.

6. Eshelman L J (ed) 1995 *Proc. 6th Int. Conf. on Genetic Algorithms (Pittsburgh, PA,
    July 1995)* (Cambridge, MA: MIT Press)

    Contains a considerable number of applications of genetic programming to
    increasingly diverse areas.

7. Angeline P J and Kinnear K E Jr (eds) 1996 *Advances in Genetic Programming II*
    (Cambridge, MA: MIT Press)

    A volume devoted exclusively to research papers on genetic programming, each
    longer and more in depth than those presented in most conference proceedings.

8. Kauffman S A 1993 *The Origins of Order: Self-Organization and Selection in
    Evolution* (New York: Oxford University Press)

A tour-de-force of interesting ideas, many of them applicable to genetic programming as well as other branches of evolutionary computation.

9. ftp.io.com pub/genetic-programming

An anonymous ftp site with considerable public domain information and implementations of genetic programming systems. This is a volunteer site, so its lifetime is unknown.

12

Learning classifier systems

Robert E Smith

12.1 Introduction

The learning classifier system (LCS) (Goldberg 1989, Holland *et al* 1986) is often referred to as the primary machine learning technique that employs genetic algorithms (GAs). It is also often described as a production system framework with a genetic algorithm as the primary rule discovery method. However, the details of LCS operation vary widely from one implementation to another. In fact, no standard version of the LCS exists. In many ways, the LCS is more of a concept than an algorithm. To explain details of the LCS concept, this article will begin by introducing the type of machine learning problem most often associated with the LCS. This discussion will be followed by a overview of the LCS, in its most common form. Final sections will introduce the more complex issues involved in LCSs.

12.2 Types of learning problem

To introduce the LCS, it will be useful to describe types of machine learning problem. Often, in the literature, machine learning problems are described in terms of cognitive psychology or animal behavior. This discussion will attempt to relate the terms used in machine learning to engineering control.

Consider the generic control problem shown in figure 12.1. In this problem, inputs from an external control system, combined with uncontrollable disturbances from other sources, change the state of the plant. These changes in state are reflected in the state information provided by the plant. Note that, in general, the state information can be incomplete and noisy.

Consider the *supervised learning* problem shown in figure 12.2 (Barto 1990). In this problem, an inverse plant model (or teacher) is available that provides errors directly in terms of control actions. Given this direct error feedback, the parameters of the control system can be adjusted by means of gradient descent, to minimize error in control actions. Note that this is the method used in the neural network backpropagation algorithm.

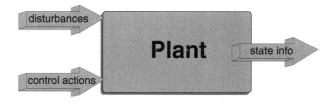

Figure 12.1. A generic control problem.

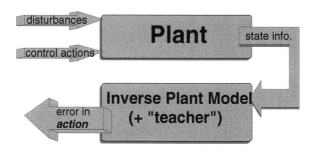

Figure 12.2. A supervised learning problem.

Now consider the *reinforcement learning* problem shown in figure 12.3 (Barto 1990). Here, no inverse plant model is available. However, a critic is available that indicates error in the state information from the plant. Because error is not directly provided in terms of control actions, the parameters of the controller cannot be directly adjusted by methods such as gradient descent.

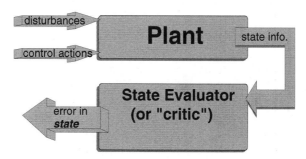

Figure 12.3. A reinforcement learning problem.

The remaining discussion will consider the control problem to operate as a *Markov decision problem*. That is, the control problem operates in discrete time steps, the plant is always in one of a finite number of discrete states, and a finite, discrete number of control actions are available. At each time step, the control action alters the probability of moving the plant from the current state to any other state. Note that deterministic environments are a specific case. Although this discussion will limit itself to discrete problems, most of the points made can be related directly to continuous problems.

A characteristic of many reinforcement learning problems is that one may need to consider a sequence of control actions and their results to determine how to improve the controller. One can examine the implications of this by associating a *reward* or *cost* with each control action. The error in state in figure 12.3 can be thought of as a cost. One can consider the long-term effects of an action formally as the *expected, infinite-horizon discounted cost*:

$$\sum_{t=0}^{t=\infty} \lambda^t c_t$$

where $0 \leq \lambda \leq 1$ is the *discount parameter*, and c_t is the cost of the action taken at time t.

To describe a strategy for picking actions, consider the following approach: for each action u associated with a state i, assign a value $Q(i, u)$. A 'greedy' strategy is to select the action associated with the best Q at every time step. Therefore, an optimum setting for the Q-values is one in which a 'greedy' strategy leads to the minimum expected, infinite-horizon discounted cost. *Q-learning* is a method that yields optimal Q-values in restricted situations. Consider beginning with random settings for each Q-value, and updating each Q-value on-line as follows:

$$Q_{t+1}(i, u_t) = Q_t(i, u)(1 - \alpha) + \alpha \left[c_i(u_t) + \lambda \min Q(j, u_{t+1}) \right]$$

where $\min Q(j, u_{t+1})$ is the minimum Q available in state j, which is the state arrived in after action u_t is taken in state i (Barto *et al* 1991, Watkins 1989). The parameter α is a learning rate parameter that is typically set to a small value between zero and one. Arguments based on *dynamic programming* and *Bellman optimality* show that if each state–action pair is tried an infinite number of times, this procedure results in optimal Q-values. Certainly, it is impractical to try every state–action pair an infinite number of times. With finite exploration, Q-values can often be arrived at that are approximately optimal. Regardless of the method employed to update a strategy in a reinforcement learning problem, this exploration–exploitation dilemma always exists.

Another difficulty in the Q-value approach is that it requires storage of a separate Q-value for each state–action pair. In a more practical approach, one could store a Q-value for a group of state–action pairs that share the same characteristics. However, it is not clear how state–action pairs should be grouped. In many ways, the LCS can be thought of as a GA-based technique for grouping state–action pairs.

12.3 Learning classifier system introduction

Consider the following method for representing a state–action pair in a reinforcement learning problem: encode a state in binary, and couple it to an action, which is also encoded in binary. In other words, the string

$$0\ 1\ 1\ 0\ /\ 0\ 1\ 0$$

represents one of 16 states and one of eight actions. This string can also be seen as a rule that says 'IF in state 0 1 1 0, THEN take action 0 1 0'. In an LCS, such a rule is called a classifier. One can easily associate a Q-value, or other performance measures, with any given classifier.

Now consider generalizing over actions by introducing a 'don't care' character (#) into the state portion of a classifier. In other words, the string

$$\#\ 1\ 1\ \#\ /\ 0\ 1\ 0$$

is a rule that says 'IF in state 0 1 1 0 OR state 0 1 1 1 OR state 1 1 1 0 OR state 1 1 1 1, THEN take action 0 1 0'. The introduction of this generality allows an LCS to represent clusters of states and associated actions. By using the genetic algorithm to search for such strings, one can search for ways of clustering states together, such that they can be assigned joint performance statistics, such as Q-values.

Note, however, that Q-learning is not the most common method of credit assignment in LCSs. The most common method is called the *bucket brigade*

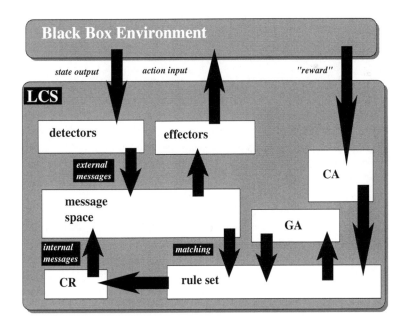

Figure 12.4. The structure of an LCS.

algorithm for updating a classifier performance statistic called *strength*. Details of the bucket brigade algorithm will be introduced later in this section.

The structure of a typical LCS is shown in figure 12.4. This is what is known as a *stimulus–response* LCS, since no internal messages are used as memory. Details of internal message posting in LCSs will be discussed later. In this system, detectors encode state information from an environment into binary messages, which are matched against a list of rules called classifiers. The classifiers used are of the form

IF (condition) THEN (action).

The operational cycle of this LCS is:

(i)   Detectors post environmental messages on the message list.

(ii)   All classifiers are matched against all messages on the message list.

(iii)   Fully matched classifiers are selected to act.

(iv)   A conflict resolution (CR) mechanism narrows the list of active classifiers to eliminate contradictory actions.

(v)   The message list is cleared.

(vi)   The CR-selected classifiers post their messages.

(vii)   Effectors read the messages from the list, and take appropriate actions in the environment.

(viii) If a reward (or cost) signal is received, it is used by a credit allocation (CA) system to update parameters associated with the individual classifiers (such as the traditional strength measure, Q-like values, or other measures (Booker 1989, Smith 1991)).

12.4 'Michigan' and 'Pitt' style learning classifier systems

There are two methods of using the genetic algorithm in LCSs. One is for each genetic algorithm population member to represent an entire set of rules for the problem at hand. This type of LCS is typified by Smith's LS-1 which was developed at the University of Pittsburgh. Often, this type of LCS is called the 'Pitt' approach. Another approach is for each genetic algorithm population member to represent a single rule. This type of LCS is typified by the CS-1 of Holland and Reitman (1978), which was developed at the University of Michigan, and is often called the 'Michigan' approach.

In the 'Pitt' approach, crossover and other operators are often employed that change the number of rules in any given population member. The 'Pitt' approach has the advantage of evaluating a complete solution within each genetic algorithm individual. Therefore, the genetic algorithm can converge to a homogeneous population, as in an optimization problem, with the best individual located by the genetic algorithm search acting as the solution. The disadvantage is that each genetic algorithm population member must be completely evaluated as a rule set. This entails a large computational expense, and may preclude on-line learning in many situations.

In the 'Michigan' approach, one need only evaluate a single rule set, that comprised by the entire population. However, one cannot use the usual genetic algorithm procedures that will converge to a homogeneous population, since one rule is not likely to solve the entire problem. Therefore, one must *coevolve* a set of cooperative rules that jointly solve the problem. This requires a genetic algorithm procedure that yields a diverse population at steady state, in a fashion that is similar to sharing (Deb and Goldberg 1989, Goldberg and Richardson 1987), or other multimodal genetic algorithm procedures. In some cases simply dividing reward between similar classifiers that fire can yield sharing-like effects (Horn *et al* 1994).

12.5   The bucket brigade algorithm (implicit form)

As was noted earlier, the bucket brigade algorithm is the most common form of credit allocation for LCSs. In the bucket brigade, each classifier has a strength, S, which plays a role analogous to a Q-value. The bucket brigade operates as follows:

(i)   Classifier A is selected to act at time t.
(ii) Reward r_t is assigned in response to this action.
(iii) Classifier B is selected to act at time $t + 1$.
(iv) The strength of classifier A is updated as follows:

$$S_A^{t+1} = S_A^t(1 - \alpha) + \alpha\,[r_t + (\lambda S_B)]\,.$$

(v)   The algorithm repeats.

Note that this is the *implicit* form of the bucket brigade, first introduced by Wilson (Goldberg 1989, Wilson 1985).

Note that this algorithm is essentially equivalent to Q-learning, but with one important difference. In this case, classifier A's strength is updated with the strength of the classifier that actually acts (classifier B). In Q-learning, the Q-value for the rule at time t is updated with the best Q-valued rule that matches the state at time $t + 1$, whether that rule is selected to act at time $t + 1$ or not. This difference is key to the convergence properties associated with Q-learning. However, it is interesting to note that recent empirical studies have indicated that the bucket brigade (and similar procedures) may be superior to Q-learning in some situations (Rummery and Niranjan 1994, Twardowski 1993).

A wide variety of variations of the bucket brigade exits. Some include a variety of *taxes*, which degrade strength based on the number of times a classifier has matched and fired and the number of generations since the classifier's creation. or other features. Some variations include a variety of methods for using classifier strength in conflict resolution through strength-based *bidding procedures* (Holland *et al* 1986). However, how these techniques fit into the broader context of machine learning, through similar algorithms such as Q-learning, remains a topic of research.

In many LCSs, strength is used as fitness in the genetic algorithm. However, a promising recent study indicates that other measures of classifier utility may be more effective (Wilson 1995).

12.6   Internal messages

The LCS discussed to this point has operated entirely in stimulus–response mode. That is, it possesses no internal memory that influences which rule fires. In a more advanced form of the LCS, the action sections of the rule are internal messages that are posted on the message list. Classifiers have a condition that matches environmental messages (those which are posted by the environment) and a condition that matches internal messages (those posted by other classifiers). Some internal messages will cause effectors to fire (causing actions in the environment), and others simply act as internal memory for the LCS.

The operational cycle of a LCS with internal memory is as follows:

(i)   Detectors post environmental messages on the message list.
(ii)   All classifiers are matched against all messages on the message list.
(iii) Fully matched classifiers are selected to act.
(iv)   A conflict resolution (CR) mechanism narrows the list of active classifiers to eliminate contradictory actions, and to cope with restrictions on the number of messages that can be posted.
(v)   The message list is cleared.
(vi)   The CR-selected classifiers post their messages.
(vii) Effectors read the messages from the list, and take appropriate actions in the environment.
(viii) If a reward (or cost) signal is received, it updates parameters associated with the individual classifiers.

In LCSs with internal messages, the bucket brigade can be used in its original, explicit form. In this form, the next rule that acts is 'linked' to the previous rule through an internal message. Otherwise, the mechanics are similar to those noted above. Once classifiers are linked by internal messages, they can form *rule chains* that express complex sequences of actions.

12.7   Parasites

The possibility of rule chains introduced by internal messages, and by 'payback' credit allocation schemes such as the bucket brigade or Q-learning, also introduces the possibility of rule *parasites*. Simply stated, parasites are rules that obtain fitness through their participation in a rule chain or a sequence of LCS actions, but serve no useful purpose in the problem environment. In some cases, parasite rules can prosper, while actually degrading overall system performance.

A simple example of parasite rules in LCSs is given by Smith (1994). In this study, a simple problem is constructed where the only performance objective is to exploit internal messages as internal memory. Although fairly effective rule sets were evolved in this problem, parasites evolved that exploited the bucket brigade, and the existing rule chains, but that were incorrect for overall system performance. This study speculates that such parasites may be an inevitable consequence in systems that use temporal credit assignment (such as the bucket brigade) and evolve internal memory processing.

12.8 Variations of the learning classification system

As was stated earlier, this article only outlines the basic details of the LCS concept. It is important to note that many variations of the LCS exist. These include:

- *Variations in representation and matching procedures.* The {1, 0, #} representation is by no means defining to the LCS approach. For instance, Valenzuela-Rendón (1991) has experimented with a fuzzy representation of classifier conditions, actions, and internal messages. Higher-cardinality alphabets are also possible. Other variations include simple changes in the procedures that match classifiers to messages. For instance, sometimes partial matches between messages and classifier conditions are allowed (Booker 1982, 1985). In other systems, classifiers have multiple environmental or internal message conditions. In some suggested variations, multiple internal messages are allowed on the message list at the same time.

- *Variations in credit assignment.* As was noted above, a variety of credit assignment schemes can be used in LCSs. The examination of such schemes is the subject of much broader research in the reinforcement learning literature. Alternate schemes for the LCS prominently include *epochal* techniques, where the history of reward (or cost) signals is recorded for some period of time, and classifiers that act during the epoch are updated simultaneously.

- *Variations in discovery operators.* In addition to various versions of the genetic algorithm, LCSs often employ other discovery operators. The most common nongenetic discovery operators are those which create new rules to match messages for which no current rules exist. Such operators are often called *create, covering,* or *guessing* operators (Wilson 1985). Other covering operators are suggested that create new rules that suggest actions not accessible in the current rule set (Riolo 1986, Robertson and Riolo 1988).

12.9 Final comments

As was stated in section 12.1, the LCS remains a concept, more than a specific algorithm. Therefore, some of the details discussed here are necessarily sketchy. However, recent research on the LCS is promising. For a particularly clear examination of a simplified LCS, see a recent article by Wilson (1994). This article also recommends clear avenues for LCS research and development. Interesting LCS applications are also appearing in the literature (Smith and Dike 1995).

Given the robust character of evolutionary computation algorithms, the machine learning techniques suggested by the LCS concept indicate a powerful avenue of future evolutionary computation application.

References

Barto A G 1990 *Some Learning Tasks from a Control Perspective* COINS Technical Report 90-122, University of Massachusetts

Barto A G, Bradtke S J and Singh S P 1991 *Real-time Learning and Control using Asynchronous Dynamic Programming* COINS Technical Report 91-57, University of Massachusetts

Booker L B 1982 Intelligent behavior as an adaptation to the task environment *Dissertations Abstracts Int.* **43** 469B; University Microfilms 8214966

——1985 Improving the performance of genetic algorithms in classifier systems *Proc. Int. Conf. on Genetic Algorithms and Their Applications* pp 80–92

——1989 Triggered rule discovery in classifier systems *Proc. 3rd Int. Conf. on Genetic Algorithms (Fairfax, VA, June 1989)* ed J D Schaffer (San Mateo, CA: Morgan Kaufmann) pp 265–74

Deb K and Goldberg D E 1989 An investigation of niche and species formation in genetic function optimization *Proc. 3rd Int. Conf. on Genetic Algorithms (Fairfax, VA, June 1989)* ed J D Schaffer (San Mateo, CA: Morgan Kaufmann) pp 42–50

Goldberg D E 1989 *Genetic Algorithms in Search, Optimization, and Machine Learning* (Reading, MA: Addison-Wesley)

Goldberg D E and Richardson J 1987 Genetic algorithms with sharing for multimodal function optimization *Proc. 2nd Int. Conf. on Genetic Algorithms (Cambridge, MA, 1987)* ed J J Grefenstette (Hillsdale, NJ: Erlbaum) pp 41–9

Holland J H, Holyoak K J, Nisbett R E and Thagard P R 1986 *Induction: Processes of Inference, Learning, and Discovery* (Cambridge, MA: MIT Press)

Holland J H and Reitman J S 1978 Cognitive systems based on adaptive algorithms *Pattern Directed Inference Systems* ed D A Waterman and F Hayes-Roth (New York: Academic) pp 313–29

Horn J, Goldberg D E and Deb K 1994 Implicit niching in a learning classifier system: Nature's way *Evolutionary Comput.* **2** 37–66

Riolo R L 1986 *CFS-C: a Package of Domain Independent Subroutines for Implementing Classifier Systems in Arbitrary User-defined Environments* University of Michigan, Logic of Computers Group, Technical Report

Robertson G G and Riolo R 1988 A tale of two classifier systems *Machine Learning* **3** 139–60

Rummery G A and Niranjan M 1994 *On-line Q-Learning using Connectionist Systems* Cambridge University Technical Report CUED/F-INFENG/TR 166

Smith R E 1991 *Default Hierarchy Formation and Memory Exploitation in Learning Classifier Systems* University of Alabama TCGA Report 91003; PhD Dissertation; University Microfilms 91-30 265

Smith R E and Dike B A 1995 Learning novel fighter combat maneuver rules via genetic algorithms *Int. J. Expert Syst.* **8** 84–94

Twardowski K 1993 Credit assignment for pole balancing with learning classifier systems *Proc. 5th Int. Conf. on Genetic Algorithms (Urbana-Champaign, IL, July 1993)* ed S Forrest (San Mateo, CA: Morgan Kaufmann) pp 238–45

Valenzuela-Rendón M 1991 The fuzzy classifier system: a classifier system for continuously varying variables *Proc. 4th Int. Conf. on Genetic Algorithms (San Diego, CA, July 1991)* ed R Belew and L Booker (San Mateo, CA: Morgan Kaufmann) pp 346–53

Watkins J C H 1989 Learning with delayed rewards

Wilson S W 1985 Knowledge growth in an artificial animal *Proc. Int. Conf. on Genetic Algorithms and Their Applications* pp 16–23

——1994 ZCS: a zeroth level classifier system *Evolutionary Comput.* **2** 1–18

——1995 Classifier fitness based on accuracy *Evolutionary Comput.* **3** 149–76

13

Hybrid methods

Zbigniew Michalewicz

There is some experimental evidence (Davis 1991, Michalewicz 1993) that the enhancement of evolutionary methods by some additional (problem-specific) heuristics, domain knowledge, or existing algorithms can result in a system with outstanding performance. Such enhanced systems are often referred to as *hybrid* evolutionary systems.

Several researchers have recognized the potential of such hybridization of evolutionary systems. Davis (1991, p 56) wrote:

> When I talk to the user, I explain that my plan is to hybridize the genetic algorithm technique and the current algorithm by employing the following three principles:
>
> - *Use the Current Encoding.* Use the current algorithm's encoding technique in the hybrid algorithm.
> - *Hybridize Where Possible.* Incorporate the positive features of the current algorithm in the hybrid algorithm.
> - *Adapt the Genetic Operators.* Create crossover and mutation operators for the new type of encoding by analogy with bit string crossover and mutation operators. Incorporate domain-based heuristics as operators as well.
>
> [...] I use the term *hybrid genetic algorithm* for algorithms created by applying these three principles.

The above three principles emerged as a result of countless experiments of many researchers, who tried to 'tune' their evolutionary algorithms to some problem at hand, that is, to create 'the best' algorithm for a particular class of problems. For example, during the last 15 years, various application-specific variations of evolutionary algorithms have been reported (Michalewicz 1996); these variations included variable-length strings (including strings whose elements were *if–then–else* rules), richer structures than binary strings, and experiments with modified genetic operators to meet the needs of particular applications. Some researchers (e.g. Grefenstette 1987) experimented with incorporating problem-specific knowledge into the initialization routine of an evolutionary system; if a (fast) heuristic algorithm provides individuals of the initial population for an evolutionary system, such a hybrid evolutionary system is guaranteed to do no worse than the heuristic algorithm which was used for the initialization.

Usually there exist several (better or worse) heuristic algorithms for a given problem. Apart from incorporating them for the purpose of initialization, some of these algorithms transform one solution into another by imposing a change in the solution's encoding (e.g. 2-opt step for the traveling salesman problem). One can incorporate such transformations into the operator set of evolutionary system, which usually is a very useful addition.

Note also (see Chapters 14 and 31) that there is a strong relationship between encodings of individuals in the population and operators, hence the operators of any evolutionary system must be chosen carefully in accordance with the selected representation of individuals. This is a responsibility of the developer of the system; again, we would cite Davis (1991, p 58):

> Crossover operators, viewed in the abstract are operators that combine subparts of two parent chromosomes to produce new children. The adopted encoding technique should support operators of this type, but it is up to you to combine your understanding of the problem, the encoding technique, and the function of crossover in order to figure out what those operators will be. [...]
>
> The situation is similar for mutation operators. We have decided to use an encoding technique that is tailored to the problem domain; the creators of the current algorithm have done this tailoring for us. Viewed in the abstract, a mutation operator is an operator that introduces variations into the chromosome. [...] these variations can be global or local, but they are critical to keeping the genetic pot boiling. You will have to combine your knowledge of the problem, the encoding technique, and the function of mutation in a genetic algorithm to develop one or more mutation operators for the problem domain.

Very often hybridization techniques make use of local search operators, which are considered as 'intelligent mutations'. For example, the best evolutionary algorithms for the traveling salesman problem use 2-opt or 3-opt procedures to improve the individuals in the population (see e.g. Mühlenbein *et al* 1988). It is not unusual to incorporate gradient-based (or hill-climbing) methods as ways for a local improvement of individuals. It is also not uncommon to combine simulated annealing techniques with some evolutionary algorithms (Adler 1993).

The class of hybrid evolutionary algorithms described so far consists of systems which extend evolutionary paradigm by incorporating additional features (local search, problem-specific representations and operators, and the like). This class also includes also so-called morphogenic evolutionary techniques (Angeline 1995), which include mappings (development functions) between representations that evolve (i.e. evolved representations) and representations which constitutes the input for the evaluation function (i.e. evaluated representations). However, there is another class of evolutionary hybrid methods, where the evolutionary algorithm acts as a separate component of a larger system. This is often the case for various scheduling systems, where the evolutionary algorithm is just responsible for ordering particular items. This is also the case for fuzzy systems, where the evolutionary algorithms may control the membership function, or of neural systems, where evolutionary algorithms may optimize the topology or weights of the network.

References

Adler D 1993 Genetic algorithms and simulated annealing: a marriage proposal *Proc. IEEE Int. Conf. on Neural Networks* pp 1104–9

Angeline P J 1995 Morphogenic evolutionary computation: introduction, issues, and examples *Proc. 4th Ann. Conf. on Evolutionary Programming (San Diego, CA, March 1995)* ed J R McDonnell, R G Reynolds and D B Fogel (Cambridge, MA: MIT Press) pp 387–401

Davis L 1991 *Handbook of Genetic Algorithms* (New York: Van Nostrand Reinhold)

Grefenstette J J 1987 Incorporating problem specific knowledge into genetic algorithms *Genetic Algorithms and Simulated Annealing* ed L Davis (Los Altos, CA: Morgan Kaufmann) pp 42–60

Michalewicz Z 1993 A hierarchy of evolution programs: an experimental study *Evolutionary Comput.* **1** 51–76

——1996 *Genetic Algorithms + Data Structures = Evolution Programs* 3rd edn (New York: Springer)

Mühlenbein H, Gorges-Schleuter M and Krämer O 1988 Evolution algorithms in combinatorial optimization *Parallel Comput.* **7** 65–85

14

Introduction to representations

Kalyanmoy Deb

14.1 Solutions and representations

Every search and optimization algorithm deals with solutions, each of which
represents an instantiation of the underlying problem. Thus, a solution must
be such that it can be completely realized in practice; that is, either it can
be fabricated in a laboratory or in a workshop or it can be used as a control
strategy or it can be used to solve a puzzle, and so on. In most engineering
problems, a solution is a real-valued vector specifying dimensions to the key
parameters of the problem. In control system problems, a solution is a time-
or frequency-dependent functional variation of key control parameters. In game
playing and some artificial-intelligence-related problems, a solution is a strategy
or an algorithm for solving a particular task. Thus, it is clear that the meaning
of a solution is inherent to the underlying problem.

As the structure of a solution varies from problem to problem, a solution
of a particular problem can be represented in a number of ways. Usually, a
search method is most efficient in dealing with a particular representation and
is not so efficient in dealing with other representations. Thus, the choice of an
efficient representation scheme depends not only on the underlying problem but
also on the chosen search method. The efficiency and complexity of a search
algorithm largely depends on how the solutions have been represented and how
suitable the representation is in the context of the underlying search operators.
In some cases, a difficult problem can be made simpler by suitably choosing a
representation that works efficiently with a particular algorithm.

In a classical search and optimization method, all decision variables are
usually represented as vectors of real numbers and the algorithm works on
one vector of solution to create a new vector of solution (Deb 1995, Reklaitis
et al 1983). Different EC methods use different representation schemes
in their search process. Genetic algorithms (GAs) have been mostly used
with a binary string representing the decision variables. Evolution strategy
and *evolutionary programming* studies have used a combination of real-

valued decision variables and a set of strategy parameters as a solution vector. In genetic programming, a solution is a LISP code representing a strategy or an algorithm for solving a task. In permutation problems solved using an EC method, a series of node numbers specifying a complete permutation is commonly used as a solution. In the following subsection, we describe a number of important representations used in EC studies.

14.2 Important representations

In most applications of GAs, decision variables are coded in binary strings of 1s and 0s. Although the variables can be integer or real valued, they are represented by binary strings of a specific length depending on the required accuracy in the solution. For example, a real-valued variable x_i bounded in the range (a, b) can be coded in five-bit strings with the strings (00000) and (11111) representing the real values a and b, respectively. Any of the other 30 strings represents a solution in the range (a, b). Note that, with five bits, the maximum attainable accuracy is only $(b - a)/(2^5 - 1)$. Binary coding is discussed further in Chapter 15. Although binary string coding has been most popular in GAs, a number of researchers prefer to use *Gray* coding to eliminate the Hamming cliff problem associated with binary coding (Schaffer *et al* 1989). In Gray coding, the number of bit differences between any two consecutive strings is one, whereas in binary strings this is not always true. However, as in the binary strings, even in Gray-coded strings a bit change in any arbitrary location may cause a large change in the decoded integer value. Moreover, the decoding of the Gray-coded strings to the corresponding decision variable introduces an artificial nonlinearity in the relationship between the string and the decoded value.

The coding of the variables in string structures make the search space discrete for GA search. Therefore, in solving a continuous search space problem, GAs transform the problem into a discrete programming problem. Although the optimal solutions of the original continuous search space problem and the derived discrete search space problem may be marginally different (with large string lengths), the obtained solutions are usually acceptable in most practical search and optimization problems. Moreover, since GAs work with a discrete search space, they can be conveniently used to solve discrete programming problems, which are usually difficult to solve using traditional methods.

The coding of the decision variables in strings allows GAs to exploit the similarities among various strings in a population to guide the search. The similarities in string positions are represented by ternary strings (with 1, 0, and $*$, where a $*$ matches a 1 or a 0) known as schema. The power of GA search is considered to lie in the implicit parallel schema processing.

Although string codings have been mostly used in GAs, there have been some studies with direct *real-valued* vectors in GAs (Deb and Agrawal 1995, Chaturvedi *et al* 1995, Eshelman and Schaffer 1993, Wright 1991). In those applications, decision variables are directly used and modified genetic operators are used to make a successful search. A detailed discussion of the real-valued vector representations is given in Chapter 16.

In evolution strategy (ES) and evolutionary programming (EP) studies, a natural representation of the decision variables is used where a real-valued solution vector is used. The numerical values of the decision variables are immediately taken from the solution vector to compute the objective function value. In both ES and EP studies, the crossover and mutation operators are used variable by variable. Thus, the relative positioning of the decision variables in the solution vector is not an important matter. However, in recent studies of ES and EP, in addition to the decision variables, the solution vector includes a set of strategy parameters specifying the variance of search mutation for each variable and variable combinations. For n decision variables, both methods use an additional number between one and $n(n + 1)/2$ such strategy parameters, depending on the degree of freedom the user wants to provide for the search algorithm. These adaptive parameters control the search of each variable, considering its own allowable variance and covariance with other decision variables. We discuss these representations in Section 16.2.

In permutation problems, the solutions are usually a vector of node identifiers representing a permutation. Depending on the problem specification, special care is taken in creating valid solutions representing a valid permutation. In these problems, the absolute positioning of the node identifiers is not as important as the relative positioning of the node identifiers. The representation of permutation problems is discussed further in Chapter 17.

In early EP works, finite-state machines were used to evolve intelligent algorithms which were operated on a sequence of symbols so as to produce an output symbol which would maximize the algorithm's performance. Finite-state representations were used as solutions to the underlying problem. The input and output symbols were taken from two different finite-state alphabet sets. A solution is represented by specifying both input and output symbols to each link connecting the finite states. The finite-state machine tranforms a sequence of input symbols to a sequence of output symbols. The finite-state representations are discussed in Chapter 18.

In genetic programming studies, a solution is usually a LISP program specifying a strategy or an algorithm for solving a particular task. Functions and terminals are used to create a valid solution. The syntax and structure of each function are maintained. Thus, if an OR function is used in the solution, at least two arguments are assigned from the terminal set to make a valid OR operation. Usually, the depth of nestings used in any solution is restricted to a specified upper limit. In recent applications of genetic programming, many special features are used in representing a solution. As the iterations progress, a part of the solution is frozen and defined as a metafunction with specified arguments. We shall discuss these features further in Chapter 19.

As mentioned earlier, the representation of a solution is important in the working of a search algorithm, including evolutionary algorithms. In EC studies, although a solution can be represented in a number of ways, the efficacy of a representation scheme cannot be judged alone; instead it depends largely on the chosen recombination operators. In the context of schema processing and the building block hypothesis, it can be argued that a representation that allows good yet important combinations of decision variables to propagate by the action of the search operators is likely to perform well. Radcliffe (1993) outlines a number of properties that a recombination operator must have in order to properly propagate good building blocks. Kargupta *et al* (1992) have shown that the success of GAs in solving a permutation problem coded by three different representations strongly depends on the appropriate recombination operator used. Thus, the choice of a representation scheme must not be made alone, but must be made in conjuction with the choice of the search operators. Guidelines for a suitable representation of decision variables are discussed in Chapter 20.

14.3   Combined representations

In many search and optimization problems, the solution vector may contain different types of variable. For example, in a mixed-integer programming problem (common to many engineering and decision-making problems) some of the decision variables could be real valued and some could be integer valued. In an engineering gear design problem, the number of teeth in a gear and the thickness of the gear could be two important design variables. The former variable is an integer variable and the latter is a real-valued variable. If the integer variable is coded in five-bit binary strings and the real variable is coded in real numbers, a typical mixed-string representation of the above gear design problem may look like (10011 23.5), representing 19 gear teeth and a thickness of 23.5 mm. Sometimes, the variables could be of different types. In a typical civil engineering truss structure problem, the topology of the truss (the connectivity of the truss members represented as presence or absence of members) and the member cross-sectional areas (real valued) are usually the design decision variables. These combined problems are difficult to solve using traditional methods, simply because the search rule in those algorithms does not allow mixed representations. Although there exists a number of mixed-integer programming algorithms such as the branch-and-bound method or the *penalty function* method, these algorithms treat the discrete variables as real valued and impose an artificial pressure for these solutions to move towards the desired discrete values. This is achieved either by adding a set of additional constraints or by penalizing infeasible solutions. These algorithms, in general, require extensive computations. However, the string representation of variables in GAs and the flexibility of using a discrete probability distribution for creating solutions in ES and EP studies allow them to be conveniently used to solve such combined problems. In these problems, a solution vector can be formed by concatenating substrings or numerical values representing or specifying each type of variable, as depicted in the above gear design problem representation. Each part of the solution vector is then operated by a suitable recombination and mutation operator. In the above gear design problem representation using GAs, a binary crossover operator may be used for the integer variable represented by the binary string and a real-coded crossover can be used for the continuous variable. Thus, the recombination operator applied to these mixed representations becomes a collection of a number of operators suitable for each type of variable (Deb 1997). Similar mixed schemes for mutation operators need also to be used for such combined representations. These representations are discussed further in Chapter 26.

References

Deb K 1995 *Optimization for Engineering Design: Algorithms and Examples* (New Delhi: Prentice-Hall)

——1997 A robust optimal design technique for mechanical component design *Evolutionary Algorithms in Engineering Applications* ed D Dasgupta and Z Michalewicz (Berlin: Springer) in press

Deb K and Agrawal R 1995 Simulated binary crossover for continuous search space *Complex Syst.* **9** 115–48

Chaturvedi D, Deb K and Chakrabarty S K 1995 Structural optimization using real-coded genetic algorithms *Proc. Symp. on Genetic Algorithms (Dehradun)* ed P K Roy and S D Mehta (Dehradun: Mahendra Pal Singh) pp 73–82

Eshelman L J and Schaffer J D 1993 Real-coded genetic algorithms and interval schemata *Foundations of Genetic Algorithms II (Vail, CO)* ed D Whitley (San Mateo, CA: Morgan Kaufmann) pp 187–202

Kargupta H, Deb K and Goldberg D E 1992 Ordering genetic algorithms and deception *Parallel Problem Solving from Nature II (Brussels)* ed R Manner and B Manderick (Amsterdam: North-Holland) pp 47–56

Radcliffe N J 1993 Genetic set recombination *Foundations of Genetic Algorithms II (Vail, CO)* ed D Whitley (San Mateo, CA: Morgan Kaufmann) pp 203–19

Reklaitis G V, Ravindran A and Ragsdell K M 1983 *Engineering Optimization: Methods and Applications* (New York: Wiley)

Schaffer J D, Caruana R A, Eshelman L J and Das R 1989 A study of control parameters affecting online performance of genetic algorithms *Proc. 3rd Int. Conf. on Genetic Algorithms (Fairfax, WA, 1989)* ed J D Schaffer (San Mateo, CA: Morgan Kaufmann) pp 51–60

Wright A 1991 Genetic algorithms for real parameter optimization *Foundations of Genetic Algorithms (Bloomington, IN)* ed G J E Rawlins (San Mateo, CA: Morgan Kaufmann) pp 205–20

15

Binary strings

Thomas Bäck

The classical representation used in so-called canonical genetic algorithms consists of binary vectors (often called bitstrings or binary strings) of fixed length ℓ; that is, the individual space I is given by $I = \{0, 1\}^\ell$ and individuals $a \in I$ are denoted as binary vectors $a = (a_1, \ldots, a_\ell) \in \{0, 1\}^\ell$ (see the book by Goldberg (1989)). The mutation operator (Section 32.1) then typically manipulates these vectors by randomly inverting single variables a_i with small probability, and the crossover operator (Section 33.1) exchanges segments between two vectors to form offspring vectors.

This representation is often well suited to problems where potential solutions have a canonical binary representation, i.e. to so-called pseudo-Boolean optimization problems of the form $f : \{0, 1\}^\ell \to \mathbb{R}$. Some examples of such combinatorial optimization problems are the maximum-independent-set problem in graphs, the set covering problem, and the knapsack problem, which can be represented by binary vectors simply by including (excluding) a vertex, set, or item i in (from) a candidate solution when the corresponding entry $a_i = 1$ ($a_i = 0$).

Canonical genetic algorithms, however, also emphasize the binary representation in the case of problems $f : S \to \mathbb{R}$ where the search space S fundamentally differs from the binary vector space $\{0, 1\}^\ell$. The most prominent example of this is given by the application of canonical genetic algorithms for continuous parameter optimization problems $f : \mathbb{R}^n \to \mathbb{R}$ as outlined by Holland (1975) and empirically investigated by De Jong (1975). The mechanisms of encoding and decoding between the two different spaces $\{0, 1\}^\ell$ and $\mathbb{R}^n$ then require us to restrict the continuous space to finite intervals $[u_i, v_i]$ for each variable $x_i \in \mathbb{R}$, to divide the binary vector into n segments of (in most cases) equal length ℓ_x, such that $\ell = n\ell_x$, and to interpret a subsegment $(a_{(i-1)\ell_x+1}, \ldots, a_{i\ell_x})$ ($i = 1, \ldots, n$) as the binary encoding of the variable x_i. Decoding then either proceeds according to the standard binary decoding function $\Gamma^i : \{0, 1\}^\ell \to [u_i, v_i]$, where (see Bäck 1996)

$$\Gamma^i(a_1, \ldots, a_\ell) = u_i + \frac{v_i - u_i}{2^{\ell_x} - 1} \left(\sum_{j=0}^{\ell_x-1} a_{i\,\ell_x - j}\, 2^j \right) \tag{15.1}$$

or by using a Gray code interpretation of the binary vectors, which ensures that adjacent integer values are represented by binary vectors with Hamming distance one (i.e. they differ by one entry only). For the Gray code, equation (15.1) is extended by a conversion of the Gray code representation to the standard code, which can be done for example according to

$$\Gamma^i(a_1, \ldots, a_{\ell_x}) = u_i + \frac{v_i - u_i}{2^{\ell_x} - 1} \left(\sum_{j=0}^{\ell_x-1} \left(\bigoplus_{k=1}^{\ell_x-j} a_{(i-1)\ell_x+k} \right) 2^j \right) \qquad (15.2)$$

where $\oplus$ denotes addition modulo two.

It is clear that this mapping from the representation space $I = \{0, 1\}^\ell$ to the search space $S = \prod_{i=1}^{n}[u_i, v_i]$ is injective but not surjective, i.e. not all points of the search space are represented by binary vectors, such that the genetic algorithm performs a grid search and, depending on the granularity of the grid, might fail to locate an optimum precisely (notice that ℓ_x and the range $[u_i, v_i]$ determine the distance of two grid points in dimension i according to $\Delta x_i = (v_i - u_i)/(2^{\ell_x} - 1)$). Moreover, both decoding mappings given by equations (15.1) and (15.2) introduce additional nonlinearities to the overall objective function $f' : \{0, 1\}^\ell \to \mathbb{R}$, where $f'(a) = (f \circ \times_{i=1}^{n}\Gamma^i)(a)$, and the standard code according to equation (15.1) might cause the problem f' to become harder than the original optimization problem f (see the work of Bäck (1993, 1996, ch 6) for a more detailed discussion).

While parameter optimization is still the dominant field where canonical genetic algorithms are applied to problems in which the search space is fundamentally different from the binary space $\{0, 1\}^\ell$, there are other examples as well, such as the traveling salesman problem (Bean 1993) and job shop scheduling (Nakano and Yamada 1991). Here, rather complex mappings from $\{0, 1\}^\ell$ to the search space were defined—to improve their results, Yamada and Nakano (1992) later switched to a more canonical integer representation space, giving a further indication that the problem characteristics should determine the representation and not vice versa.

The reasons why a binary representation of individuals in genetic algorithms is favored by some researchers can probably be split into historical and schema-theoretical aspects. Concerning the history, it is important to notice that Holland (1975, p 21) does not define adaptive plans to work on binary variables (alleles) $a_i \in \{0, 1\}$, but allows arbitrary but finite individual spaces $I = A_1 \times \ldots \times A_\ell$, where $A_i = \{\alpha_{i_1}, \ldots, \alpha_{i_{k_i}}\}$. Furthermore, his notion of schemata (certain subsets of I characterized by the fact that all members—so-called instances of a schema—share some similarities) does not require binary variables either, but is based on extending the sets A_i defined above by an additional 'don't care' symbol (Holland 1975, p 68). For the application example of parameter optimization, however, he chooses a binary representation (Holland 1975, pp 57, 70), probably because this is the canonical way to map the continuous object variables to the discrete allele sets A_i defined in his adaptive plans, which in turn are likely to be discrete because they aim at modeling the adaptive capabilities of natural evolution on the genotype level.

Interpreting a genetic algorithm as an algorithm that processes schemata, Holland (1975, p 71) then argues that the number of schemata available under a certain representation is maximized by using binary variables; that is, the maximum number of schemata is processed by the algorithm if $a_i \in \{0, 1\}$. This result can be derived by noticing that, when the cardinality of an alphabet A for the allele values is $k = |A|$, the number of different schemata is $(k + 1)^\ell$ (i.e. 3^ℓ in the case of binary variables). For binary alleles, 2^ℓ different solutions can be represented by vectors of length ℓ, and in order to encode the same number of solutions by a k-ary alphabet, a vector of length

$$\ell' = \ell \frac{\ln 2}{\ln k} \tag{15.3}$$

is needed. Such a vector, however, is an instance of $(k+1)^{\ell'}$ schemata, a number that is always smaller than 3^ℓ for $k > 2$; that is, fewer schemata exist for an alphabet of higher cardinality, if the same number of solutions is represented.

Goldberg (1989, p 80) weakens the general requirement for a binary alphabet by proposing the so-called *principle of minimal alphabets*, which states that 'The user should select the smallest alphabet that permits a natural expression of the problem' (presupposing, however, that the binary alphabet permits a natural expression of continuous parameter optimization problems and is no worse than a real-valued representation (Goldberg 1991)). Interpreting this strictly, the requirement for binary alphabets can be dropped, as many practitioners (e.g. Davis 1991 and Michalewicz 1996) who apply (noncanonical) genetic algorithms to industrial problems have already done, using nonbinary, problem-adequate representations such as real-valued vectors (Chapter 16), integer lists representing permutations (Chapter 17), finite-state machine representations (Chapter 18), and parse trees (Chapter 19).

At present, there are neither clear theoretical nor empirical arguments that a binary representation should be used for arbitrary problems other than those that have a canonical representation as pseudo-Boolean optimization problems. From an optimization point of view, where the quality of solutions represented by individuals in a population is to be maximized, the interpretation of genetic algorithms as schema processors and the corresponding implicit parallelism and schema theorem results are of no practical use. From our point of view, the decoding function $\Gamma : \{0, 1\}^\ell \to S$ that maps the binary representation to the canonical search space a problem is defined on plays a much more crucial role than the schema processing aspect, because, depending on the properties of Γ, the overall pseudo-Boolean optimization problem $f' = f \circ \Gamma$ might become more complex than the original search problem $f : S \to \mathbb{R}$. Consequently, one might propose the requirement that, if a mapping between representation space and search space is used at all, it should be kept as simple as possible and obey some structure preserving conditions that still need to be formulated as a guideline for finding a suitable encoding.

References

Bäck T 1993 Optimal mutation rates in genetic search *Proc. 5th Int. Conf. on Genetic Algorithms (Urbana-Champaign, IL, July 1993)* ed S Forrest (San Mateo, CA: Morgan Kaufmann) pp 2–8

——1996 *Evolutionary Algorithms in Theory and Practice* (New York: Oxford University Press)

Bean J C 1993 *Genetics and Random Keys for Sequences and Optimization* Technical Report 92–43, University of Michigan Department of Industrial and Operations Engineering

Davis L 1991 *Handbook of Genetic Algorithms* (New York: Van Nostrand Reinhold)

De Jong K A 1975 *An Analysis of the Behaviour of a Class of Genetic Adaptive Systems* PhD Thesis, University of Michigan

Goldberg D E 1989 *Genetic Algorithms in Search, Optimization and Machine Learning* (Reading, MA: Addison Wesley)

——1991 The theory of virtual alphabets *Parallel Problem Solving from Nature—Proc. 1st Workshop, PPSN I (Lecture Notes in Computer Science 496)* ed H-P Schwefel and R Männer (Berlin: Springer) pp 13–22

Holland J H 1975 *Adaptation in Natural and Artificial Systems* (Ann Arbor, MI: University of Michigan Press)

Michalewicz Z 1996 *Genetic Algorithms + Data Structures = Evolution Programs* (Berlin: Springer)

Nakano R and T Yamada 1991 Conventional genetic algorithms for job shop problems *Proc. 4th Int. Conf. on Genetic Algorithms (San Diego, CA, July 1991)* ed R K Belew and L B Booker (San Mateo, CA: Morgan Kaufmann) pp 474–9

Yamada T and R Nakano 1992 A genetic algorithm applicable to large-scale job-shop problems *Parallel Problem Solving from Nature, 2 (Proc. 2nd Int. Conf. on Parallel Problem Solving from Nature, Brussels, 1992)* ed R Männer and B Manderick (Amsterdam: Elsevier) pp 281–90

16

Real-valued vectors

David B Fogel

16.1 Object variables

When posed with a real-valued function optimization problem of the form 'find the vector x such that $F(x) : \mathbb{R}^n \to \mathbb{R}$ is minimized (or maximized)', evolution strategies (Bäck and Schwefel 1993) and evolutionary programming (Fogel 1995, pp 75–84, 136–7) typically operate directly on the real-valued vector x (with the components of x identified as *object parameters*). In contrast, traditional genetic *algorithms* operate on a coding (often binary) of the vector x (Goldberg 1989, pp 80–4). The choice to use a separate coding rather than operating on the parameters themselves relies on the fundamental belief that it is useful to operate on subsections of a problem and try to optimize these subsections (i.e. building blocks) in isolation, and then subsequently recombine them so as to generate improved solutions. More specifically, Goldberg (1989, p 80) recommends

> The user should select a coding so that short, low-order schemata are relevant to the underlying problem and relatively unrelated to schemata over other fixed positions.

> The user should select the smallest alphabet that permits a natural expression of the problem.

Although the smallest alphabet generates the greatest implicit parallelism, there is no empirical evidence to indicate that binary codings allow for greater effectiveness or efficiency in solving real-valued optimization problems (see the tutorial by Davis (1991, p 63) for a commentary on the ineffectiveness of binary codings).

Evolution strategies and evolutionary programming are not generally concerned with the recombination of building blocks in a solution and do not consider schema processing. Instead, solutions are viewed in their entirety, and no attempt is made to decompose whole solutions into subsections and assign credit to these subsections.

With the belief that maximizing the number of schemata being processed is not necessarily useful, or may even be harmful (Fogel and Stayton 1994), there is no compelling reason in a real-valued optimization problem to act on anything except the real values of the vector x themselves. Moreover, there has been a general trend away from binary codings within genetic algorithm research (see e.g. Davis 1991, Belew and Booker 1991, Forrest 1993, and others). Michalewicz (1992, p 82) indicated that for real-valued numerical optimization problems, floating-point representations outperform binary representations because they are more consistent and more precise and lead to faster execution. This trend may reflect a growing rejection of the building block hypothesis as an explanation for how genetic algorithms act as optimization procedures.

With evolution strategies and evolutionary programming, the typical method for searching a real-valued solution space is to add a multivariate zero-mean Gaussian random variable to each parent involved in the creation of offspring (see Section 32.2). In consequence, this necessitates the setting of the covariance matrix for the Gaussian perturbation. If the covariances between parameters are ignored, only a vector of standard deviations in each dimension is required. There are a variety of methods for setting these standard deviations. Section 32.2 offers a variety of procedures for mutating real-valued vectors.

16.2 Object variables and strategy parameters

It has been recognized since 1967 (Rechenberg 1994, Reed *et al* 1967) that it is possible for each solution to possess an auxiliary vector of parameters that determine how the solution will be changed. Two general procedures for adjusting the object parameters via Gaussian mutation have been proposed (Schwefel 1981, Fogel *et al* 1991) (see Section 32.2). In each case, a vector of *strategy parameters* for self-adaptation is included with each solution and is subject to random variation. The vector may include covariance or rotation information to indicate how mutations in each parameter may covary. Thus the representation consists of two or three vectors:

$$(x, \sigma, \alpha)$$

where x is the vector of object parameters $(x_1, \ldots, x_n)$, σ is the vector of standard deviations, and α is the vector of rotation angles corresponding to the covariances between mutations in each dimension, and may be omitted if these covariances are set to be zero. The vector σ may have n components $(\sigma_1, \ldots, \sigma_n)$ where each entry corresponds to the standard deviation in the ith dimension, $i = 1, \ldots, n$. The vector σ may also degenerate to a scalar σ in which case this single value is used as the standard deviation in all dimensions. Intermediate numbers of standard deviations are also possible, although such implementation is uncommon (this also applies to the rotation angles α). Very recent efforts by Ostermeier *et al* (1994) offer a variation on the methods of Schwefel (1981) and further study is required to determine the general effectiveness of this new technique (see Section 32.2).

Recent efforts in genetic algorithms have also included self-adaptive procedures (see e.g. Spears 1995) and these may incorporate similar real-valued coding for variation operators including crossover and point mutations, on both real-valued or binary or otherwise coded object parameters.

References

Bäck T and Schwefel H-P 1993 An overview of evolutionary algorithms for parameter optimization *Evolutionary Comput.* **1** 1–24

Belew R K and Booker L B (eds) 1991 *Proc. 4th Int. Conf. on Genetic Algorithms (San Diego, CA, July 1991)* (San Mateo, CA: Morgan Kaufmann)

Davis L 1991 A genetic algorithms tutorial IV. Hybrid genetic algorithms *Handbook of Genetic Algorithms* ed L Davis (New York: Van Nostrand Reinhold)

Fogel D B 1995 *Evolutionary Computation: Toward a New Philosophy of Machine Intelligence* (Piscataway, NJ: IEEE)

Fogel D B, Fogel L J and Atmar J W 1991 Meta-evolutionary programming *Proc. 25th Asilomar Conf. on Signals, Systems, and Computers* ed R R Chen (San Jose, CA: Maple) pp 540–5

Fogel D B and Stayton L C 1994 On the effectiveness of crossover in simulated evolutionary optimization *BioSystems* **32** 171–82

Forrest S (ed) 1993 *Proc. 5th Int. Conf. on Genetic Algorithms (Urbana-Champaign, IL, July 1993)* (San Mateo, CA: Morgan Kaufmann)

Goldberg D E 1989 *Genetic Algorithms in Search, Optimization and Machine Learning* (Reading, MA: Addison-Wesley)

Michalewicz Z 1992 *Genetic Algorithms + Data Structures = Evolution Programs* (Berlin: Springer)

Ostermeier A, Gawelczyk A and Hansen N 1994 A derandomized approach to self-adaptation of evolution strategies *Evolutionary Comput.* **2** 369–80

Rechenberg I 1994 Personal communication

Reed J, Toombs R and Barricelli N A 1967 Simulation of biological evolution and machine learning *J. Theor. Biol* **17** 319–42

Schwefel H-P 1981 *Numerical Optimization of Computer Models* (Chichester: Wiley)

Spears W M 1995 Adapting crossover in evolutionary algorithms *Evolutionary Programming IV: Proc. 4th Ann. Conf. on Evolutionary Programming (San Diego, CA, March 1995)* ed J R McDonnell, R G Reynolds and D B Fogel (Cambridge, MA: MIT Press) pp 367–84

17

Permutations

Darrell Whitley

17.1 Introduction

To quote Knuth (1973), 'A permutation of a finite set is an arrangement of its elements into a row.' Given n unique objects, $n!$ permutations of the objects exist. There are various properties of permutations that are relevant to the manipulation of permutation representations by evolutionary algorithms, both from a representation point of view and from an analytical perspective.

As researchers began to apply evolutionary algorithms to applications that are naturally represented as permutations, it became clear that these problems pose different coding challenges than traditional parameter optimization problems. First, for some types of problem there are multiple equivalent solutions. When a permutation is used to represent a cycle, as in the traveling salesman problem (TSP), then all shifts of the permutation are equivalent solutions. Furthermore, all reversals of a permutation are also equivalent solutions. Such symmetries can pose problems for evolutionary algorithms that rely on recombination.

Another problem is that permutation problems cannot be processed using the same general recombination and mutation operators which are applied to parameter optimization problems. The use of a permutation representation may in fact mask very real differences in the underlying combinatorial optimization problems. An example of these differences is evident in the description of classic problems such as the TSP and the problem of resource scheduling.

The traveling salesman problem is the problem of visiting each vertex (i.e. city) in a full connected graph exactly once while minimizing a cost function defined with respect to the edges between adjacent vertices. In simple terms, the problem is to minimize the total distance traveled while visiting all the cities and returning to the point of origin. The TSP is closely related to the problem of finding a Hamiltonian circuit in an arbitrary graph. The Hamiltonian circuit is a set of edges that form a cycle which visits every vertex exactly once. It is relatively easy to show that the problem of finding a set of Boolean values that yield an evaluation of 'true' for a three-conjunction normal form Boolean expression is directly *polynomial-time reducible* to the problem of finding a Hamiltonian circuit in a specific type of graph (Cormen *et al* 1990). The Hamiltonian circuit problem in turn is reducible to the TSP. All of these problems have a nondeterministic polynomial-time (NP) solution but have no known polynomial-time solution. These problems are also members of the set of hardest problems in NP, and hence are NP complete.

Permutations are also important for scheduling applications, variants of which are also often NP complete. Some scheduling problems are directly related to the TSP. Consider minimizing setup times between a set of N jobs, where the function Setup(A, B) is the cost of switching from job A to job B. If Setup(A, B) = Setup(B, A) this is a variant of the symmetric TSP, except that the solution may be a path instead of a cycle through the graph (i.e. it visits every vertex, but does not necessarily return to the origin.) The TSP and setup minimization problem may also be nonsymmetric: the cost of going from vertex A to B may not be equal to the cost of going from vertex B to A.

Other types of scheduling problem are different from the TSP. Assume that one must schedule service times for a set of customers. If this involves access to a critical resource, then those customers that are scheduled early may have access to resources that are unavailable to later customers. If one is scheduling appointments, for example, later customers will have less choice with respect to which time slots are available to them. In either case, access to limited resources is critical. We would like to optimize the match between resources and customers. This could allow us to give more customers what they want in terms of resources, or the goal might be to increase the number of customers who can be serviced. In either case, permutations over the set of customers can be used as a priority queue for scheduling. While there are various classic problems in the scheduling literature, the term *resource scheduling* is used here to refer to scheduling applications where resources are consumed.

Permutations are also sometimes used to represent multisets. A multiset is also sometimes referred to as a *bag*, which is analogous to a set except that a bag may contain multiple copies of identical elements. In sets, the duplication of elements is not significant, so that

$$\{a, a, b\} = \{a, b\}.$$

However, in the following multiset,

    M = {a, a, b, b, b, c, d, e, e, f, f}

there are two a's, three b's, one c, one d, two e's and two f's, and duplicates are significant. In scheduling applications that map jobs to machines, it may be necessary to schedule two jobs of type a, three jobs of type b, and so on. Note that it is not necessary that all jobs of type a be scheduled contiguously. While M in the above illustration contains 11 elements, there are not 11! unique permutations. Rather, the number of unique permutations is given by

$$\frac{11!}{2!3!1!1!2!2!}$$

and in general

$$\frac{n!}{n_1! n_2! n_3! \ldots}$$

where n is the number of elements in the multiset and n_i is the number of elements of type i (Knuth 1973). Radcliffe (1993) considers the application of genetic and evolutionary operators when the solution is expressed as a set or multiset (bag).

Before looking in more detail at the relationship between permutations and evolutionary algorithms, some general properties of permutations are reviewed that are both interesting and useful.

17.2 Mapping integers to permutations

The set of $n!$ permutations can be mapped onto the set of integers in various ways. Whitley and Yoo (1995) give the following algorithm which converts an integer X into the corresponding permutation.

(i)   Choose some ordering of the permutation which is defined to be sorted.
(ii) Sort and index the N elements ($N \geq 1$) of the permutation from 1 to N. Pick an index X for a specific permutation such that $0 \leq X < N!$.
(iii) If $X = 0$, pick all remaining elements in the sorted permutation list in the sequence in which they occur and stop.
(iv) IF $X < (N - 1)!$ pick the first element of the remaining list; GOTO (vi). Otherwise, continue.
(v)   Find Y such that $(Y - 1)(N - 1)! \leq X < Y(N - 1)!$. The Yth element of the sort list is the next element of the permutation. $X = X - (Y - 1)((N - 1)!)$.
(vi) Delete the chosen element from the list of sorted elements; $N = N - 1$; GOTO (iii).

This algorithm can also be inverted to map integers to permutations. For permutations of length three this generates the following correspondance:

```
X = 0 indexes 123    X = 3 indexes 231
X = 1 indexes 132    X = 4 indexes 312
X = 2 indexes 213    X = 5 indexes 321.
```

17.3 The inverse of a permutation

One important property of a permutation is that it has a well-defined *inverse* (Knuth 1973). Let $a_1 a_2 \ldots a_n$ be a permutation of the set $\{1, 2, \ldots, n\}$. This can be written in a two-line form

$$\begin{pmatrix} 1 & 2 & 3 & \ldots & n \\ a_1 & a_2 & a_3 & \ldots & a_n \end{pmatrix}.$$

The inverse is obtained by reordering both rows such that the second row is transformed into the sequence $123 \ldots n$; the reordering of the first that occurs as a consequence of reordering the second row yields the inverse of permutation $a_1 a_2 a_3 \ldots a_n$. The inverse is denoted $a_1' a_2' a_3' \ldots a_n'$. Knuth (1973) gives the following example of a permutation, 5 9 1 8 2 6 4 7 3, and shows that its inverse can be obtained as follows:

$$\left(\begin{array}{c} 591826473 \\ 123456789 \end{array} \right) = \left(\begin{array}{c} 123456789 \\ 359716842 \end{array} \right)$$

which yields the inverse 3 5 9 7 1 6 8 4 2. Knuth also points out that $a_j' = k$ if and only if $a_k = j$. The inverse can be used as part of a function for mapping permutations to a canonical form, which in turn makes it easier to model problems with permutation representations.

17.4   The mapping function

When modeling evolutionary algorithms it is often useful to compute a transmission function $r_{i,j}(k)$ which yields the probability of recombining strings i and j and obtaining an arbitrary string k. Whitley and Yoo (1995) explain how to compute the transmission function for a single string k and then to generalize the results to all other strings. In this case, the strings represent permutations and the remapping function, denoted @, functions as follows:

$$r_{3421,1342}(3124) = r_{3421@3124,1342@3124}(1234).$$

The computation $Y = A @ X$ behaves as follows. Let any permutation X be represented by $x_1 x_2 x_3 \ldots x_n$. Then $a_1 a_2 a_3 \ldots a_n @ x_1 x_2 x_3 \ldots x_n$ yields $Y = y_1 y_2 y_3 \ldots y_n$ where $y_i = j$ when $a_i = x_j$. Thus, $(3421 @ 3124)$ yields (1432) since $(a_1 = 3 = x_1) \Rightarrow (y_1 = 1)$. Next, $(a_2 = 4 = x_4) \Rightarrow (y_2 = 4)$, $(a_3 = 2 = x_3) \Rightarrow (y_3 = 3)$, and $(a_4 = 1 = x_2) \Rightarrow (y_4 = 2)$. This mapping function is analogous to the bitwise addition (mod 2) used to reorder the vector s for binary strings. However, note that $A @ X \neq X @ A$. Furthermore, for permutation recombination operators it is *not generally true* that $r_{i,j} = r_{j,i}$.

This allows one to compute the transmission function with respect to a canonical permutation, in this case 1234, and generalize this mapping to all other permutations. This mapping can be achieved by simple element substitution. First, the function r can be generalized as follows:

$$r_{3421,1342}(3124) = r_{wzyx,xwzy}(wxyz)$$

where w, x, y, and z are variables representing the elements of the permutation (e.g. $w = 3, x = 1, y = 2, z = 4$). If $wxyz$ now represents the canonically ordered permutation 1234,

$$r_{wzyx,xwzy}(wxyz) = r_{1432,2143}(1234) = r_{3421@3124,1342@3124}(1234).$$

We can also relate this mapping operator to the process of finding an inverse. The permutations in the expression

$$r_{3421,1342}(3124) = r_{1432,2143}(1234)$$

are included as rows in an array. To map the left-hand side of the preceding expression to the terms in the right-hand side, first compute the inverses for each of the terms in the left-hand side:

$$\begin{pmatrix} 3421 \\ 1234 \end{pmatrix} = \begin{pmatrix} 1234 \\ 4312 \end{pmatrix}$$

$$\begin{pmatrix} 1342 \\ 1234 \end{pmatrix} = \begin{pmatrix} 1234 \\ 1423 \end{pmatrix}$$

$$\begin{pmatrix} 3124 \\ 1234 \end{pmatrix} = \begin{pmatrix} 1234 \\ 2314 \end{pmatrix}.$$

Collect the three inverses into a single array. We also then add 1 2 3 4 to the array and inverse the permutation 2 3 1 4, at the same time rearranging all the other permutations in the array:

$$\begin{pmatrix} 4312 \\ 1423 \\ 2314 \\ 1234 \end{pmatrix} = \begin{pmatrix} 1432 \\ 2143 \\ 1234 \\ 3124 \end{pmatrix}.$$

This yields the permutations 1432, 2143, and 1234 which represent the desired canonical form as it relates to the notion of substitution into a symbolic canonical form. One can also reverse the process to find the permutations p_i and p_j in the following context:

$$r_{1432,2143}(1234) = r_{p_i,p_j}(3124) = r_{p_i@3124,p_j@3124}(1234).$$

17.5 Matrix representations

When comparing the TSP to the problem of resource scheduling, in one case adjacency is important (the TSP) and in the other case relative order is important (resource scheduling). One might also imagine problems where absolute position is important. One way in which the differences between adjacency and relative order can be illustrated is to use a matrix representation of the information contained in a permutation.

When we discuss adjacency in the TSP, we typically are referring to a symmetric TSP: the distance of going from city A to B is the same as going from B to A. Thus, when we define a matrix representing a tour, there will be two edges in every row of the matrix, where a row–column entry of 1 represents an edge connecting the two cities. Thus, the matrices for the tours [A B C D E F] and [C D E B F A] (left and right, respectively) are as follows:

	A	B	C	D	E	F
A	0	1	0	0	0	1
B	1	0	1	0	0	0
C	0	1	0	1	0	0
D	0	0	1	0	1	0
E	0	0	0	1	0	1
F	1	0	0	0	1	0

	A	B	C	D	E	F
A	0	0	1	0	0	1
B	0	0	0	0	1	1
C	1	0	0	1	0	0
D	0	0	1	0	1	0
E	0	1	0	1	0	0
F	1	1	0	0	0	0

One thing that is convenient about the matrix representation is that it is easy to extract information about where common edges occur. This can also be expressed in the form of a matrix, where a zero or one respectively is placed in the matrix where there is agreement in the two parent structures. If the values in the parent matrices conflict, we will place a # in the matrix. Using the two above structures as parents, the following common information is obtained:

	A	B	C	D	E	F
A	0	#	#	0	0	1
B	#	0	#	0	#	#
C	#	#	0	1	0	0
D	0	0	1	0	1	0
E	0	#	0	1	0	#
F	1	#	0	0	#	0

This matrix can be interpreted in the following way. If we convert the # symbols to * symbols, then (in the notation typically used by the genetic algorithm community) a hyperplane is defined in this binary space in which both of the parents reside. If a recombination operator is applied, the offspring should also reside in this same subspace (this is the concept of *respect*, as used by Radcliffe (1991); note mutation can still be applied after recombination).

This matrix representation does bring out one feature rather well: the common subtour information can automatically and easily be extracted and passed on to the offspring during recombination.

The matrix defining the common hyperplane information also defines those offspring that represent a recombination of the information contained in the parent structures. In fact, any assignment of 0 or 1 bits to the locations occupied by the # symbols could be considered valid recombinations, *but not all are feasible solutions to the TSP,* because not all recombinations result in a Hamiltonian circuit. We would like to have an offspring that is not only a valid recombination, but also a feasible solution.

The matrix representation can also make explicit relative order information. Consider the same two parents: [A B C D E F] and [C D E B F A]. Relative order can be represented as follows. Each row will be the relative order information for a particular element of a permutation. The columns will be all permutation elements in some canonical order. If A is the first element in a permutation, then a one will be placed in every column (except column A; the diagonal will again be zero) to indicate A precedes all other cities. This representation is given by Fox and McMahon (1991). Thus, the matrices for [A B C D E F] and [C D E B F A] are as follows (left and right, respectively):

	A	B	C	D	E	F
A	0	1	1	1	1	1
B	0	0	1	1	1	1
C	0	0	0	1	1	1
D	0	0	0	0	1	1
E	0	0	0	0	0	1
F	0	0	0	0	0	0

	A	B	C	D	E	F
A	0	0	0	0	0	0
B	1	0	0	0	0	1
C	1	1	0	1	1	1
D	1	1	0	0	1	1
E	1	1	0	0	0	1
F	1	0	0	0	0	0

In this case, the lower triangle of the matrix flags *inversions*, which should not be confused with an *inverse*. If $a_1a_2a_3 \ldots a_n$ is a permutation of the canonically ordered set $1, 2, 3, \ldots, n$ then the pair (a_i, a_j) is an *inversion* if $i < j$ and $a_i > a_j$ (Knuth 1973). Thus, the number of 1 bits in the lower triangles of the above matrices is also a count of the number of inversions (which should also not be confused with the *inversion* operator used in simple genetic algorithms, see Holland 1975, p 106, Goldberg 1989, p 166).

The common information can also extracted as before. This produces the following matrix:

	A	B	C	D	E	F
A	0	#	#	#	#	#
B	#	0	#	#	#	1
C	#	#	0	1	1	1
D	#	#	0	0	1	1
E	#	#	0	0	0	1
F	#	0	0	0	0	0

Note that this binary matrix is again symmetric around the diagonal, except that the lower triangle and upper triangle have complementary bit values. Thus only $N(N - 1)/2$ elements are needed to represent relative order information.

There have been few studies of how recombination crossover operators generate offspring in this particular representation space. Fox and McMahon (1991) offer some work of this kind and also define several operators that work directly on these binary matrices for relative order.

While matrices may not be the most efficient form of implementation, they do provide a tool for better understanding sequence recombination operators designed to exploit relative order. It is clear that adjacency and relative order relationships are different and are best expressed by different binary matrices. Likewise, absolute position information also has a different matrix representation (for example, rows could represent cities and the columns represent positions). Cycle crossover (Section 33.3.6; see Starkweather *et al* 1991, Oliver *et al* 1987) appears to be a good absolute position operator, although it is hard to find problems in the literature where absolute position is critical.

17.6 Alternative representations

Let P be an arbitrary permutation and P_j be the jth element of the permutation. One notable alternative representation of a permutation is to define some canonical ordering, C, over the elements in the permutation and then define a vector of integers, I, such that the integer in position j corresponds to the position in which element C_j appears in P. Such a vector I can then serve as a representation of a permutation. More precisely,

$$P_{(I_j)} = C_j.$$

To illustrate:

$$C = a\ b\ c\ d\ e\ f\ g\ h$$

$$I = 6\ 2\ 5\ 3\ 8\ 7\ 1\ 4 \quad \text{which represents} \quad P = g\ b\ d\ h\ c\ a\ f\ e.$$

This may seem like a needless indirection, but consider that I can be generalized to allow a larger number of possible values than there are permutation elements. I can also be generalized to allow all real values (although for computer implementations the distinction is somewhat artificial since all digital representations of real values are discrete and finite). We now have a parameter-based presentation of the permutation such that we can generate random vectors I representing permutations. If the number of values for which elements in I are defined is dramatically larger than the number of elements in the permutation, then duplicate values in randomly generated vectors will occur with very small probability.

This representation allows a permutation problem to be treated as if it were a more traditional parameter optimization problem with the constraint that no two elements of vector I should be equal, or that there is a well defined way to resolve ties. Evolutionary algorithm techniques normally used for parameter optimization problems can thus be applied to permutation problems using this representation.

This idea has been independently invented on a couple of occasions. The first use of this coding method was by Steve Smith of Thinking Machines. A version of this coding was used by the ARGOT Strategy (Shaefer 1987) and the representation was picked up by Syswerda (1989) and by Schaffer *et al* (1989) for the TSP. More recently, a similar idea was introduced by Bean (1994) under the name *random keys*.

17.7   Ordering schemata and other metrics

Goldberg and Lingle (1985) built on earlier work by Franz (1972) to describe similarity subsets between different permutations. Franz's calculations were related to the use of inversion operators for traditional genetic algorithm binary representations. The use of inversion operators is very much relevant to the topic of permutations, since in order to apply inversion the binary alleles must be tagged in some way and inversion acts in the space of all possible permutations of allele orderings. Thus,

```
((6 0)(3 1)(2 0)(8 1)(1 0)(5 1)(7 0)(4 0))
```

is equivalent to

```
((1 0)(2 0)(3 1)(4 0)(5 1)(6 0)(7 0)(8 1))
```

which represents the binary string 00101001 in a position-independent fashion (Holland 1975).

Goldberg and Lingle were more directly concerned with problems where the permutation was itself the problem representation, and, in particular, they present early results for the TSP. They also introduced the partially mapped crossover (PMX) operator and the notion of *ordering schemata*, or *o-schemata*. For o-schemata, the symbol ! acts as a wild card match symbol. Thus, the template

!   !   1 !   !   7 3 !

represents all permutations with a one as the third element, a seven as the sixth element, and a three as the seventh element. Given o selected positions in a permutation of length l, there are $(l - o)!$ permutations that match an o-schemata. One can also count the number of possible o-schema. There are clearly $\binom{l}{o}$ ways to choose o fixed positions; there are also $\binom{l}{o}$ ways to pick the permutation elements that fill the slots, and $o!$ ways of ordering the elements (i.e. the number of permutations over the chosen combination of subelements). Thus, Goldberg (1989, Goldberg and Lingle 1985) notes that the total number of o-schemata, n_{os}, can be calculated by

$$n_{os} = \sum_{j=0}^{l} \frac{l!}{(l - j)! j!} \frac{l!}{(l - j)!}.$$

Note that in this definition of the o-schemata, relative order is not accounted for. In other words, if relative order is important then all of the following shifted o-schemata,

1   !   !   7   3   !   !   !
!   1   !   !   7   3   !   !
!   !   1   !   !   7   3   !
!   !   !   1   !   !   7   3

could be viewed as equivalent. Such schemata may or may not 'wrap around'. Goldberg discusses o-schemata which have an absolute fixed position (o-schemata, type a) and those with relative position which are shifts of a specified template (o-schemata, type r).

This work on o-schemata predates the distinctions between relative order permutation problems, absolute position problems, and adjacency-based problems. Thus, o-schemata appear to be better for understanding resource scheduling applications than for the TSP. In subsequent work, Kargupta *et al* (1992) attempt to use ordering schemata to construct deceptive functions for ordering problems—that is, problems where the average fitness values of the o-schemata provide misleading information. Note that such problems are constructed to mislead simple genetic algorithms and may or may not be difficult with respect to other types of algorithm. (For a discussion of deception see the article by Goldberg (1987) and Whitley (1991) and for another perspective see the article by Grefenstette (1993).) The analysis of Kargupta *et al* (1992) considers PMX, a uniform ordering crossover operator (UOX), and a relative ordering crossover operator (ROX).

An alternative way of constructing relative order problems and of comparing the similarity of permutations is given by Whitley and Yoo (1995). Recall that a relative order matrix has a 1 bit in position (X, Y) if row element X appears before column element Y in a permutation. Note that the matrix representation yields a unique binary representation for each permutation. Using this representation one can also define the Hamming distance between two permutations P1 and P2; Hamming distance is denoted by HD(index(P1), index(P2)), where the permutations are represented by their integer index. In the following examples, the Hamming distance is computed with respect to the lower triangle (i.e. it is a count of the number of 1 bits in the lower triangle):

```
                 A B C D
                 ---------
             A | 0 1 1 1
    A B C D  B | 0 0 1 1        HD(0,0) = 0
             C | 0 0 0 1
             D | 0 0 0 0


                 A B C D
                 ---------
             A | 0 0 0 0
    B D C A  B | 1 0 1 1        HD(0,11) = 4
             C | 1 0 0 0
             D | 1 0 1 0


                 A B C D
                 ---------
             A | 0 0 0 0
    D C B A  B | 1 0 0 0        HD(0,23) = 6
             C | 1 1 0 0
             D | 1 1 1 0
```

Whitley and Yoo (1995) point out that this representation is not perfect. Since $2^{(N^2)} > N!$, certain binary strings are undefined. For example, consider the following upper triangle:

```
    1 1 1
      0 1
        0
```

Element 1 occurs before 2, 3, and 4, which poses no problem, but 2 occurs after 3, 2 occurs before 4, and 4 occurs before 3. Using > to denote relative order, this implies a nonexistent ordering such that

```
3 > 2 > 4    but   4 > 3
```

Thus, not all matrices correspond to permutations. Nevertheless, the binary representation does afford a metric in the form of Hamming distance and suggests an alternative way of constructing deceptive ordering problems, since once a binary representation exists several methods for constructing misleading problems could be employed. Deb and Goldberg (1992) explain how to construct trap functions. Whitley (1991) also discusses the construction of deceptive binary functions.

While the topic of deception has been the focus of some controversy (cf Grefenstette 1993), there are few tools for understanding the difficulty or complexity of permutation problems. Whitley and Yoo found that simulation results failed to provide clear evidence that *deceptive* functions built using o-schema fitness averages really were misleading or difficult for simple genetic algorithms.

Aside from the fact that many problems with permutation-based representations are known to be NP complete problems, there is little work which characterizes the complexity of specific instances of these problems, especially from a genetic algorithm perspective. One can attempt to estimate the size and depth of basins of attraction, but such methods must presuppose the use of a particular search methods. The use of different local search operators can induce different numbers of local optima and different sized basins of attraction. Changing representations can have the same effect.

17.8   Operator descriptions and local search

Section 32.3, on mutation for permutations, also provides information on local search operators, the best known of which is 2-opt. Information on the most commonly used forms of recombination for permutation-based representations is found in Section 33.3. For a general discussion of permutations, see the books by Niven (1965) and Knuth (1973). Whitley and Yoo (1995) present methods for constructing infinite-population models for simple genetic algorithms applied to permutation problems which can be easily converted into finite-population Markov models.

References

Bean J C 1994 Genetic algorithms and random keys for sequencing and optimization *ORSA J. Comput.* **6**

Cormen T, Leiserson C and Rivest R 1990 *Introduction to Algorithms* (Cambridge, MA: MIT Press)

Deb K and Goldberg D 1993 Analyzing deception in trap functions *Foundations of Genetic Algorithms 2* ed D Whitley (San Mateo, CA: Morgan Kaufmann)

Fox B R and McMahon M B 1991 *Genetic Operators for Sequencing Problems Foundations of Genetic Algorithms* ed G J E Rawlins (San Mateo, CA: Morgan Kaufmann) pp 284–300

Franz D R 1972 *Non-Linearities in Genetic Adaptive Search* PhD Dissertation, University of Michigan

Goldberg D 1987 Simple genetic algorithms and the minimal, deceptive problem *Genetic Algorithms and Simulated Annealing* ed L Davis (Boston, MA: Pitman)

——1989 *Genetic Algorithms in Search, Optimization and Machine Learning* (Reading, MA: Addison-Wesley)

Goldberg D and Lingle R Jr 1985 Alleles, loci, and the traveling salesman problem *Proc. 1st Int. Conf. on Genetic Algorithms and Their Applications* ed J J Grefenstette (Hillsdale, NJ: Erlbaum)

Grefenstette J 1993 Deception considered harmful *Foundations of Genetic Algorithms 2* ed D Whitley (San Mateo, CA: Morgan Kaufmann)

Holland J 1975 *Adaptation In Natural and Artificial Systems* (University of Michigan Press)

Kargupta H, Deb K and Goldberg D 1992 Ordering genetic algorithms and deception *Parallel Problems Solving from Nature, 2* ed R Manner and B Manderick (Amsterdam: Elsevier) pp 47–56

Knuth R M 1973 *The Art of Computer Programming Volume 3: Sorting and Searching* (Reading, MA: Addison-Wesley)

Niven I 1965 *Mathematics of Choice, or How to Count without Counting* (Mathematical Association of America)

Oliver I, Smith D and Holland J 1987 A study of permutation crossover operators on the traveling salesman problem *2nd Int. Conf. on Genetic Algorithms (Pittsburgh, PA, 1987)* ed J J Grefenstette (Hillsdale, NJ: Erlbaum) pp 224–30

Radcliffe N 1991 Forma analysis and random respectful recombination *Fourth Int. Conf. on Genetic Algorithms (San Diego, CA, July 1991)* ed R Belew and L Booker (San Mateo, CA: Morgan Kaufmann) pp 222–9

——1993 Genetic set recombination *Foundations of Genetic Algorithms 2* ed D Whitley (San Mateo, CA: Morgan Kaufmann) pp 203–19

Schaffer J D, Caruana R A, Eshelman L J and Das R 1989 A study of control parameters affecting online performance of genetic algorithms for function optimization *Proc. 3rd Int. Conf. on Genetic Algorithms (Fairfax, VA, 1989)* ed J D Schaffer (San Mateo, CA: Morgan Kaufmann) pp 51–60

Shaefer C 1987 The ARGOT strategy: adaptive representation genetic optimizer technique *2nd Int. Conf. on Genetic Algorithms (Pittsburgh, PA, 1987)* ed J J Grefenstette (Hillsdale, NJ: Erlbaum) pp 50–8

Starkweather T, McDaniel S, Mathias K, Whitley D and Whitley C 1991 A comparison of genetic sequencing operators *Proc. 4th Int. Conf. on Genetic Algorithms (San Diego, CA, 1991)* ed R K Belew and L B Booker (San Mateo, CA: Morgan Kauffman)

Syswerda G 1989 Uniform crossover in genetic algorithms *Proc. 3rd Int. Conf. on Genetic Algorithms (Fairfax, VA, 1989)* ed J D Schaffer (San Mateo, CA: Morgan Kaufmann) pp 2–9

Whitley D 1991 Fundamental principles of deception in genetic search *Foundations of Genetic Algorithms* ed G J E Rawlins (San Mateo, CA: Morgan Kaufmann) pp 221–41

Whitley D and Yoo N-W 1995 Modeling simple genetic algorithms for permutation problems *Foundations of Genetic Algorithms 3* ed D Whitley and M Vose (San Mateo, CA: Morgan Kaufmann) pp 163–84

18

Finite-state representations

David B Fogel

18.1 Introduction

A finite-state machine is a mathematical logic. It is essentially a computer program: it represents a sequence of instructions to be executed, each depending on a current state of the machine and the current stimulus. More formally, a finite-state machine is a 5-tuple

$$M = (Q, \tau, \rho, s, o)$$

where Q is a finite set, the set of states, τ is a finite set, the set of input symbols, ρ is a finite set, the set of output symbols, $s : Q \times \tau \rightarrow Q$ is the next state function, and $o : Q \times \tau \rightarrow \rho$ is the next output function.

Any 5-tuple of sets and functions satisfying this definition is to be interpreted as the mathematical description of a machine that, if given an input symbol x while it is in state q, will output the symbol $o(q, x)$ and transition to state $s(q, x)$. Only the information contained in the current state describes the behavior of the machine for a given stimulus. The entire set of states serves as the 'memory' of the machine. Thus a finite-state machine is a transducer that can be stimulated by a finite alphabet of input symbols, that can respond in a finite alphabet of output symbols, and that possesses some finite number of different internal states. The corresponding input–output symbol pairs and next-state transitions for each input symbol, taken over every state, specify the behavior of any finite-state machine, given any starting state. For example, a three-state machine is shown in figure 18.1. The alphabet of input symbols are elements of the set $\{0, 1\}$, whereas the alphabet of output symbols are elements of the set $\{\alpha, \beta, \gamma\}$ (input symbols are shown to the left of the slash, output symbols are shown to the right). The finite-state machine transforms a sequence of input symbols into a sequence of output symbols. Table 18.1 indicates the response of the machine to a given string of symbols, presuming that the machine is found in state C. It is presumed that the machine acts when each input symbol is perceived and the output takes place before the next input symbol arrives.

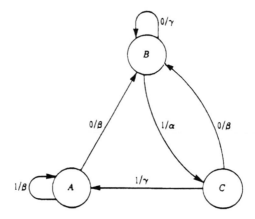

Figure 18.1. A three-state finite machine. Input symbols are shown to the left of the slash. Output symbols are to the right of the slash. Unless otherwise specified, the machine is presumed to start in state A. (After Fogel *et al* 1966, p 12).

Table 18.1. The response of the finite-state machine shown in figure 18.1 to a string of symbols. In this example, the machine starts in state C.

Present state	C	B	C	A	A	B
Input symbol	0	1	1	1	0	1
Next state	B	C	A	A	B	C
Output symbol	β	α	γ	β	β	α

18.2 Applications

Finite-state representations are often convenient when the required solutions to a particular problem of interest require the generation of a sequence of symbols having specific meaning. For example, consider the problem offered by Fogel *et al* (1966) of predicting the next symbol in a sequence of symbols taken from some alphabet A (here, $\tau = \rho = A$). A population of finite-state machines is exposed to the environment, that is, the sequence of symbols that have been observed up to the current time. For each parent machine, as each input symbol is offered to the machine, each output symbol is compared with the next input symbol. The worth of this prediction is then measured with respect to the given payoff function (e.g. all–none, absolute error, squared error, or any other expression of the meaning of the symbols). After the last prediction is made, a function of the payoff for each symbol (e.g. average payoff per symbol) indicates the fitness of the machine. Offspring machines are created through mutation (Section 32.4) and/or recombination (Section 33.4). The machines that provide the greatest payoff are retained to become parents of the next generation. This process is iterated until an actual prediction of the next symbol (as yet

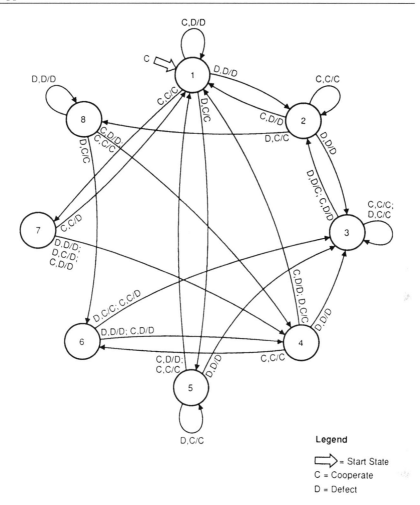

Figure 18.2. A finite-state machine evolved in prisoner's dilemma experiments detailed by Fogel (1995b, p 215). The input symbols form the set $\{(C, C), (C, D), (D, C), (D, D)\}$ and the output symbols form the set $\{C, D\}$. The machine also has an associated first move indicated by the arrow; here the machine cooperates initially then proceeds into state 6.

inexperienced) in the environment is required. The best machine generates this prediction, the new symbol is added to the experienced environment, and the process is repeated.

There is an inherent versatility in such a procedure. The payoff function can be arbitrarily complex and can possess temporal components; there is no requirement for the classical squared-error criterion or any other smooth function. Further, it is not required that the predictions be made with a one-step look ahead. Forecasting can be accomplished at an arbitrary length of time into the future. Multivariate environments can be handled, and the environmental process need not be stationary because the simulated evolution will adapt to changes in the transition statistics.

For example, Fogel (1991, 1993, 1995a) has used finite-state machines to describe behaviors in the iterated prisoner's dilemma. The input alphabet was selected as the set $\{(C, C), (C, D), (D, C), (D, D)\}$ where C corresponds to a move for cooperation and D corresponds to a move for defection. The ordered pair (X, Y) indicates that the machine played X in the last move, while its opponent played Y. The output alphabet was the set $\{C, D\}$ and corresponded to the next move of the machine based on the previous pair of moves and the current state of the machine (see figure 18.2).

Other applications of finite-state representations have been offered. For example, Jefferson *et al* (1991), Angeline and Pollack (1993), and others employed a finite-state machine to describe the behavior of a simulated ant on a trail placed on a grid. The input alphabet was the set $\{0, 1\}$, where 0 indicated that the ant did not see a trail cell ahead and 1 indicated that it did see such a cell ahead. The output alphabet was $\{M, L, R, N\}$ where M indicated a move forward, L indicated a turn to the left without moving, R indicated a turn to the right without moving, and N indicated a condition to do nothing. The task was to evolve a finite-state machine that would generate a sequence of moves to traverse the trail in the shortest number of time steps.

References

Angeline P J and Pollack J B 1993 Evolutionary module acquisition *Proc. 2nd Ann. Conf. on Evolutionary Programming (San Diego, CA)* ed D B Fogel and W Atmar (La Jolla, CA: Evolutionary Programming Society) pp 154–63

Fogel D B 1991 The evolution of intelligent decision-making in gaming *Cybernet. Syst.* **22** 223–36

——1993 Evolving behaviors in the iterated prisoner's dilemma *Evolut. Comput.* **1** 77–97

——1995a On the relationship between the duration of an encounter and the evolution of cooperation in the iterated prisoner's dilemma *Evolut. Comput.* **3** 349–63

——1995b *Evolutionary Computation: Toward a New Philosophy of Machine Intelligence* (Piscataway, NJ: IEEE)

Fogel L J, Owens A J and Walsh M J 1966 *Artificial Intelligence Through Simulated Evolution* (New York: Wiley)

Jefferson D, Collins R, Cooper C, Dyer M, Flowers M, Korf R, Taylor C and Wang A 1991 Evolution as a theme in artificial life: the Genesys/Tracker system *Artificial Life II* ed C G Langton, C Taylor, J D Farmer and S Rasmussen (Reading, MA: Addison-Wesley) pp 549–77

19

Parse trees

Peter J Angeline

When an executable structure such as a program or a function is the object of an evolutionary computation, representation plays a crucial role in determining the ultimate success of the system. If a traditional, syntax-laden programming language is chosen to represent the evolving programs, then manipulation by simple evolutionary operators will most likely produce syntactically invalid offspring. A more beneficial approach is to design the representation to ensure that only syntactically correct programs are created. This reduces the ultimate size of the search space considerably. One method for ensuring syntactic correctness of generated programs is to evolve the desired program's parse tree rather than an actual, unparsed, syntax-laden program. Use of the parse tree representation completely removes the 'syntactic sugar' introduced into a programming language to ensure human readability and remove parsing ambiguity.

Cramer (1985), in the first use of a parse tree representation in a genetic algorithm, described two distinct representations for evolving sequential computer programs based on a simple algorithmic language and emphasized the need for offspring programs to remain syntactically correct after manipulation by the genetic operators. To accomplish this, Cramer investigated two encodings of the language into fixed-length integer representations.

Cramer (1985) first represented a program as an ordered collection of statements. Each statement consisted of N integers; the first integer identified the command to be executed and the remainder specified the arguments to the command. If the command required fewer than $N - 1$ arguments, then the trailing integers in the statement were ignored. Depending on the syntax of the statement's command, an integer argument could identify a variable to be manipulated or a statement to be executed. Consequently, even though the program was stored as a sequence it implicitly encoded an execution tree that could be reconstructed by replacing all arguments referring to program statements with the actual statement. Cramer (1985) noted that this representation was not suitable for manipulation by genetic operators and occasionally resulted in infinite loops when two auxiliary statements referred to each other.

The second representation for simple programs reviewed by Cramer (1985) alleviated some of the deficiencies of the first by making the implicit tree representation explicit. Instead of evolving a sequence of statements with arguments that referred to other statements, this representation replaces these arguments with the actual statement. For instance, an encoded program would have the form (0 (3 5) (1 3 (1 4 (4 5)))) where a matching set of parentheses denotes a single complete statement. Note that in the language used by Cramer (1985), a subtree argument does not return a value to the calling statement but only designates a command to be executed.

Probably the best known use of the parse tree representation is that by Koza (1992), an example of which is shown in figure 19.1. The only difference between the representations used in genetic programming (Koza 1992) and the explicit parse tree representation (Cramer 1985) is that the subtree arguments in genetic programming return values to their calling statements. This provides a direct mechanism for the communication of intermediate values to other portions of the parse tree representation and fortifies a subtree as an independent computational unit. The variety of problems investigated by Koza (1992) demonstrates the flexibility and applicability of this representational paradigm.

An appealing aspect of the parse tree representation is its natural recursive definition, which allows for dynamically sized structures. All parse tree representations investigated to date have included an associated restriction on the size of the evolving programs. Without such a restriction, the natural dynamics of evolutionary systems would continually increase the size of the evolving programs, eventually swamping the available computational resources. Size restrictions take on two distinct forms. Depth limitation restricts the size of evolving parse trees based on a user-defined maximal depth parameter. Node limitation places a limit on the total number of nodes available for an individual parse tree. Node limitation is the preferred method of the two since it encodes fewer restrictions on the structural organization of the evolving programs (Angeline 1996).

In a parse tree representation, the primitive language—the contents of the parse tree—determines the power and suitability of the representation. Sometimes the elements of this language are taken from existing programming languages, but typically it is more prudent to design the primitive language so that it takes into consideration as much domain-specific knowledge as available. Failing to select language primitives tailored to the task at hand may prevent the acquisition of solutions. For instance, if the objective is to evolve a function that has a particular periodic behavior, it is important to include base language primitives that also have periodic behavior, such as the mathematical functions $\sin x$ and $\cos x$.

Due to the acyclic structure of parse trees, iterative computations are often not naturally represented. It is often difficult for an evolutionary computation to correctly identify appropriate stopping criteria for loop constructs introduced into the primitive language. To compensate, the evolved function often is evaluated within an implied 'repeat until done' loop that reexecutes the evolved function until some predetermined stopping criterion is satisfied. For instance, Koza (1992) describes evolving a controller for an artificial ant for which the fitness function repeatedly applies its program until a total of 400 commands are executed or the ant completes the task. Numerous examples of such implied loops can be found in the genetic programming literature (e.g. Koza 1992, pp 147, 329, 346, Teller 1994, Reynolds 1994, Kinnear 1993).

Often it is necessary to include constants in the primitive language, especially when mathematical expressions are being evolved. The general practice is to include as a potential terminal of the language a special symbol that denotes a constant. When a new individual is created and this symbol is selected to be a terminal, rather than enter the symbol into the parse tree, a numerical constant is inserted drawn uniformly from a user-defined range (Koza 1992). Figure 19.1 shows a number of numerical constants that would be inserted into the parse tree in this manner.

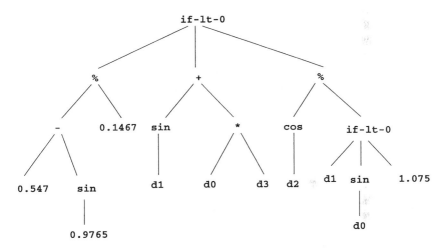

Figure 19.1. An example parse tree representation for a complex numerical function. The function if-lt-0 is a numerical conditional that returns the value of its second argument if its first argument evaluates to a negative number and otherwise returns the value of its third argument. The function % denotes a protected division operator that returns a value of 1.0 if the second argument (the denominator) is zero.

Typically, the language defined for a parse tree representation is syntactically homogenous, meaning that the return values of all functions and terminals are the same computational type, (e.g. integer). Montana (1995) has investigated the evolution of multityped parse trees and shown that extra syntactic considerations do not drastically increase the complexity of the associated genetic operators. Koza (1992) also investigates constrained parse tree representations.

Given the recursive nature of parse trees, they are a natural representation in which to investigate issues concerning induction of modular structures. Currently, three methods for inducing modular parse trees have been proposed. Angeline and Pollack (1994) add two mutation operators to their Genetic Program Builder (GLiB) system, which dynamically form and destroy modular components out of the parse tree. The *compress* mutation operation, which bears some resemblance to the *encapsulate* operator of Koza (1992), selects a subtree and makes it a new representational primitive in the language. The *expand* mutation operation reverses the actions of the compress mutation by selecting a compressed subtree in the individual and replacing it with the original subtree. Angeline and Pollack (1994) claim that the natural evolutionary dynamics of the genetic program automatically discover effective modularizations of the evolving programs. Rosca and Ballard (1996) with their Adaptive Representation method use a set of heuristics to evaluate the usefulness of all subtrees in the population and then create subroutines from the ones that are most useful. Koza (1994) describes a third method for creating modular programs, called automatically defined functions (ADFs) (see Chapter 11), which allow the user to determine the number of subroutines to which the main program can refer. During evolution, the definitions of both the main routine and all of its subroutines are evolved in parallel. Koza and Andre (1996) have more recently included a number of mutations to dynamically modify various aspects of ADFs in order to reduce the amount of prespecification required by the user.

References

Angeline P J 1996 Genetic programming's continued evolution *Advances in Genetic Programming* vol 2, ed P J Angeline and K Kinnear (Cambridge, MA: MIT Press) pp 1–20

Angeline P J and Pollack J B 1994 Co-evolving high-level representations *Artificial Life III* ed C G Langton (Reading, MA: Addison-Wesley) pp 55–71

Cramer N L 1985 A representation for the adaptive generation of simple sequential programs *Proc. 1st Int. Conf. on Genetic Algorithms (Pittsburgh, PA, July 1985)* ed J J Grefenstette (Hillsdale, NJ: Erlbaum) pp 183–7

Kinnear K E 1993 Generality and difficulty in genetic programming: evolving a sort *Proc. 5th Int. Conf. on Genetic Algorithms (Urbana-Champaign, IL, July 1993)* ed S Forrest (San Mateo, CA: Morgan Kaufmann) pp 287–94

Koza J R 1992 *Genetic Programming: on the Programming of Computers by Means of Natural Selection* (Cambridge, MA: MIT Press)

——1994 *Genetic Programming II: Automatic Discovery of Reusable Programs* (Cambridge, MA: MIT Press)

Koza J R and Andre D 1996 Classifying protein segments as transmembrane domains using architecture-altering operations in genetic programming *Advances in Genetic Programming* vol 2, ed P J Angeline and K Kinnear (Cambridge, MA: MIT Press) pp 155–76

Montana D J 1995 Strongly typed genetic programming *Evolutionary Comput.* **3** 199–230

Reynolds C W 1994 Evolution of obstacle avoidance behavior: using noise to promote robust solutions *Advances in Genetic Algorithms* ed K Kinnear (Cambridge, MA: MIT Press) pp 221–43

Rosca J P and Ballard D H 1996 Discovery of subroutines in genetic programming *Advances in Genetic Programming* vol 2, ed P J Angeline and K Kinnear (Cambridge, MA: MIT Press) pp 177–202

Teller A 1994 The evolution of mental models *Advances in Genetic Algorithms* ed K Kinnear (Cambridge, MA: MIT Press) pp 199–220

20

Guidelines for a suitable encoding

David B Fogel and Peter J Angeline

In any evolutionary computation application to an optimization problem, the human operator determines at least four aspects of the approach: representation, variation operators, method of selection, and objective function. It could be argued that the most crucial of these four is the objective function because it defines the purpose of the operator in quantitative terms. Improperly specifying the objective function can lead to generating the right answer to the wrong problem. However it should be clear that the selections made for each of these four aspects depend in part on the choices made for all the others. For example, the objective function cannot be specified in the absence of a problem representation. The choice for appropriate representation, however, cannot be made in the absence of anticipating the variation operators, the selection function, and the mathematical formulation of the problem to be solved. Thus, an iterative procedure for adjusting the representation and search and selection procedures in light of a specified objective function becomes necessary in many applications of evolutionary computation. This section focuses on selecting the representation for a problem, but it is important to remain cognizant of the interdependent nature of these operations within any evolutionary computation.

There have been proposals that the most suitable encoding for any problem is a binary encoding (Chapter 15) because it maximizes the number of schemata being searched implicitly (Holland 1975, Goldberg 1989), but there have been many examples in the evolutionary computation literature where alternative representations have provided for algorithms with greater efficiency and optimization effectiveness when compared with identical problems (see e.g. the articles by Bäck and Schwefel (1993) and Fogel and Stayton (1994) among others). Davis (1991) and Michalewicz (1996) comment that in many applications real-valued (Chapter 16) or other representations may be chosen to advantage over binary encodings. There does not appear to be any general benefit to maximizing implicit parallelism in evolutionary algorithms, and, therefore, forcing problems to fit binary representation is not recommended.

The close relationship between representation and other facets of evolutionary computation suggests that, in many cases, the appropriate choice of representation arises from the operator's ability to visualize the dynamics of the resulting search on an adaptive landscape. For example, consider the problem of finding the minimum of the quadratic surface

$$f(x, y) = x^2 + y^2 \qquad x, y \in \mathbb{R}.$$

Immediately, it is easy to visualize this function as shown in figure 20.1. If an evolutionary approach to the problem were to be taken, an intuitive representation suggested by the surface is to operate on the real values of (x, y) directly (rather than recoding these values into some other alphabet). Accordingly, a reasonable choice of variation operator would be the imposition of a continuous random perturbation to each dimension (x, y) (perhaps a zero-mean Gaussian perturbation as is common in evolution strategies and evolutionary programming). This would be followed by a hard selection against all but the best solution in the current population, given that the function is strongly convex. With even slight experience, the resulting population dynamics of this approach can be visualized without executing a single line of code. In contrast, for this problem other representational choices and variation operators (e.g. mapping the real numbers into binary and then applying crossover operators to the binary encoding) are contrived, difficult to visualize, and appear more likely to be ineffectual (see Schraudolph and Belew 1992, Fogel and Stayton 1994).

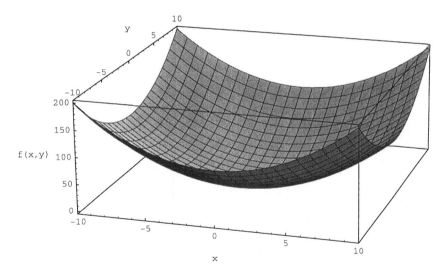

Figure 20.1. A quadratic bowl in two dimensions. The shape of the response surface suggests a natural approach for optimization. The intuitive choice is to use real-valued encodings and continuous variation operators. The shape of a response surface can be useful in suggesting choices for suitable encodings.

Thus, the basic recommendation for choosing a suitable encoding is that the representation should be suggested from the problem at hand. If a traveling salesman problem is to be investigated, obvious natural choices for the encoding are a list of cities to be visited in order, or a corresponding list of edges. For a discrete-symbol time-series prediction problem, finite-state machines (Chapter 18)may be especially appropriate. For continuous time-series prediction, other model forms (e.g. neural networks, ARMA, or Box–Jenkins) appear better suited. In nonstationary environments, that is, fitness functions that are dynamic rather than static, it is often necessary to include some form of memory in the representation. Diplodic representations—representations that include two alleles per gene—have been used to model cyclic environments (Goldberg and Smith 1987, Ng and Wong 1995). The most natural choice for representation is a subjective choice, and it will differ across investigators, although, like a suitable scientific model, a suitable representation should be as complex as necessary (and no more so) and should 'explain' the phenomena investigated, which here means that the resulting search should be visualizable or imaginable to some extent.

References

Bäck T and Schwefel H-P 1993 An overview of evolutionary algorithms for parameter optimization *Evolutionary Comput.* **1** 1–24

Davis L (ed) 1991 *Handbook of Genetic Algorithms* (New York: Van Nostrand Reinhold)

Fogel D B and Stayton L C 1994 On the effectiveness of crossover in simulated evolutionary optimization *BioSystems* **32** 171–82

Goldberg D E 1989 *Genetic Algorithms in Search, Optimization, and Machine Learning* (Reading, MA: Addison-Wesley)

Goldberg D E and Smith R E 1987 Nonstationary function optimization using genetic algortihms with dominance and diploidy *Proc. 2nd Int. Conf. on Genetic Algorithms (Cambridge, MA, 1987)* ed J J Grefenstette (Hillsdale, NJ: Erlbaum) pp 59–68

Holland J H 1975 *Adaptation in Natural and Artificial Systems* (Ann Arbor, MI: University of Michigan Press)

Michalewicz Z 1996 *Genetic Algorithms + Data Structures = Evolution Programs* 3rd edn (Berlin: Springer)

Ng K P and Wong K C 1995 A new diploid scheme and dominance change mechanism for non-stationary function optimization *Proc. 6th Int. Conf. on Genetic Algorithms (Pittsburgh, PA, July 1995)* ed L J Eshelman (San Mateo, CA: Morgan Kaufmann) pp 159–66

Schraudolph N N and Belew R K 1992 Dynamic parameter encoding for genetic algorithms *Machine Learning* **9** 9–21

21

Other representations

Peter J Angeline and David B Fogel

21.1 Mixed-integer structures

Many real-world applications suggest the use of representations that are hybrids of the canonical representations. One common instance is the simultaneous use of discrete and continuous object variables, with a general formulation of the global optimization problem as follows (Bäck and Schütz 1995):

$$\min\{f(x, d)|x \in M, R^n \supseteq M, d \in N, Z^{n_d} \supseteq N\}.$$

Within evolution strategies and evolutionary programming, the common representation is simply the real-integer vector pair (i.e. no effort is made to encode these vectors into another representation such as binary). Sections 32.6 and 33.6 offer methods for mutating and recombining the above representations.

Mixed representations also occur in the application of evolutionary algorithms to neural networks or fuzzy logic systems, where real-world parameters are used to define weights or shapes of membership functions and integer values are used to define the number of nodes and their connections, or the number of membership functions (see e.g. Fogel 1995, Angeline *et al* 1994, McDonnell and Waagen 1994, Haffner and Sebald 1993).

21.2 Introns

In contrast to the above hybridization of different forms of representation, another 'nontraditional' approach has involved the inclusion of noncoding regions (introns) within a solution (see e.g. Levenick 1991, Golden *et al* 1995, Wu and Lindsay 1995). Solutions are represented in the form

$$x_1|\text{intron}|x_2|\text{intron}|\ldots|\text{intron}|x_n$$

where there are n components to vector x. Introns have been hypothesized to allow for greater efficiency in recombining building blocks (see Section 33.6).

In the standard genetic algorithm representation, the semantics of an allele value (how the allele is interpreted) is usually tied to its position in the fixed-length n-ary string. For instance, in a binary string representation, each position signifies the presence or absence of a specific feature in the genome being decoded. The difficulty with such a representation is that with positions in the string representation that are semantically linked but separated by a large number of intervening positions in the string crossover has a high probability of disrupting beneficial settings for these two positions. Goldberg *et al* (1989) describe a representation for a genetic algorithm that embodies one approach to addressing this problem. In their messy genetic algorithm (mGA), each allele value is represented as a pair of values, one specifying the actual allele value and one specifying the position the allele occupies. Messy GAs are defined to be of variable length, and Goldberg *et al* (1989) describe appropriate methods for resolving underdetermined or overdetermined genomes. In this representation it is important to note that the semantics are literally carried along with the allele value in the form of the allele's string position.

21.3   Diploid representations

Diploid representations, representations that include multiple allele values for each position in the genome, have been offered as mechanisms for modeling cyclic environments. In a diploid representation, a method for determining which allele value for a gene will be expressed is required to adjudicate when the allele values do not agree. Building on earlier investigations (see e.g. Bagley 1967, Hollstein 1971, Brindle 1981), Goldberg and Smith (1987) demonstrate that an evolving dominance map allows quicker adaptation to cyclical environment changes than either a haploid representation or a diploid representation using a fixed dominance mapping. Goldberg and Smith (1987) use a triallelic representation from Hollstein (1971): 1, i, and 0. Both 1 and i map to the allele value of '1', while 0 maps to the allele value of '0' with 1 dominating both i and 0 and 0 dominating i. Thus, the dominance of a 1 over a 0 allele value could be altered via mutation by altering the value to an i. Ng and Wong (1995) extend the multiallele approach to dominance computation by adding a fourth value for a recessive 0. Thus 1 dominates 0 and o while 0 dominates i and o. When both allele values for a gene are dominant or recessive, then one of the two values is chosen randomly to be the dominant value. Ng and Wong (1995) also suggest that the dominance of all of the components in the genome should be reversed when the fitness value of an individual falls by 20% or more between generations.

References

Angeline P J, Saunders G M and Pollack J B 1994 An evolutionary algorithm that constructs recurrent neural networks *IEEE Trans. Neural Networks* **NN-5** 54–65

Bäck T and Schütz M 1995 Evolution strategies for mixed-integer optimization of optical multilayer systems *Proc. 4th Ann. Conf. on Evolutionary Programming (San Diego, CA, March 1995)* ed J R McDonnell, R G Reynolds and D B Fogel (Cambridge, MA: MIT Press) pp 33–51

Bagley J D 1967 *The Behavior of Adaptive Systems which Employ Genetic and Correlation Algorithms* Doctoral Dissertation, University of Michigan; University Microfilms 68-7556

Brindle A 1981 *Genetic Algorithms for Function Optimization* Doctoral Dissertation, University of Alberta

Cobb H G and Grefenstette J J 1993 Genetic algorithms for tracking changing environments *Proc. 5th Int. Conf. on Genetic Algorithms (Urbana-Champaign, IL, July 1993)* ed S Forrest (San Mateo, CA: Morgan Kaufmann) pp 523–30

Fogel D B 1995 *Evolutionary Computation: Toward a New Philosophy of Machine Intelligence* (Piscataway, NJ: IEEE)

Goldberg D E, Korb D E and Deb K 1989 Messy genetic algorithms: motivation, analysis, and first results *Complex Syst.* **3** 493–530

Goldberg D E and Smith R E 1987 Nonstationary function optimization using genetic algorithms with dominance and diploidy *Proc. 2nd Int. Conf. on Genetic Algorithms (Cambridge, MA, July 1987)* ed J J Grefenstette (Hillsdale, NJ: Erlbaum) pp 59–68

Golden J B, Garcia E and Tibbetts C 1995 Evolutionary optimization of a neural network-based signal processor for photometric data from an automated DNA sequencer *Proc. 4th Ann. Conf. on Evolutionary Programming (San Diego, CA, March 1995)* ed J R McDonnell, R G Reynolds and D B Fogel (Cambridge, MA: MIT Press) pp 579–601

Haffner S B and Sebald A V 1993 Computer-aided design of fuzzy HVAC controllers using evolutionary programming *Proc. 2nd Ann. Conf. on Evolutionary Programming (San Diego, CA, 1993)* ed D B Fogel and W Atmar (La Jolla, CA: Evolutionary Programming Society) pp 98–107

Hollstein R B 1971 *Artificial Genetic Adaptation in Computer Control Systems* Doctoral Dissertation, University of Michigan; University Microfilms 71-23, 773

Levenick J R 1991 Inserting introns improves genetic algorithm success rate: taking a cue from biology *Proc. 4th Int. Conf. on Genetic Algorithms (San Diego, CA, July 1991)* ed R K Belew and L B Booker (San Mateo, CA: Morgan Kaufmann) pp 123–27

McDonnell J R and Waagen D 1994 Evolving recurrent perceptrons for time-series modeling *IEEE Trans. Neural Networks* **NN-5** 24–38

Ng K P and Wong K C 1995 A new diploid scheme and dominance change mechanism for non-stationary function optimization *Proc. 6th Int. Conf. on Genetic Algorithms (Pittsburgh, PA, July 1995)* ed L J Eshelman (San Mateo, CA: Morgan Kaufmann) pp 159–66

Wu A S and Lindsay R K 1995 Empirical studies of the genetic algorithm with noncoding segments *Evolutionary Comput.* **3** 121–48

22

Introduction to selection

Kalyanmoy Deb

22.1 Working mechanisms

Selection is one of the main operators used in evolutionary algorithms. The primary objective of the selection operator is to *emphasize* better solutions in a population. This operator does not create any new solution, instead it selects relatively good solutions from a population and deletes the remaining, not-so-good, solutions. Thus, the selection operator is a mix of two different concepts—reproduction and selection. When one or more copies of a good solution are reproduced, this operation is called reproduction. Multiple copies of a solution are placed in a population by deleting some inferior solutions. This concept is known as selection. Although some EC studies use both these concepts simultaneously, some studies use them separately.

The identification of good or bad solutions in a population is usually accomplished according to a solution's fitness. The essential idea is that a solution having a better fitness must have a higher probability of selection. However, selection operators differ in the way the copies are assigned to better solutions. Some operators sort the population according to fitness and deterministically choose the best few solutions, whereas some operators assign a probability of selection to each solution according to fitness and make a copy using that probability distribution. In the probabilistic selection operator, there is some finite, albeit small, probability of rejecting a good solution and choosing a bad solution. However, a selection operator is usually designed in a way so that the above is a low-probability event. There is, of course, an advantage of allowing this stochasticity (or flexibility) in the evolutionary algorithms. Due to a small initial population or an improper parameter choice or in solving a complex nonlinear fitness function, the best few individuals in a finite population may sometimes represent a suboptimal region. If a deterministic selection operator is used, these seemingly good individuals in the population will be emphasized and the population may finally converge to a wrong solution. However, if a stochastic selection operator is used, diversity in the population will be maintained by occasionally choosing not-so-good solutions. This event may prevent EC algorithms from making a hasty decision in converging to a wrong solution.

In the following, we present a pseudocode for the selection operator and then discuss briefly some of the popular selection operators.

22.2 Pseudocode

Some EC algorithms (specifically, genetic algorithms (GAs) and *genetic programming* (GP)) usually apply the selection operator first to select good solutions and then apply the recombination and mutation operators on these good solutions to create a hopefully better set of solutions. Other EC algorithms (specifically, evolution strategies (ES) and *evolutionary programming* (EP)) prefer using the recombination and mutation operator first to create a set of solutions and then use the selection operator to choose a good set of solutions. The selection operator in $(\mu + \lambda)$ ES and EP techniques chooses the offspring solutions from a combined population of parent solutions and solutions obtained after recombination and mutation. In the case of EP, this is done statistically. However, the selection operator in (μ, λ) ES chooses the offspring solutions only from the solutions obtained after the recombination and mutation operators. Since the selection operators are different in different EC studies, it is difficult to present a common code for all selection operators. However, the following pseudocode is a generic for most of the selection operators used in EC studies.

The parameters μ and λ are the numbers of parent solutions and offspring solutions after recombination and mutation operators, respectively. The parameter q is a parameter related to the operator's selective pressure, a matter we discuss later in this section. The population at iteration t is denoted by $P(t) = \{a_1, a_2, \ldots\}$ and the population obtained after the recombination and mutation operators is denoted by $P'(t) = \{a'_1, a'_2, \ldots\}$. Since GAs and GP techniques use the selection operator first, the population $P'(t)$ before the selection operation is an empty set, with no solutions. The fitness function is represented by $F(t)$.

Input: μ, λ, q, $P(t) \in I^{\mu}$, $P'(t) \in I^{\lambda}$, $F(t)$

Output: $P''(t) = \{a''_1, a''_2 \ldots, a''_{\mu}\} \in I^{\mu}$

1   **for** $i \leftarrow 1$ **to** μ
       $a''_i(t) \leftarrow s_{\text{selection}}(P(t), P'(t), F(t), q)$;
2   **return**$(\{a''_1(t), \ldots, a''_{\mu}(t)\})$;

Detailed discussions of some of the selection operators are presented in the subsequent sections. Here, we outline a brief introduction to some of the popular selection schemes, mentioned as $s_{\text{selection}}$ in the above pseudocode.

In the *proportionate* selection operator, the expected number of copies a solution receives is assigned proportionally to its fitness. Thus, a solution having twice the fitness of another solution receives twice as many copies. The simplest form of the proportionate selection scheme is known as the roulette-wheel selection, where each solution in the population occupies an area on the roulette wheel proportional to its fitness. Then, conceptually, the roulette wheel is spun as many times as the population size, each time selecting a solution marked by the roulette-wheel pointer. Since the solutions are marked proportionally to their fitness, a solution with a higher fitness is likely to receive more copies than a solution with a low fitness. There exists a number of variations to this simple selection scheme, which are discussed in Chapter 23. However, one limitation of the proportionate selection scheme is that since copies are assigned proportionally to the fitness values, negative fitness values are not allowed. Also, this scheme cannot handle minimization problems directly. (Minimization problems must be transformed to an equivalent maximization problem in order to use this operator.) Selecting solutions proportional to their fitness has two inherent problems. If a population contains a solution having exceptionally better fitness than the rest of the solutions in the population, this so-called *supersolution* will occupy most of the roulette-wheel area. Thus, most spinning of the roulette wheel is likely to choose the same supersolution. This may cause the population to lose genetic diversity and cause the algorithm to prematurely converge to a suboptimal solution. The second inherent difficulty may arise later in a simulation run, when most of the population members have more or less the same fitness. In this case, the roulette wheel is marked almost equally for each solution in the population and every solution becomes equally likely to be selected. This has the effect of a random selection. Both these inherent difficulties can be avoided by using a *scaling* scheme, where every solution fitness is linearly mapped between a lower and an upper bound before marking the roulette wheel (Goldberg 1989). This allows the selection operator to assign a controlled number of copies, thereby eliminating both the above problems of too large and random assignments. We discuss this scaling scheme further in the next section. Although this selection scheme has been mostly used with GAs and GP applications, in principle it can also be used with both multimembered ES and EP techniques.

In the *tournament* selection operator, both the scaling problems mentioned above are eliminated by playing tournaments among a specified number of parent solutions according to fitness of solutions. In a tournament of q solutions, the best solution is selected either deterministically or probabilistically. After the tournament is played, there are two options—either all participating q solutions are replaced into the population for the next tournament or they are not replaced until a certain number of tournaments have been played. In its simplest form (called the binary tournament selection), two solutions are picked and the better solution is chosen. One advantage of this selection method is that this scheme can handle both minimization and maximization problems without any structural change in the fitness function. Only the solution having either the highest or the lowest objective function value need to be chosen depending on whether the problem is a maximization or a minimization problem. Moreover, it has no restriction on negative objective function values. An added advantage of this scheme is that it is ideal for a parallel implementation. Since only a few solutions are required to be compared at a time without resorting to calculation of the population average fitness or any other population statistic, all solutions participating in a tournament can be sent to one processor. Thus, tournaments can be played in parallel on multiple processors and the complete selection process may be performed quickly. Because of these properties, tournament selection is fast becoming a popular selection scheme in most EC studies. Tournament selection is discussed in detail in Chapter 24.

The *ranking* selection operator is similar to proportionate selection except that the solutions are ranked according to descending or ascending order of their fitness values depending on whether it is a maximization or minimization problem. Each solution is assigned a ranked fitness based on its rank in the population. Thereafter, copies are allocated with the resulting selection probabilities of the solutions calculated using the ranked fitness values. Like tournament selection, this selection scheme can also handle negative fitness values. There exists a number of other schemes based on the concept of the ranking of solutions; these are discussed in Chapter 25.

In the *Boltzmann* selection operator, a modified fitness is assigned to each solution based on a Boltzmann probability distribution: $\mathcal{F}_i = 1/(1+\exp(F_i/T))$, where T is a parameter analogous to the temperature term in the Boltzmann distribution. This parameter is reduced in a predefined manner in successive iterations. Under this selection scheme, a solution is selected based on the above probability distribution. Since a large value of T is used initially, almost any solution is equally likely to be selected, but, as the iterations progress, the parameter T becomes small and only good solutions are selected. We discuss this selection scheme further in Chapter 26.

In the $(\mu + \lambda)$ ES, the selection operator selects μ best solutions deterministically from a pool of all μ parent solutions and λ offspring solutions. Since all parent and offspring solutions are compared, if performed deterministically, this selection scheme guarantees preservation of the best solution found in any iteration.

On the other hand, in the (μ, λ) ES, the selection operator chooses μ best solutions from λ (usually $\lambda > \mu$) offspring solutions obtained by the recombination and mutation operators. Unlike the $(\mu + \lambda)$ ES selection scheme, the best solution found in any iteration is not guaranteed to be preserved throughout a simulation. However, since many offspring solutions are created in this scheme, the search is more exhaustive than that in the $(\mu + \lambda)$ ES scheme. In most applications of the (μ, λ) ES selection scheme, a deterministic selection of best μ solutions is adopted.

In modern variants of the EP technique, a slightly different selection scheme is used. In a pool of parent (of size μ) and offspring solutions (of size the same as the parent population size), each solution is first assigned a score depending on how many solutions it is better than from a set of random solutions (of size

q) chosen from the pool. The complete pool is then sorted in descending order of this score and the first μ solutions are chosen deterministically. Thus, this selection scheme is similar to the ($\mu + \mu$) ES selection scheme with a tournament selection of q tournament size. Bäck *et al* (1994) analyzed this selection scheme as a combination of ($\mu + \mu$) ES and tournament selection schemes, and found some convergence characteristics of this operator.

Goldberg and Deb (1991) have compared a number of popular selection schemes in terms of their convergence properties, selective pressure, takeover times, and growth factors, all of which are important in the understanding of the power of different selection schemes used in GA and GP studies. Similar studies have also been performed by Bäck *et al* (1994) for selection schemes used in ES and EP studies. A detailed discussion of some analytical as well as experimental comparisons of selection schemes is presented in Chapter 29. In the following section, we briefly discuss the theory of selective pressure and its importance in choosing a suitable selection operator for a particular application.

22.3 Theory of selective pressure

Selection operators are characterized by a parameter known as the *selective pressure*, which relates to the takeover time of the selection operator. The takeover time is defined as the speed at which the best solution in the initial population would occupy the complete population by repeated application of the selection operator alone (Bäck 1994, Goldberg and Deb 1991). If the takeover time of a selection operator is *large* (that is, the operator takes a large number of iterations for the best solution to take over the population), the selective pressure of the operator is *small*, and vice versa. Thus, the selective pressure or the takeover time is an important parameter for successful operation of an EC algorithm (Bäck 1994, Goldberg *et al* 1993). This parameter gives an idea of how greedy the selection operator is in terms of making the population uniform with one particular solution. If a selection operator has a large selective pressure, the population loses diversity in the population quickly. Thus, in order to avoid premature convergence to a wrong solution, either a large population is required or highly disruptive recombination and mutation operators are needed. However, a selection operator with a small selection pressure makes a slow convergence and permits the recombination and mutation operators enough iterations to properly search the space. Goldberg and Deb (1991) have calculated takeover times of a number of selection operators used in GAs and GP studies and Bäck (1994) has calculated the takeover time for a number of selection operators used in ES, EP, and GA studies. The former study has also introduced two other parameters—early and late growth rate—characterizing the selection operators.

The growth rate is defined as the ratio of the number of the best solutions in two consecutive iterations. Since most selection operators have different growth rates as the iterations progress, two different growth rates—early and late growth rates—are defined. The early growth rate is calculated initially, when the proportion of the best solution in the population is negligible. The late growth rate is calculated later, when the proportion of the best solution in the population is large (about 0.5). The early growth rate is important, especially if a quick near-optimizer algorithm is desired, whereas the late growth rate can be a useful measure if precision in the final solution is important. Goldberg and Deb (1991) have calculated these growth rates for a number of selection operators used in GAs. A comparison of different selection schemes based on some of the above criteria is given in Chapter 29.

The above discussion suggests that, for a successful EC simulation, the required selection pressure of a selection operator depends on the recombination and mutation operators used. A selection scheme with a large selection pressure can be used, but only with highly disruptive recombination and mutation operators. Goldberg *et al* (1993) and later Thierens and Goldberg (1993) have found functional relationships between the selective pressure and the probability of crossover for successful working of selectorecombinative GAs. These studies show that a large selection pressure can be used but only with a large probability of crossover. However, if a reasonable selection pressure is used, GAs work successfully for a wide variety of crossover probablities. Similar studies can also be performed with ES and EP algorithms.

References

Bäck T 1994 Selective pressure in evolutionary algorithms: a characterization of selection mechanisms *Proc. 1st IEEE Conf. on Evolutionary Computation (Orlando, FL, 1994)* (Piscataway, NJ: IEEE) pp 57–62

Bäck T, Rudolph G and Schwefel H-P 1994 Evolutionary programming and evolution strategies: similarities and differences *Proc. 2nd Ann. Conf. on Evolutionary Programming (San Diego, CA, July 1994)* ed D B Fogel and W Atmar (La Jolla, CA: Evolutionary Programming Society)

Goldberg D E 1989 *Genetic Algorithms in Search, Optimization, and Machine Learning* (Reading, MA: Addison-Wesley)

Goldberg D E and Deb K 1991 A comparison of selection schemes used in genetic algorithms *Foundations of Genetic Algorithms (Bloomington, IN)* ed G J E Rawlins (San Mateo, CA: Morgan Kaufmann) pp 69–93

Goldberg D E, Deb K and Theirens D 1993 Toward a better understanding of mixing in genetic algorithms *J. SICE* **32** 10–6

Thierens D and Goldberg D E 1993 Mixing in genetic algorithms *Proc. 5th Int. Conf. on Genetic Algorithms (Urbana-Champaign, IL, July 1993)* ed S Forrest (San Mateo, CA: Morgan Kaufmann) pp 38–45

23

Proportional selection and sampling algorithms

John Grefenstette

23.1   Introduction

Selection (Chapter 22) is the process of choosing individuals for reproduction in an evolutionary algorithm. One popular form of selection is called *proportional selection*. As the name implies, this approach involves creating a number of offspring in proportion to an individual's fitness. This approach was proposed and analyzed by Holland (1975) and has been used widely in many implementations of evolutionary algorithms.

Besides having some interesting mathematical properties, proportional selection provides a natural counterpart in artificial evolutionary systems to the usual practice in population genetics of defining an individual's fitness in terms of its number of offspring.

For clarity of discussion, it is convenient to decompose the selection process into distinct steps, namely:

(i)   map the objective function to fitness,
(ii)   create a probability distribution proportional to fitness, and
(iii)   draw samples from this distribution.

The first three sections of this article discuss these steps. The final section discusses some results in the theory of proportional selection, including the schema theorem and the impact of the fitness function, and two characterizations of selective pressure.

23.2   Fitness functions

The evaluation process of individuals in an evolutionary algorithm begins with the user-defined *objective function*,

$$f : A_x \rightarrow \mathbb{R}$$

where A_x is the object variable space. The objective function typically measures some cost to be minimized or some reward to be maximized. The definition of the objective function is, of course, application dependent. The characterization of how well evolutionary algorithms perform on different classes of objective functions is a topic of continuing research. However, a few general design principles are clear when using an evolutionary algorithm.

(i) The objective function must reflect the relevant measures to be optimized. Evolutionary algorithms are notoriously opportunistic, and there are several known instances of an algorithm optimizing the stated objective function, only to have the user realize that the objective function did not actually represent the intended measure.

(ii) The objective function should exhibit some regularities over the space defined by the selected representation.

(iii) The objective function should provide enough information to drive the selective pressure of the evolutionary algorithm. For example, 'needle-in-a-haystack' functions, i.e. functions that assign nearly equal value to every candidate solution except the optimum, should be avoided.

The *fitness function*

$$\Phi : A_x \rightarrow \mathbb{R}_+$$

maps the raw scores of the objective function to a non-negative interval. The fitness function is often a composition of the objective function and a scaling function g:

$$\Phi(a_i(t)) = g(f(a_i(t)))$$

where $a_i(t) \in A_x$. Such a mapping is necessary if the goal is to minimize the objective function, since higher fitness values correspond to lower objective values in this case. For example, one fitness function that might be used when the goal is to minimize the objective function is

$$\Phi(a_i(t)) = f_{\max} - f(a_i(t))$$

where $f_{\max}$ is the maximum value of the objective function. If the global maximum value of the objective function is unknown, an alternative is

$$\Phi(a_i(t)) = f_{\max}(t) - f(a_i(t))$$

where $f_{\max}(t)$ is the maximum observed value of the objective function up to time t. There are many other plausible alternatives, such as

$$\Phi(a_i(t)) = \frac{1}{1 + f(a_i(t)) - f_{\min}(t)}$$

where $f_{\min}(t)$ is the minimum observed value of the objective function up to time t. For maximization problems, this becomes

$$\Phi(a_i(t)) = \frac{1}{1 + f_{\max}(t) - f(a_i(t))}.$$

Note that the latter two fitness functions yield a range of $(0, 1]$.

23.2.1 Fitness scaling

As an evolutionary algorithm progresses, the population often becomes dominated by high-performance individuals with a narrow range of objective values. In this case, the fitness functions described above tend to assign similar fitness values to all members of the population, leading to a loss in the selective pressure toward the better individuals. To address this problem, *fitness scaling* methods that accentuate small differences in objective values are often used in order to maintain a productive level of selective pressure.

One approach to fitness scaling (Grefenstette 1986) is to define the fitness function as a time-varying linear transformation of the objective value, for example

$$\Phi(a_i(t)) = \alpha f(a_i(t)) - \beta(t)$$

where α is $+1$ for maximization problems and -1 for minimization problems, and $\beta(t)$ represents the worst value seen in the last few generations. Since $\beta(t)$ generally improves over time, this scaling method provides greater selection pressure later in the search. This method is sensitive, however, to 'lethals', poorly performing individuals that may occasionally arise through crossover or mutation. Smoother scaling can be achieved by defining $\beta(t)$ as a recency-weighted running average of the worst observed objective values, for example

$$\beta(t) = \delta\beta(t-1) + (1-\delta)(f_{\text{worst}}(t))$$

where δ is an update rate of, say, 0.1, and $f_{\text{worst}}(t)$ is the worst objective value in the population at time t.

Sigma scaling (Goldberg 1989) is based on the distribution of objective values within the current population. It is defined as follows:

$$\Phi(a_i(t)) = \begin{cases} f(a_i(t)) - (\bar{f}(t) - c\sigma_f(t)) & \text{if } f(a_i(t)) > (\bar{f}(t) - c\sigma_f(t)) \\ 0 & \text{otherwise} \end{cases}$$

where $\bar{f}(t)$ is the mean objective value of the current population, $\sigma_f(t)$ is the (sample) standard deviation of the objective values in the current population, and c is a constant, say $c = 2$. The idea is that $\bar{f}(t) - c\sigma_f(t)$ represents the least acceptable objective value for any reproducing individual. As the population improves, this statistic tracks the improvement, yielding a level of selective pressure that is sensitive to the spread of performance values in the population.

Fitness scaling methods based on power laws have also been proposed. A fixed transformation of the form

$$\Phi(a_i(t)) = f(a_i(t))^k,$$

where k is a problem-dependent parameter, is used by Gillies (1985). *Boltzmann selection* (de la Maza and Tidor 1993) is a power-law-based scaling method that draws upon techniques used in simulated annealing. The fitness function is a time-varying transformation given by

$$\Phi(a_i(t)) = \exp\left(f(a_i(t))/T\right)$$

where the parameter T can be used to control the level of selective pressure during the course of the evolution. It is suggested by de la Maza and Tidor (1993) that, if T decreases with time as in a simulated annealing procedure, then a higher level of selective pressure results than with proportional selection without fitness scaling.

23.3 Selection probabilities

Once the fitness values are assigned, the next step in proportional selection is to create a probability distribution such that the probability of selecting a given individual for reproduction is proportional to the individual's fitness. That is,

$$\Pr_{\text{prop}}(i) = \frac{\Phi(i)}{\sum_{i=1}^{\mu} \Phi(i)}.$$

23.4 Sampling

In an incremental, or steady-state, algorithm, the probability distribution can be used to select one parent at a time. This procedure is commonly called the *roulette wheel* sampling algorithm, since one can think of the probability distribution as defining a roulette wheel on which each slice has a width corresponding to the individual's selection probability, and the sampling can be envisioned as spinning the roulette wheel and testing which slice ends up at the top. The pseudocode for this is shown below:

Input: probability distribution Pr
Output: n, the selected parent

```
1    roulette wheel (Pr):
2        n ← 1;
3        sum ← Pr(n);
4        sample u ~ U(0, 1);
5        while sum < u do
                n ← (n + 1);
                sum ← sum + Pr(n);
         od
6        return (n);
```

In a generational algorithm, the entire population is replaced during each generation, so the probability distribution is sampled μ times. This could be implemented by μ independent calls to the roulette wheel procedure, but such an implementation may exhibit a high variance in the number of offspring assigned to each individual. For example, it is possible that the individual with the largest selection probability may be assigned no offspring in a particular generation. Baker (1987) developed an algorithm called *stochastic universal sampling* (SUS) that exhibits less variance than repeated calls to the roulette wheel algorithm. The idea is to make a single draw from a uniform distribution, and use this to determine how many offspring to assign to all parents. The pseudocode for SUS follows:

Input: a probability distribution, Pr; the total number of children to assign, λ.
Output: $c = (c_1, \ldots, c_\mu)$, where c_i is the number of children assigned to individual a_i, and $\sum c_i = \lambda$.

```
1    SUS(Pr, λ):
2        sample u ~ U(0, 1/λ);
3        sum ← 0.0;
4        for i = 1 to μ do
5            c_i ← 0;
6            sum ← sum + Pr(i);
7            while u < sum do
8                c_i ← c_i + 1;
9                u ← u + 1/λ;
             od
         od
10       return c;
```

Note that the pseudocode allows for any number $\lambda > 0$ of children to be specified. If $\lambda = 1$, SUS behaves like the roulette wheel function. For generational algorithms, SUS is usually invoked with $\lambda = \mu$.

In can be shown that the expected number of offspring that SUS assigns to individual i is $\lambda \Pr(i)$, and that on each invocation of the procedure, SUS assigns either $\lfloor \lambda \Pr(i) \rfloor$ or $\lceil \lambda \Pr(i) \rceil$ offspring to individual i. Finally, SUS is optimally efficient, making a single pass over the individuals to assign all offspring.

23.5   Theory

The section presents some results from the theory of proportional selection. First, the schema theorem is described, following by a discussion of the effects of the fitness function on the allocation of trials to schemata. The selective pressure of proportional selection is characterized in two ways. First, the selection differential describes the effects of selection on the mean population fitness. Second, the takeover time describes the convergence rate of population toward the optimal individual, in the absence of other genetic operators.

23.5.1 The schema theorem

In the above description, $\Pr_{\text{prop}}(i)$ is the probability of selecting individual i for reproduction. In a generational evolutionary algorithm, the entire population is replaced, so the expected number of offspring of individual i is $\mu \Pr_{\text{prop}}(i)$. This value is called the *target sampling rate,* $\text{tsr}(a_i, t)$ of the individual (Grefenstette 1991). For any selection algorithm, the allocation of offspring to individuals induces a corresponding allocation to hyperplanes represented by the individuals:

$$\text{tsr}(H, t) =_{\text{def}} \sum_{i=1}^{m(H,t)} \frac{\text{tsr}(a_i, t)}{m(H, t)}$$

where $a_i \in H$ and $m(H, t)$ denotes the number of representatives of hyperplane H in population $P(t)$. In the remainder of this discussion, we will refer to $\text{tsr}(H, t)$ as the *target sampling rate* of H at time t.

For proportional selection, we have

$$\text{tsr}(a_i, t) = \frac{\Phi(a_i)}{\bar{\Phi}(t)}$$

where Φ is the fitness function and $\bar{\Phi}(t)$ denotes the average fitness of the individuals in $P(t)$. The most important feature of proportional selection is that it induces the following target sampling rates for all hyperplanes in the population:

$$\begin{aligned}
\text{tsr}(H, t) &= \sum_{i=1}^{m(H,t)} \frac{\text{tsr}(a_i, t)}{m(H, t)} \\
&= \sum_{i=1}^{m(H,t)} \frac{\Phi(a_i)}{\bar{\Phi}(t)\, m(H, t)} \\
&= \frac{\Phi(H, t)}{\bar{\Phi}(t)}
\end{aligned} \tag{23.1}$$

where $\Phi(H, t)$ is simply the average fitness of the representatives of H in $P(t)$. This result is the heart of the schema theorem (Holland 1975), which has been called the *fundamental theorem of genetic algorithms* (Goldberg 1989).

Schema theorem. In a genetic algorithm using a proportional selection algorithm, the following holds for each hyperplane H represented in $P(t)$:

$$M(H, t+1) \geq M(H, t) \left(\frac{\Phi(H, t)}{\bar{\Phi}(t)} \right) (1 - p_{\text{disr}}(H, t))$$

where $M(H, t)$ is the expected number of representatives of hyperplane H in $P(t)$, and $p_{\text{disr}}(H, t)$ is the probability of disruption due to genetic operators such as crossover and mutation.

Holland provides an analysis of the disruptive effects of various genetic operators, and shows that hyperplanes with short defining lengths, for example, have a small chance of disruption due to one-point crossover and mutation operators. Others have extended this analysis to many varieties of genetic operators.

The main thrust of the schema theorem is that trials are allocated *in parallel* to a large number of hyperplanes (i.e. the ones with short definition lengths) according to the sampling rate (23.1), with minor disruption from the recombination operators. Over succeeding generations, the number of trials allocated to extant short-definition-length hyperplanes with persistently above-average observed fitness is expected to grow rapidly, while trials to those with below-average observed fitness generally decline rapidly.

23.5.2   Effects of the fitness function

In his early analysis of genetic algorithms, Holland implicitly assumes a nonnegative fitness and does not explicitly address the problem of mapping from the objective function to fitness in his brief discussion of function optimization (Holland 1975, ch 3). Consequently, many of the schema analysis results in the literature use the symbol f to refer to the *fitness* and not to *objective function* values. The methods mentioned above for mapping the objective function to the fitness values must be kept in mind when interpreting the schema theorem. For example, consider two genetic algorithms that both use proportional selection but that differ in that one uses the fitness function

$$\Phi_1(x) = \alpha f(x) + \beta$$

and the other uses the fitness function

$$\Phi_2(x) = \Phi_1(x) + \gamma$$

where $\gamma \neq 0$. Then for any hyperplane H represented in a given population $P(t)$, the target sampling rate for H in the first algorithm is

$$\text{tsr}_1(H, t) = \frac{\Phi_1(H, t)}{\bar{\Phi}_1(t)}$$

while the target sampling rate for H in the second algorithm is

$$\mathrm{tsr}_2(H, t) = \frac{\Phi_2(H, t)}{\bar{\Phi}_2(t)}$$

$$= \frac{\Phi_1(H, t) + \gamma}{\bar{\Phi}_1(t) + \gamma}.$$

Even though both genetic algorithms behave according to the schema theorem, they clearly allocate trials to hyperplane H at different rates, and thus produce entirely different sequences of populations. The relationship between the schema theorem and the objective function becomes even more complex if the fitness function Φ is dynamically scaled during the course of the algorithm. Clearly, the allocation of trials described by schema theorem depends on the precise form of the fitness function used in the evolutionary algorithm. And of course, crossover and mutation will also interact with selection.

23.5.3 Selection differential

Drawing on the terminology of selective breeding, Mühlenbein and Schlierkamp-Voosen (1993) define the *selection differential* $S(t)$ of a selection method as the difference between the mean fitness of the selected parents and the mean fitness of the population at time t. For proportional selection, they show that the selection differential is given by

$$S(t) = \frac{\sigma_p^2(t)}{\bar{\Phi}(t)}$$

where $\sigma_p^2(t)$ is the fitness variance of the population at time t. From this formula, it is easy to see that, without dynamic fitness scaling, an evolutionary algorithm tends to stagnate over time since $\sigma_p^2(t)$ tends to decrease and $\bar{\Phi}(t)$ tends to increase. The fitness scaling techniques described above are intended to mitigate this effect. In addition, operators which produce random variation (e.g. mutation) can also be used to reduce stagnation in the population.

23.5.4 Takeover time

Takeover time refers to the number of generations required for an evolutionary algorithm operating under selection alone (i.e. no other operators such as mutation or crossover) to converge to a population consisting entirely of instances of the optimal individual, starting from a population that contains a single instance of the optimal individual. Goldberg and Deb (1991) show that, assuming $\Phi = f$, the takeover time τ in a population of size μ for proportional selection is

$$\tau_1 = \frac{\mu \ln \mu - 1}{c}$$

for $f_1(x) = x^c$, and

$$\tau_2 = \frac{\mu \ln \mu}{c}$$

for $f_2(x) = \exp(cx)$. Goldberg and Deb compare these results with several other selection mechanisms and show that the takeover time for proportional selection (without fitness scaling) is larger than for many other selection methods.

References

Bäck T 1994 Selective pressure in evolutionary algorithms: a characterization of selection mechanisms *Proc. 1st IEEE Int. Conf. on Evolutionary Computation (Orlando, FL, June 1994)* (Piscataway, NJ: IEEE) pp 57–62

Baker J E 1987 Reducing bias and inefficiency in the selection algorithm *Proc. 2nd Int. Conf. on Genetic Algorithms (Cambridge, MA, 1987)* ed J Grefenstette (Hillsdale, NJ: Erlbaum) pp 14–21

de la Maza M and Tidor B 1993 An analysis of selection procedures with particular attention paid to proportional and Boltzmann selection *Proc. 5th Int. Conf. on Genetic Algorithms (Urbana-Champaign, IL, July 1993)* ed S Forrest (San Mateo, CA: Morgan Kaufmann) pp 124–31

Gillies A M 1985 *Machine Learning Procedures for Generating Image Domain Feature Detectors* Doctoral Dissertation, University of Michigan, Ann Arbor

Goldberg D E 1989 *Genetic Algorithms in Search, Optimization, and Machine Learning* (Reading, MA: Addison-Wesley)

Goldberg D and Deb K 1991 A comparative analysis of selection schemes used in genetic algorithms *Foundations of Genetic Algorithms* ed G Rawlins (San Mateo, CA: Morgan Kaufmann) pp 69–93

Grefenstette J 1986 Optimization of control parameters for genetic algorithms *IEEE Trans. Syst. Man Cybernet.* **SMC-16** 122–8

——1991 Conditions for implicit parallelism *Foundations of Genetic Algorithms* ed G Rawlins (San Mateo, CA: Morgan Kaufmann) pp 252–61

Holland J H 1975 *Adaptation in Natural and Artificial Systems* (Ann Arbor, MI: University of Michigan Press)

Mühlenbein H and Schlierkamp-Voosen D 1993 Predictive models for the breeder genetic algorithm *Evolut. Comput.* **1** 25–49

24

Tournament selection

Tobias Blickle

24.1 Working mechanism

In tournament selection a group of q individuals is randomly chosen from the population. They may be drawn from the population with or without replacement. This group takes part in a *tournament*; that is, a winning individual is determined depending on its fitness value. The best individual having the highest fitness value is usually chosen deterministically though occasionally a stochastic selection may be made. In both cases only the winner is inserted into the next population and the process is repeated λ times to obtain a new population. Often, tournaments are held between two individuals (binary tournament). However, this can be generalized to an arbitrary group size q called *tournament size*.

The following description assumes that the individuals are drawn with replacement and the winning individual is deterministically selected.

> **Input:** Population $P(t) \in I^\lambda$, tournament size $q \in \{1, 2, \ldots, \lambda\}$
> **Output:** Population after selection $P(t)'$
> 1 tournament(q, $a_1, \ldots, a_\lambda$):
> 2 **for** $i \leftarrow 1$ **to** λ **do**
> 3 $a_i' \leftarrow$ best fit individual from q randomly chosen individuals from $\{a_1, \ldots, a_\lambda\}$;
>   **od**
> 4 **return** $\{a_1', \ldots, a_\lambda'\}$.

Tournament selection can be implemented very efficiently and has the time complexity $\mathcal{O}(\lambda)$ as no sorting of the population is required. However, the above algorithm leads to high variance in the expected number of offspring as λ independent trials are carried out.

Tournament selection is translation and scaling invariant (de la Maza and Tidor 1993). This means that a scaling or translation of the fitness value does not affect the behavior of the selection method. Therefore, scaling techniques as used for proportional selection are not necessary, simplifying the application of the selection method.

Furthermore, tournament selection is well suited for parallel evolutionary algorithms. In most selection schemes global calculations are necessary to compute the reproduction rates of the individuals. For example, in proportional selection the mean of the fitness values in the population is required, and in ranking selection and truncation selection a sorting of the whole population is necessary. However, in tournament selection the tournaments can be performed independently of each other such that only groups of q individuals need to communicate.

24.2 Parameter settings

$q = 1$ corresponds to no selection at all (the individuals are randomly picked from the population). Binary tournament is equivalent to linear ranking selection with $\eta^- = 1/\lambda$ (Blickle and Thiele 1995a), where η^- gives the expected number of offspring of the worst individual. With increasing q the selection pressure increases (for a quantitative discussion of selection pressure see below). For many applications in genetic programming values $q \in \{6, \ldots, 10\}$ have been recommended.

24.3 Formal description

Tournament selection has been well studied (Goldberg and Deb 1991, Bäck 1994, 1995, Blickle and Thiele 1995a, b, Miller and Goldberg 1995). The following description is based on the fitness distribution of the population.

Let $\gamma(P)$ denote the number of unique fitness values in the population. Then $\rho(P) = (\rho_{F_1(P)}, \rho_{F_2(P)}, \ldots, \rho_{F_{\gamma(P)}(P)}) \in [0, 1]^{\gamma(P)}$ is the fitness distribution of the population P, with $F_1(P) < F_2(P) < \cdots < F_{\gamma(P)}(P)$. $\rho_{F_i(P)}$ gives the proportion of individuals with fitness value $F_i(P)$ in the population P. Furthermore the cumulative fitness distribution is denoted by $R(P) = (R_{F_1(P)}, R_{F_2(P)}, \ldots, R_{F_{\gamma(P)}(P)}) \in [0, 1]^{\gamma(P)}$. $R_{F_i(P)}$ gives the number of individuals with fitness value $F_i(P)$ or less in the population P, i.e. $R_{F_i(P)} = \sum_{j=1}^{j=i} \rho_{F_j(P)}$ and $R_{F_0(P)} := 0$.

With these definitions, the selection operator s can be viewed as an operator on fitness distributions (Blickle and Thiele 1995b). The expected fitness distribution after tournament selection with tournament size q is $s_{\text{tour}}(q)$: $\mathbb{R}^{\gamma(P)} \mapsto \mathbb{R}^{\gamma(P)}$, $s_{\text{tour}}(q)(\rho(P)) = (\rho'_{F_1(P)}, \rho'_{F_2(P)}, \ldots, \rho'_{F_\gamma(P)})$, where

$$\rho'_{F_i(P)} = (R_{F_i(P)})^q - (R_{F_{i-1}(P)})^q. \tag{24.1}$$

The expected number of occurrences of an individual with fitness value $F_i(P)$ is given by $\rho'_{F_i(P)}/\rho_{F_i(P)}$. Consequently, stochastic universal sampling (Baker 1987) (see Chapter 23) can also be used for tournament selection. This almost completely reduces the usually high variance in the expected number of offspring. However, the time complexity of the selection algorithm increases to $\mathcal{O}(\lambda \ln \lambda)$ as calculation of the fitness distribution is required.

For the analytical analysis it is advantageous to use continuous fitness distributions. The continuous form of (24.1) is given by

$$\bar{\rho}'(F) = q\bar{\rho}(F) \left(\bar{R}(F)\right)^{q-1} \tag{24.2}$$

where $\bar{\rho}(F)$ is the continuous form of $\rho(P)$, $\bar{R}(F) = \int_{F_0(P)}^{F} \bar{\rho}(x)\,dx$ is the cumulative continuous fitness distribution and $F_0(P) < F \leq F_{\gamma(P)}(P)$ the range of the distribution function $\bar{\rho}(F)$.

24.4 Properties

24.4.1 Concatenation of tournaments

An interesting property of tournament selection is the concatenation of several selection phases. Assuming an arbitrary population with a fitness distribution $\bar{\rho}$, tournament selection with tournament size q_1 is applied followed by tournament selection with tournament size q_2 on the resulting population and no recombination in between. The obtained expected fitness distribution is the same as if only a single tournament selection with tournament size $q_1 q_2$ were applied to the initial distribution $\bar{\rho}$ (Blickle and Thiele 1995b):

$$s_{\text{tour}}(q_2)(s_{\text{tour}}(q_1)(\bar{\rho})) = s_{\text{tour}}(q_1 q_2)(\bar{\rho}). \tag{24.3}$$

24.4.2 Takeover time

The takeover time was introduced by Goldberg and Deb (1991) to describe the selection pressure of a selection method. The takeover time τ^* is the number of generations needed under pure selection for a initial single best-fit individual to fill up the whole population. The takeover time can, for example, be calculated combining (24.1) and (24.3) as follows. Only the best individual is considered and its expected proportion ρ'_{best} after tournament selection can be obtained as $\rho'_{\text{best}} = 1 - (1 - 1/\lambda)^q$, which is a special case of (24.1) using $\rho_{\text{best}} = 1/\lambda$ and $R_{\text{best}} = 1$. Performing τ such tournaments subsequently with no recombination in between leads to $\hat{\rho}_{\text{best}} = 1 - (1 - 1/\lambda)^{q^\tau}$ by repeatedly applying (24.3). Goldberg and Deb (1991) solved this equation for τ and gave the following approximation for the takeover time:

$$\tau^*_{\text{tour}}(q) \approx \frac{1}{\ln q}(\ln \lambda + \ln(\ln \lambda)). \tag{24.4}$$

Figure 24.1 shows the dependence of the takeover time on the tournament size q. For scaling purposes an artificial population size of $\lambda = e$ is assumed, such that (24.4) simplifies to $\tau^*_{\text{tour}}(q) \approx 1/\ln q$.

24.4.3 Selection intensity

The selection intensity is another measure for the strength of selection which is borrowed from population genetics. The *selection intensity* S is the change in the average fitness of the population due to selection divided by the mean variance of the population before selection σ, that is, $S = (u^* - u)/\sigma$, with u average fitness before selection, and u^* average fitness after selection. To eliminate the dependence of the selection intensity on the initial distribution one usually assumes a Gaussian-distributed initial population (Mühlenbein and Schlierkamp-Voosen 1993). Under this assumption, the selection intensity of tournament selection is determined by

$$S_{\text{tour}}(q) = \int_{-\infty}^{\infty} qx \, \frac{1}{(2\pi)^{1/2}} \, e^{-x^2/2} \left(\int_{-\infty}^{x} \frac{1}{(2\pi)^{1/2}} \, e^{-y^2/2} \, dy \right)^{q-1} dx. \quad (24.5)$$

The dependence of the selection intensity on the tournament size is shown in figure 24.1.

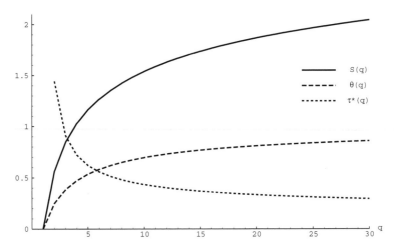

Figure 24.1. The selection intensity S, the loss of diversity θ, and the takeover time τ^* (for $\lambda = e$) of tournament selection in dependence on the tournament size q.

The known exact solutions of the integral equation (24.5) are given in table 24.1. These values can also be obtained using the results of the order statistics theory (Bäck 1995). The following formula was derived by Blickle and Thiele (1995b) and approximates the selection intensity with a relative error of less than 1% for tournament sizes of $q > 5$:

$$S_{\text{tour}}(q) \approx (2(\ln(q) - \ln((4.14 \, \ln(q))^{1/2})))^{1/2}.$$

Table 24.1. Known exact values for the selection intensity of tournament selection.

q	1	2	3	4	5
$S_{\text{tour}}(q)$	0	$\frac{1}{\pi^{1/2}}$	$\frac{3}{2\pi^{1/2}}$	$\frac{6}{\pi\pi^{1/2}}\tan^{-1}2^{1/2}$	$\frac{10}{\pi^{1/2}}\left(\frac{3}{2\pi}\tan^{-1}2^{1/2}-\frac{1}{4}\right)$

24.4.4 Loss of diversity

During every selection phase bad individuals are replaced by copies of better ones. Thereby a certain amount of 'genetic material' contained in the bad individuals is lost. The *loss of diversity* θ is the proportion of the population that is not selected for the next population (Blickle and Thiele 1995b). Baker (1989) introduces a similar measure called 'reproduction rate, RR'. RR gives the percentage of individuals that is selected to reproduce, hence RR $= 100(1-\theta)$.

For tournament selection this value computes to (Blickle and Thiele 1995b)

$$\theta_{\text{tour}}(q) = q^{-1/(q-1)} - q^{-q/(q-1)}.$$

It is interesting to note that the loss of diversity is independent of the initial fitness distribution $\bar{\rho}$. Furthermore, a relatively moderate tournament size of $q = 5$ leads to a loss of diversity of almost 50% (see figure 24.1).

References

Bäck T 1994 Selective pressure in evolutionary algorithms: a characterization of selection mechanisms *Proc. 1st IEEE Conf. on Evolutionary Computation (Orlando, FL, June 1994)* (Piscataway, NJ: IEEE) pp 57–62
——1995 Generalized convergence models for tournament- and (μ, λ)-selection *Proc. 6th Int. Conf. on Genetic Algorithms (Pittsburg, PA, July 1995)* ed L J Eshelman (San Mateo, CA: Morgan Kaufmann) pp 2–8
Baker J E 1987 Reducing bias and inefficiency in the selection algorithm *Proc. 2nd Int. Conf. on Genetic Algorithms (Cambridge, MA, 1987)* ed J J Grefenstette (Hillsdale, NJ: Erlbaum) pp 14–21
——1989 *An Analysis of the Effects of Selection in Genetic Algorithms* PhD Thesis, Graduate School of Vanderbilt University, Nashville, TN
Blickle T and Thiele L 1995a *A Comparison of Selection Schemes used in Genetic Algorithms* Technical Report 11, Computer Engineering and Communication Networks Lab (TIK), Swiss Federal Institute of Technology (ETH) Zurich
——1995b A mathematical analysis of tournament selection *Proc. 6th Int. Conf. on Genetic Algorithms (Pittsburg, PA, July 1995)* ed L J Eshelman (San Mateo, CA: Morgan Kaufmann) pp 9–16
de la Maza M and Tidor B 1993 An analysis of selection procedures with particular attention paid to proportional and Boltzmann selection *Proc. 5th Int. Conf. on Genetic Algorithms (Urbana-Champaign, IL, July 1993)* ed S Forrest (San Mateo, CA: Morgan Kaufmann) pp 124–31

Goldberg D E and Deb K 1991 A comparative analysis of selection schemes used in genetic algorithms *Foundations of Genetic Algorithms* ed G Rawlins (San Mateo, CA: Morgan Kaufmann) pp 69–93

Miller B L and Goldberg D E 1995 *Genetic Algorithms, Tournament Selection, and the Effects of Noise* Technical Report 95006, Illinois Genetic Algorithm Laboratory, University of Urbana-Champaign

Mühlenbein H and Schlierkamp-Voosen D 1993 Predictive models for the breeder genetic algorithm *Evolut. Comput.* **1** 25–49

25

Rank-based selection

John Grefenstette

25.1 Introduction

Selection is the process of choosing individuals for reproduction or survival in an evolutionary algorithm. *Rank-based selection* or *ranking* means that only the rank ordering of the fitness of the individuals within the current population determines the probability of selection.

As discussed in Chapter 23, the selection process may be decomposed into distinct steps:

(i) Map the objective function to fitness.
(ii) Create a probability distribution based on fitness.
(iii) Draw samples from this distribution.

Ranking simplifies step (i), the mapping from the objective function f to the fitness function Φ. All that is needed is

$$\Phi(a_i) = \delta f(a_i)$$

where δ is $+1$ for maximization problems and -1 for minimization problems.

Ranking also eliminates the need for fitness scaling (see Section 23.1), since selection pressure is maintained even if the objective function values within the population converge to a very narrow range, as often happens as the population evolves.

This section discusses step (ii), the creation of the selection probability distribution based on fitness. The final step (iii) is independent of the selection method, and the stochastic universal sampling algorithm (see Section 23.4) is an appropriate sampling procedure.

Besides its simplicity, other motivations for using rank-based selection include:

(i) Under proportional selection, a 'super' individual, i.e. an individual with vastly superior objective value, might completely take over the population in a single generation unless an artificial limit is placed on the maximum number of offspring for any individual. Ranking helps prevent premature convergence due to 'super' individuals, since the best individual is always assigned the same selection probability, regardless of its objective value.

(ii) Ranking may be a natural choice for problems in which it is difficult to precisely specify an objective function, e.g. if the objective function involves a person's subjective preference for alternative solutions. For such problems it may make little sense to pay too much attention to the exact values of the objective function, if exact values exist at all.

The following sections describe various forms of linear and nonlinear ranking algorithms. The final section presents some of the theory of rank-based selection.

25.2   Linear ranking

Linear ranking assigns a selection probability to each individual that is proportional to the individual's rank (where the rank of the least fit is defined to be zero and the rank of the most fit is defined to be $\mu - 1$, given a population of size μ). For a generational algorithm, linear ranking can be implemented by specifying a single parameter, β_{rank}, the expected number of offspring to be allocated to the best individual during each generation. The selection probability for individual i is then defined as follows:

$$\Pr_{\text{lin\_rank}}(i) = \frac{\alpha_{\text{rank}} + [\text{rank}(i)/(\mu - 1)](\beta_{\text{rank}} - \alpha_{\text{rank}})}{\mu}$$

where α_{rank} is the number of offspring allocated to the worst individual. The sum of the selection probabilities is then

$$\sum_{i=0}^{\mu-1} \frac{\alpha_{\text{rank}} + [\text{rank}(i)/(\mu - 1)](\beta_{\text{rank}} - \alpha_{\text{rank}})}{\mu} = \alpha_{\text{rank}} + \frac{\beta_{\text{rank}} - \alpha_{\text{rank}}}{\mu(\mu - 1)} \sum_{i=0}^{\mu-1} i$$

$$= \alpha_{\text{rank}} + \tfrac{1}{2}(\beta_{\text{rank}} - \alpha_{\text{rank}})$$

$$= \tfrac{1}{2}(\beta_{\text{rank}} + \alpha_{\text{rank}}).$$

It follows that $\alpha_{\text{rank}} = 2 - \beta_{\text{rank}}$, and $1 \leq \beta_{\text{rank}} \leq 2$. That is, the expected number of offspring of the best individual is no more than twice that of the population average. This shows how ranking can avoid premature convergence caused by 'super' individuals.

25.3   Nonlinear ranking

Nonlinear ranking assigns selection probabilities that are based on each individual's rank, but are not proportional to the rank. For example, the selection probabilities might be proportional to the square of the rank:

$$\Pr_{\text{sq\_rank}}(i) = \frac{\alpha + [\text{rank}(i)^2/(\mu - 1)^2](\beta - \alpha)}{c}$$

where $c = (\beta - \alpha)\mu(2\mu - 1)/6(\mu - 1) + \mu\alpha$ is a normalization factor. This version has two parameters, α and β, where $0 < \alpha < \beta$, such that the selection probabilities range from α/c to β/c.

Even more aggressive forms of ranking are possible. For example, one could assign selection probabilities based on a geometric distribution:

$$\mathrm{Pr}_{\mathrm{geom\_rank}} = \alpha(1 - \alpha)^{\mu - 1 - \mathrm{rank}(i)}.$$

This distribution arises if selection occurs as a result of independent Bernoulli trials over the individuals in rank order, with the probability of selecting the next individual equal to α, and was introduced in the GENITOR system (Whitley and Kauth 1988, Whitley 1989).

Another variation that provides exponential probabilities based on rank is

$$\mathrm{Pr}_{\mathrm{exp\_rank}}(i) = \frac{1 - e^{-\mathrm{rank}(i)}}{c} \tag{25.1}$$

for a suitable normalization factor c. Both of the latter methods strongly bias the selection toward the best few individuals in the population, perhaps at the cost of premature convergence.

25.4 (μ, λ), $(\mu + \lambda)$ and threshold selection

The (μ, λ) and $(\mu + \lambda)$ methods used in evolution strategies (see Chapter 9 and Schwefel 1977) are deterministic rank-based selection methods. In (μ, λ) selection, $\lambda = k\mu$ for some $k > 1$. The process is that k offspring are generated from each parent in the current population through mutation or possibly recombination, and the best μ offspring are selected for retention. This method is similar to the technique called *beam search* in artificial intelligence (Shapiro 1990). Experimental studies indicate that a value of $k \approx 7$ is optimal (Schwefel 1987).

In $(\mu + \lambda)$ selection, the best μ individuals are selected from the union of the μ parents and the λ offspring. Thus, $(\mu + \lambda)$ is an elitist method, since it always retains the best individuals unless they are replaced by superior individuals. According to Bäck and Schwefel (1993), the (μ, λ) method is preferable to $(\mu + \lambda)$, since it is more robust on probabilistic or changing environments.

The (μ, λ) method is closely related to methods known as *threshold selection* or *truncation selection* in the genetic algorithm literature. In threshold selection the best $T\mu$ individuals are assigned a uniform selection probability, and the rest of the population is discarded:

$$\mathrm{Pr}_{\mathrm{thresh\_rank}}(i) = \begin{cases} 0 & \text{if } \mathrm{rank}(i) < (1 - T)\mu \\ 1/T\mu & \text{otherwise.} \end{cases}$$

The parameter T is the called the *threshold*, where $0 < T \leq 1$. According to Mühlenbein and Schlierkamp-Voosen (1993), T should be chosen in the range 0.1–0.5.

Threshold selection is essentially a (μ', λ) method, with $\mu' = T\mu$ and $\lambda = \mu$, except that threshold selection is usually implemented as a probabilistic procedure using the distribution $\text{Pr}_{\text{thresh\_rank}}$, while systems using (μ, λ) are usually deterministic.

25.5 Theory

The theory of rank-based selection has received less attention than the proportional selection method, due in part to the difficulties in applying the schema theorem to ranking. The next subsection describes the issues that arise in the schema analysis of ranking, and shows that ranking does exhibit a form of implicit parallelism. Characterizations of the selective pressure of ranking are also described, including its fertility rate, selective differential, and takeover time. Finally, a simple substitution result is mentioned.

25.5.1 Ranking and implicit parallelism

The use of rank-based selection makes it difficult to relate the schema theorem to the original objective function, since the mean observed rank of a schema is generally unrelated to the mean observed objective value for that schema. As a result, the relative target sampling rates (see Section 23.5.1) of two schemata under ranking cannot be predicted based on the mean objective values of the schemata, in contrast to proportional selection. For example, consider the following case:

$$f(a_1) = 59 \qquad f(a_2) = 15 \qquad f(a_3) = 5 \qquad f(a_4) = 1 \qquad f(a_5) = 0$$

where

$$a_1, a_4, a_5 \in H_1 \qquad a_2, a_3 \in H_2.$$

Assume that the goal is to maximize the objective function f. Even though $\overline{f}(H_1) = 20 > 10 = \overline{f}(H_2)$, ranking will assign a higher target sampling rate to H_2 than to H_1.

However, ranking does exhibit a weaker form of *implicit parallelism*, meaning that it allocates search effort in a way that differentiates among a large number of competing areas of the search space on the basis of a limited number of explicit evaluations of knowledge structures (Grefenstette 1991). The following definitions assume that the goal is to maximize the objective function. A fitness function Φ is called *monotonic* if

$$\Phi(a_i) \le \Phi(a_j) \Leftrightarrow f(a_i) \le f(a_j).$$

That is, a monotonic fitness function does not reverse the sense of any pairwise ranking provided by the objective function. A fitness function is called *strictly monotonic* if it is monotonic and

$$f(a_i) < f(a_j) \Rightarrow \Phi(a_i) < \Phi(a_j).$$

A strictly monotonic fitness function preserves the relative ranking of any two individuals in the search space with distinct objective function values. Since $\Phi(a_i) = \delta f(a_i)$, ranking uses a strictly monotonic fitness function by definition.

Likewise, a selection algorithm is called *monotonic* if

$$\text{tsr}(a_i) \leq \text{tsr}(a_j) \Leftrightarrow \Phi(a_i) \leq \Phi(a_j)$$

where $\text{tsr}(a)$ is the target sampling rate, or expected number of offspring, for individual a. That is, a monotonic selection algorithm is one that respects the *survival-of-the-fittest* principle. A selection algorithm is called *strictly monotonic* if it is monotonic and

$$\Phi(a_i) < \Phi(a_j) \Rightarrow \text{tsr}(a_i) < \text{tsr}(a_j).$$

A strictly monotonic selection algorithm assigns a higher selection probability to individuals with better fitness values. Linear ranking selection and proportional selection are both strictly monotonic, whereas threshold selection is monotonic but not strict, since it may assign the same number of offspring to individuals with different fitness values.

Finally, an evolutionary algorithm is called *admissible* if its fitness function and selection algorithm are both monotonic. An evolutionary algorithm is *strict* iff its fitness function and selection algorithm are both strictly monotonic.

Now, consider two arbitrary subsets of the solution space, A and B, sorted by objective function value. By definition, B *partially dominates* A $(A \prec B)$ at time t if each representative of B is at least as good as the corresponding representative of A. The following theorem (Grefenstette 1991) partially characterizes the implicit parallelism exhibited by ranking (any many other selection methods):

Implicit parallelism of admissible evolutionary algorithms. In any admissible evolutionary algorithm, if $(A \prec B)$ then $\text{tsr}(A) \leq \text{tsr}(B)$. Furthermore, in any strict evolutionary algorithm, if $(A \prec B)$ then $\text{tsr}(A) < \text{tsr}(B)$.

One illustration of this result to rank-based selection is shown in figure 25.1. Let A be the set of points in the space with objective function values between the dotted lines. Let B be the set of points in the space with objective values above the region between the dotted lines. Then, in any population that contains points from both set A and set B, the number of offspring allocated to B by any strict evolutionary algorithm grows strictly faster than the number allocated to set A, since any subset of B dominates any subset of A. This example illustrates *implicit parallelism* because it holds no matter where the dotted lines are drawn. This result holds not only for rank-based selection, but for any fitness function and selection algorithm that satisfy the requirement of admissibility.

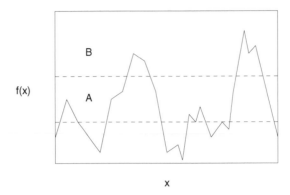

Figure 25.1. Two regions defined by range of objective values.

25.5.2 Fertility rate

The *fertility rate* $\mathcal{F}$ of a selection method is the proportion of the population that is expected to have at least one offspring as a result of the selection process. Other terms that have been used for this include *fertility factor* (Baker 1985, 1987), *reproductive rate* (Baker, 1989), and *diversity* (Blickle and Thiele, 1995).

Baker (1987, 1989) shows that, for linear ranking, the fertility rate obeys the following formula:

$$\mathcal{F} = 1 - \frac{\beta - 1}{4}$$

where β is the number of offspring allocated to the best individual, $1 \le \beta \le 2$. So $\mathcal{F}$ ranges in value from 1 (if $\beta = 1$) to 0.75 (if $\beta = 2$) for linear ranking.

25.5.3 Selection differential

Drawing on the terminology of selective breeding, Mühlenbein and Schlierkamp-Voosen (1993) define the *selection differential* $S(t)$ of a selection method as the difference between the mean fitness of the selected parents and the mean fitness of the population at time t. If the fitness values are normally distributed the selection differential for truncation selection is approximately

$$S(t) \approx I \sigma_{\mathrm{p}}$$

where σ_{p} is the standard deviation of the fitness values in the population, and I is a value called the *selection intensity*. Bäck (1995) quantifies the selection intensity for general (μ, λ) selection as follows:

$$I = \frac{1}{\mu} \sum_{i=\lambda-\mu+1}^{\lambda} E(Z_{i:\lambda})$$

where $Z_{i:\lambda}$ are order statistics based in the fitness of individuals in the current population. That is, I is the average of the expectations of the μ best samples taken from iid normally distributed random variables Z. This analysis shows that I is approximately proportional to λ/μ, and experimental studies confirm this relationship (Bäck 1995, Mühlenbein and Schlierkamp-Voosen 1993).

25.5.4 Takeover time

Takeover time refers to the number of generations required for an evolutionary algorithm operating under selection alone (i.e. no other operators such as mutation or crossover) to converge to a population consisting entirely of instances of the optimal individual, starting from a population that contains a single instance of the optimal individual. According to Goldberg and Deb (1991), the approximate takeover time τ in a population of size μ for rank-based selection is

$$\tau \approx \frac{\ln \mu + \ln(\ln \mu)}{\ln 2}$$

for linear ranking with $\beta_{\text{rank}} = 2$ and

$$\tau \approx \frac{2}{\mu - 1} \ln(\mu - 1)$$

for linear ranking with $1 < \beta_{\text{rank}} < 2$.

25.5.5 Substitution theorem

One interesting feature of rank-based selection is that it is clearly less sensitive to the objective function than proportional selection. As a result, it possible to make the following observation about evolutionary algorithms that use rank-based selection:

Substitution theorem. Let EA be an evolutionary algorithm that uses rank-based selection, along with any forms of mutation and recombination that are independent of the the objective values of individuals. If EA optimizes an objective function f then EA also optimizes the function $g \circ f$, for any monotonically increasing g.

Proof. For any monotonically increasing function g, the composition $g \circ f$ induces the same rank ordering of the search space as f. It follows that a rank-based algorithm EA produces an identical sequence of populations for objective functions f and $g \circ f$, assuming that mutation and recombination in EA are independent of the the objective values of individuals. Since f and $g \circ f$ have the same optimal solutions, the result follows.

For example, a rank-based evolutionary algorithm that optimizes a given function $f(x)$ in t steps will also optimize the function $(f(x))^n$ in t steps, for any even $n > 0$.

References

Bäck T 1995 Generalized convergence models for tournament- and (μ, λ)-selection *Proc. 6th Int. Conf. on Genetic Algorithms (Pittsburgh, PA, July 1995)* ed L J Eshelman (San Mateo, CA: Morgan Kaufmann) pp 2–8

Bäck T and H-P Schwefel 1993 An overview of evolutionary algorithms for parameter optimization *Evolut. Comput.* **1** 1–23

Baker J 1985 Adaptive selection methods for genetic algorithms *Proc. 1st Int. Conf. on Genetic Algorithms (Pittsburgh, PA, July 1985)* ed J J Grefenstette (Hillsdale, NJ: Lawrence Erlbaum) pp 101–11

——1987 Reducing bias and inefficiency in the selection algorithm *Proc. 2nd Int. Conf. on Genetic Algorithms (Cambridge, MA, 1987)* ed J J Grefenstette (Hillsdale, NJ: Erlbaum) pp 14–21

——1989 *Analysis of the Effects of Selection in Genetic Algorithms* Doctoral Dissertation, Department of Computer Science, Vanderbilt University

Blickle T and Thiele L 1995 A mathematical analysis of tournament selection *Proc. 6th Int. Conf. on Genetic Algorithms (Pittsburgh, PA, July 1995)* ed L Eshelman (San Mateo, CA: Morgan Kaufmann) pp 9–16

Goldberg D and Deb K 1991 A comparative analysis of selection schemes used in genetic algorithms *Foundations of Genetic Algorithms* ed G Rawlins (San Mateo, CA: Morgan Kaufmann) pp 69–93

Grefenstette J 1991 Conditions for implicit parallelism *Foundations of Genetic Algorithms* ed G Rawlins (San Mateo, CA: Morgan Kaufmann) pp 252–61

Mühlenbein H and Schlierkamp-Voosen D 1993 Predictive models for the breeder genetic algorithm *Evolut. Comput.* **1** 25–49

Schwefel H-P 1977 *Numerische Optimierung von Computer-Modellen mittels der Evolutionsstrategie (Interdisciplinary System Research 26)* (Basel: Birkhäuser)

——1987 Collective phenomena in evolutionary systems *Preprints of the 31st Ann. Meeting International Society for General Systems Research (Budapest)* vol 2, pp 1025–33

Shapiro S C (ed) 1990 *Encyclopedia of Artificial Intelligence* vol 1 (New York: Wiley)

Whitley D 1989 The GENITOR algorithm and selective pressure: why rank-based allocation of reproductive trials is best *Proc. 3rd Int. Conf. on Genetic Algorithms (Fairfax, VA, June 1989)* ed J Schaffer (San Mateo, CA: Morgan Kaufmann) pp 116–21

Whitley D and Kauth J 1988 GENITOR: a different genetic algorithm *Proc. Rocky Mountain Conf. on Artificial Intelligence (Denver, CO)* pp 118–30

26

Boltzmann selection

Samir W Mahfoud

26.1 Introduction

Boltzmann selection mechanisms thermodynamically control the selection pressure in an evolutionary algorithm (EA), using principles from *simulated annealing* (SA) (Kirpatrick *et al* 1983). Boltzmann selection mechanisms can be used to indefinitely prolong an EA's search, in order to locate better final solutions.

In EAs that employ Boltzmann selection mechanisms, it is often impossible to separate the selection mechanism from the rest of the EA. In fact, the mechanics of the recombination and neighborhood operators are critical to the generation of the proper temporal population distributions. Therefore, most of the following discusses *Boltzmann EAs* rather than Boltzmann selection mechanisms in isolation.

Boltzmann EAs represent parallel extensions of the inherently serial SA. In addition, theoretical proofs of asymptotic, global convergence for SA carry over to certain Boltzmann selection EAs (Mahfoud and Goldberg 1995).

The heart of Boltzmann selection mechanisms is the *Boltzmann trial*, a competition between current solution i and alternative solution j, in which i wins with logistic probability

$$\frac{1}{1 + e^{(f_i - f_j)/T}} \tag{26.1}$$

where T is temperature and f_i is the energy, cost, or objective function value (assuming minimization) of solution i. Slight variations of the Boltzmann trial exist, but all variations essentially accomplish the same thing when iterated (the winner of a trial becomes solution i for the next trial): at fixed T, given a sufficient number of Boltzmann trials, a Boltzmann distribution arises among the winning solutions (over time). The intent of the Boltzmann trial is that at high T, i and j win with nearly equal probabilities, making the system fluctuate wildly from solution to solution; at low T, the better of the two solutions nearly always wins, resulting in a relatively stable system.

Several types of Boltzmann algorithm exist, each designed for slightly different purposes. Boltzmann *tournament selection* (see Chapter 24 and Goldberg 1990, Mahfoud 1993) is designed to give the population *niching* capabilities (Mahfoud 1995), but is not able to significantly slow the population's convergence. (*Convergence* refers to a population's decrease in diversity over time, as measured by an appropriate diversity measure.) Whether any Boltzmann EA is capable of performing effective niching remains an open question.

The Boltzmann selection method of de la Maza and Tidor (1993) scales the fitnesses of population elements, following fitness assignment, according to the Boltzmann distribution. It is designed to control the convergence of traditional selection.

Parallel recombinative simulated annealing (PRSA) (Mahfoud and Goldberg 1992, 1995) allows control of EA convergence, achieves a true parallelization of SA, and inherits SA's convergence proofs. PRSA is the Boltzmann EA discussed in the remainder of this section.

26.2   Simulated annealing

SA is an optimization technique, analogous to the physical process of annealing. SA starts with a high temperature T and any initial state. A neighborhood operator is applied to the current state i to yield state j. If $f_j < f_i$, j becomes the current state. Otherwise j becomes the current state with probability $e^{(f_i - f_j)/T}$. (If j does not become the current state, i remains the current state.) The application of the neighborhood operator and the probabilistic acceptance of the newly generated state are repeated either for a fixed number of iterations or until a quasi-equilibrium is reached. The entire above-described procedure is performed repeatedly, each time starting from the current i and from a lower T.

At any given T, a sufficient number of iterations always leads to equilibrium, at which point the temporal distribution of accepted states is stationary. (This stationary distribution is Boltzmann.) The SA algorithm, as described above, is called the *Metropolis algorithm*. What distinguishes the Metropolis algorithm is the criterion by which the newly generated state is accepted or rejected. An alternative criterion is that of equation (26.1). Both criteria lead to a Boltzmann distribution.

The key to achieving good performance with SA, as well as to proving global convergence, is that a stationary distribution must be reached at each temperature, and cooling (lowering T) must proceed sufficiently slowly.

26.3   Working mechanism for parallel recombinative simulated annealing

PRSA is a population-level implementation of simulated annealing. Instead of processing one solution at a time, it processes an entire population of solutions in parallel, using a recombination operator (typically crossover, see Chapter 33) and a neighborhood operator (typically *mutation*, see Chapter 32). The combination of crossover and mutation produces a population-level neighborhood operator whose action on the entire population parallels the action of SA's neighborhood operator on a single solution. (See figure 26.1.) It is interesting to note that without crossover, PRSA would be equivalent to running μ independent SAs, where μ is population size. Without mutation, PRSA's global convergence proofs would no longer hold.

PRSA works by pairing all population elements, at random, for crossover each generation. After crossover and mutation, children compete against their parents in Boltzmann trials. Winners advance to the next generation.

In the Boltzmann trial step, many competitions are possible between two children and two parents. One possibility, *double acceptance/rejection*, allows both parents to compete as a unit against both children: the sum of the two parents' energies should be substituted for f_i in equation (26.1); the sum of the two childrens' energies, for f_j. A second possibility, *single acceptance/rejection*, holds two competitions, each time pitting one child against one parent. There are several possible single acceptance/rejection competitions. For instance, each parent can always compete against the child formed from its own right end and the other parent's left end (assuming single-point crossover). Other possibilities and their consequences are outlined by Mahfoud and Goldberg (1995).

26.4   Pseudocode for a common variation of parallel recombinative simulated annealing

The pseudocode at the top of the next page describes a common variation of PRSA that employs single acceptance/rejection competitions, a static stopping criterion, and random—without replacement—pairing of population elements for recombination. The cooling schedule is set by the two functions initialize_temperature() and adjust_temperature(). These two functions, as well as initialize_population(), are shown without arguments, because their arguments depend upon the type of cooling schedule and initialization chosen by the user. The function random() simply returns a pseudorandom real number on the interval $(0, 1)$.

26.5   Parameters and their settings

PRSA allows the use of any recombination and neighborhood operators. It performs minimization by default; maximization can be accomplished by reversing the sign of all objective function values. Population size (μ) remains constant from generation to generation. The number of generations the algorithm runs can either be fixed, as in the pseudocode, or dynamic, determined by a user-specified stopping or convergence criterion that is perhaps tied to the cooling schedule.

Input: g—number of generations to run, μ—population size
Output: $P(g)$—the final population

$P(0) \leftarrow$ initialize_population()
$T(1) \leftarrow$ initialize_temperature()
for $t \leftarrow 1$ **to** g **do**
    $P(t) \leftarrow$ shuffle($P(t-1)$)
    for $i \leftarrow 0$ **to** $\mu/2 - 1$ **do**
        $p_1 \leftarrow a_{2i+1}(t)$
        $p_2 \leftarrow a_{2i+2}(t)$
        $\{c_1, c_2\} \leftarrow$ recombine(p_1, p_2)
        $c'_1 \leftarrow$ neighborhood(c_1)
        $c'_2 \leftarrow$ neighborhood(c_2)
        if random() $> [1 + e^{[f(p_1)-f(c'_1)]/T(t)}]^{-1}$ **then** $a_{2i+1}(t) \leftarrow c'_1$ **fi**
        if random() $> [1 + e^{[f(p_2)-f(c'_2)]/T(t)}]^{-1}$ **then** $a_{2i+2}(t) \leftarrow c'_2$ **fi**
    od
    $T(t+1) \leftarrow$ adjust_temperature()
od

PRSA requires a user to select a population size, a type of competition, recombination and neighborhood operators, and a cooling schedule. Prior research offers some guidelines (Mahfoud and Goldberg 1992, 1995). A good rule of thumb for population size is to choose as large a population size as system limitations and time constraints allow. In general, smaller populations require longer cooling schedules. The type of competition previously employed is single acceptance/rejection, in which each parent competes against the child formed from its own right end and the other parent's left end (under single-point crossover).

Appropriate recombination and neighborhood operators are problem specific. For example, in optimization of traditional binary encodings, one might employ single-point crossover and mutation; in permutation problems, permutation-based crossover and inversion would be more appropriate.

Many styles of cooling schedule exist, but their discussion is beyond the scope of this section. Several studies contain thorough discussions of cooling (Aarts and Korst 1989, Azencott 1992, Ingber and Rosen 1992, Romeo and Sangiovanni-Vincentelli 1991). Perhaps the simplest type of cooling schedule is to start at a high T, and to periodically lower T through multiplication by a positive constant such as 0.95. At each T, a number of generations are performed. In general, the more generations performed at each T and the higher the multiplicative constant, the better the end result.

26.6 Global convergence theory and proofs

The most straightforward global convergence proof for any variation of PRSA shows that the variation is a special case of standard SA. This results in the transfer of SA's convergence proof to the PRSA variant. Details of PRSA's convergence proofs are given by Mahfoud and Goldberg (1995).

The variation of PRSA that we consider employs selection of parents with replacement, and double acceptance/rejection. No population element may be selected as both parents. (Self-mating is disallowed.)

Many authors have taken the viewpoint that SA is essentially an EA with a population size of one. Our proof takes the opposite viewpoint, showing an EA (PRSA) to be a special case of SA. To see this, concatenate all strings of the PRSA population in a side-by-side fashion to form one superstring. Define the fitness of this superstring to be the sum of the individual fitnesses of its component substrings (the former population elements). Let cost be the negated fitness of this superstring. The cost function will reach a global minimum only when each substring is identically at a global maximum. Thus, to maximize all elements of the former population, PRSA can search for a global minimum for the cost function assigned to its superstring.

Consider the superstring as our structure to be optimized. Our chosen variation of PRSA, as displayed graphically in figure 26.1, is now a special case of SA, in which the crossover-plus-mutation neighborhood operator is applied to selected portions of the superstring to generate new superstrings. Crossover-plus-mutation's net effect as a population-level neighborhood operator is to swap two blocks of the superstring, and then probabilistically flip bits of these swapped blocks and of two other blocks (the other halves of each parent).

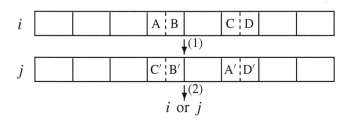

Figure 26.1. The population, after application of crossover and mutation (step 1), transitions from superstring i to superstring j. After a Boltzmann trial (step 2), either i or j becomes the current population. Individual population elements are represented as rectangles within the superstrings. Blocks A, B, C, and D represent portions of individual population elements, prior to crossover and mutation. Crossover points are shown as dashed lines. Blocks A', B', C', and D' result from applying mutation to A, B, C, and D.

As a special case of SA, the chosen variation of PRSA inherits the global convergence proof of SA, provided the population-level neighborhood operator meets certain conditions. According to Aarts and Korst (1989), two conditions on the neighborhood generation mechanism are sufficient to guarantee asymptotic global convergence. The first condition is that the neighborhood operator must be able to move from any state to a globally optimal state in a finite number of transitions. The presence of mutation satisfies this requirement. The second condition is symmetry. It requires that the probability at any temperature of generating state y from state x is the same as the probability of generating state x from state y. Symmetry holds for common crossover operators such as single-point, multipoint, and uniform crossover (Mahfoud and Goldberg 1995).

References

Aarts E and Korst J 1989 *Simulated Annealing and Boltzmann Machines: a Stochastic Approach to Combinatorial Optimization and Neural Computing* (Chichester: Wiley)

Azencott R (ed) 1992 *Simulated Annealing: Parallelization Techniques* (New York: Wiley)

de la Maza M and Tidor B 1993 An analysis of selection procedures with particular attention paid to proportional and Boltzmann selection *Proc. 5th Int. Conf. on Genetic Algorithms (Urbana-Champaign, IL, July 1993)* ed S Forrest (San Mateo, CA: Morgan Kaufmann) pp 124–31

Goldberg D E 1990 A note on Boltzmann tournament selection for genetic algorithms and population-oriented simulated annealing *Complex Syst.* **4** 445–60

Ingber L and Rosen B 1992 Genetic algorithms and very fast simulated re-annealing: a comparison *Math. Comput. Modelling* **16** 87–100

Kirpatrick S, Gelatt C D Jr and Vecchi M P 1983 Optimization by simulated annealing *Science* **220** 671–80

Mahfoud S W 1993 Finite Markov chain models of an alternative selection strategy for the genetic algorithm *Complex Syst.* **7** 155–70

——1995 *Niching Methods for Genetic Algorithms* Doctoral Dissertation and IlliGAL Report 95001, University of Illinois at Urbana-Champaign, Illinois Genetic Algorithms Laboratory; *Dissertation Abstracts Int.* **56**(9) p 49878 *(University Microfilms 9543663)*

Mahfoud S W and Goldberg D E 1992 A genetic algorithm for parallel simulated annealing *Proc. 2nd Int. Conf. on Parallel Problem Solving from Nature (Brussels, 1992)* ed R Männer and B Manderick (Amsterdam: Elsevier) pp 301–10

——1995 Parallel recombinative simulated annealing: a genetic algorithm *Parallel Comput.* **21** 1–28

Romeo F and Sangiovanni-Vincentelli A 1991 A theoretical framework for simulated annealing *Algorithmica* **6** 302–45

27

Other selection methods

David B Fogel

27.1 Introduction

In addition to the methods of selection presented in other sections of this chapter, other procedures for selecting parents of successive generations are of interest. These include the tournament selection typically used in evolutionary programming (Fogel 1995, p 137), soft brood selection offered within research in genetic programming (Altenberg 1994a, b), disruptive selection (Kuo and Hwang 1993), Boltzmann selection (de la Maza and Tidor 1993), nonlinear ranking selection (Michalewicz 1996), competitive selection (Hillis 1992, Angeline and Pollack 1993, Sebald and Schlenzig 1994), and the use of lifespan (Bäck 1996).

27.2 Tournament selection

The tournament selection typically performed in evolutionary programming allows for tuning the degree of stringency of the selection imposed. Rather than selecting on the basis of each solution's fitness or error in light of the objective function at hand, selection is made on the basis of the number of *wins* earned in a competition. Each solution is made to compete with some number, q, of randomly selected solutions from the population. In each pairing, if the first solution's score is at least as good as the randomly selected opponent, the first solution receives a win. Thus up to q wins can be earned. This competition is conducted for all solutions in the population and selection then chooses the best subset of a given size from the population based on the number of wins each solution has earned. For $q = 1$, the procedure yields essentially a random walk with very low selection pressure. For $q = \infty$, the procedure becomes selection based on objective function scores (with no probabilistic selection). For practical purposes, $q \geq 10$ is often considered relatively *hard* selection, and q in the range of three to five is considered *soft*. Soft selection allows for lower probabilities of becoming trapped at local optima for periods of time.

27.3 Soft brood selection

Soft brood selection holds a tournament (see Chapter 24) between members of a *brood* of two parents. The winner of the tournament is considered to be the offspring contributed by the mating. Soft brood selection is intended to shield the recombination operator from the costs of producing deleterious offspring. It culls such offspring, essentially testing for their viability before being placed into competition with the remainder of the population. (For further details on the effects of soft brood selection on subexpressions in tree structures, see the article by Altenberg (1994a).)

27.4 Disruptive selection

Disruptive selection can be used to select against individuals with moderate values (in contrast to stabilizing selection which acts against extreme values, or directional selection which acts to increase or decrease values). Kuo and Hwang (1993) suggested a fitness function of the form

$$u(x) = |f(x) - f(t)|$$

where $f(x)$ is the objective value of the solution x and $f(t)$ is the mean of all solutions in the population at time t. Thus a solution's fitness increases with its distance from the mean of all current solutions. The idea is to distribute more search effort to both the extremely good and extremely bad solutions. The utility of this method is certainly very problem dependent.

27.5 Boltzmann selection

Boltzmann selection (as offered by de la Maza and Tidor 1993) proceeds as

$$F_i(U(X)) = \exp(U_i(X)/T)$$

where X is a population of solutions, $U(X)$ is the problem dependent objective function, $F_i(\cdot)$ is the fitness function for the ith solution in X, $U_i(\cdot)$ is the objective function evaluated for the ith solution in X, and T is a variable temperature parameter. De la Maza and Tidor (1993) suggest that this method of assigning fitness proportional selection converges faster than traditional proportional selection. Bäck (1994), however, describes this as a 'misleading name for yet another scaling method for proportional selection'.

27.6 Nonlinear ranking selection

Nonlinear ranking selection (Michalewicz 1996, pp 60–1) is a variant of linear ranking selection. Recall that for linear ranking selection, the probability of a solution with a given rank being selected can be set as

$$P(\text{rank}) = q - (\text{rank} - 1)r$$

where q is a user-defined parameter. For each lower rank, the probability of being selected is reduced by a factor of r. The requirement that the sum of all the probabilities for each ranked solution must be equal to unity implies that

$$q = r(\text{popsize} - 1)/2 + 1/\text{popsize}$$

where popsize is the number of solutions in the population. This relationship can be made nonlinear by setting:

$$P(\text{rank}) = q(1 - q)^{\text{rank}-1}$$

where $q \in (0, 1)$ and does not depend on popsize; larger values of q imply stronger selective pressure. Bäck (1994) notes that this nonlinear ranking method fails to sum to unity and can be made practically identical to tournament selection under the choice of q.

27.7 Competitive selection

Competitive selection is implemented such that the fitness of a solution is determined by its interactions with other members of the population, or other members of a jointly evolving but separate population. Hillis (1992) used this concept to evolve sorting networks in which a population of sorting networks competed against a population of various permutations; the networks were scored in light of how well they sorted the permutations and the permutations were scored in light of how well they could defeat the sorting networks. Angeline and Pollack (1993) used a similar idea to evolve programs to play tic-tac-toe. Sebald and Schlenzig (1994) used evolutionary programming on competing populations to generate suitable blood pressure controllers for simulated patients undergoing cardiac surgery (i.e. controllers were scored on how well they maintained the patient's blood pressure while patients were scored on how well they defeated the controllers). Fogel and Burgin (1969) describe experiments in which competing evolutionary programs played a prisoner's dilemma game using finite-state machines, but insufficient detail is provided to allow for replication of the results. Axelrod (1987), and others, offered an apparently similar procedure for evolving rule sets describing alternative behaviors in the iterated prisoner's dilemma.

27.8 Variable lifespan

Finally, Bäck (1996) notes that the concept of a variable lifespan has been incorporated into the (μ, λ) selection of evolution strategies by Schwefel and Rudolph (1995) by allowing the parents to survive some number of generations. When this number is one generation, the method is the familiar comma strategy; at infinity, the method becomes a plus strategy.

References

Altenberg L 1994a Emergent phenomena in genetic programming, *Proc. 3rd Ann. Conf. on Evolutionary Programming (San Diego, CA, February 1994)* ed A V Sebald and L J Fogel (Singapore: World Scientific) pp 233–41

——1994b The evolution of evolvability in genetic programming *Advances in Genetic Programming* ed K Kinnear (Cambridge, MA: MIT Press) pp 47–74

Axelrod R 1987 The evolution of strategies in the iterated prisoner's dilemma *Genetic Algorithms and Simulated Annealing* ed L Davis (Los Altos, CA: Morgan Kaufmann) pp 32–41

Angeline P J and Pollack J B 1993 Competitive environments evolve better solutions for complex tasks *Proc. 5th Int. Conf. on Genetic Algorithms (Urbana-Champaign, IL, July 1993)* ed S Forrest (San Mateo, CA: Morgan Kaufmann) pp 264–70

Bäck T 1994 Selective pressure in evolutionary algorithms: a characterization of selection mechanisms *Proc. 1st IEEE Conf. on Evolutionary Computation (Orlando, FL, 1993)* (Piscataway, NJ: IEEE) pp 57–62

——1996 *Evolutionary Algorithms in Theory and Practice* (New York: Oxford University Press)

de la Maza M and Tidor B 1993 An analysis of selection procedures with particular attention paid to proportional and Boltzmann selection *Proc. 5th Int. Conf. on Genetic Algorithms (Urbana-Champaign, IL, July 1993)* ed S Forrest (San Mateo, CA: Morgan Kaufmann) pp 124–31

Fogel D B 1995 *Evolutionary Computation: Toward a New Philosophy of Machine Intelligence* (New York: IEEE)

Fogel L J and Burgin G H 1969 *Competitive Goal-Seeking Through Evolutionary Programming* Final Report, Contract No AF 19(628)-5927, Air Force Cambridge Research Labratories.

Hillis W D 1992 Co-evolving parasites improves simulated evolution as an optimization procedure *Artificial Life II* ed C Langton, C Taylor, J Farmer and S Rasmussen (Reading, MA: Addison-Wesley) pp 313–24

Kuo T and Hwang S-Y 1993 A genetic algorithm with disruptive selection *Proc. 5th Int. Conf. on Genetic Algorithms (Urbana-Champaign, IL, July 1993)* ed S Forrest (San Mateo, CA: Morgan Kaufmann) pp 65–9

Michalewicz Z 1996 *Genetic Algorithms + Data Structures = Evolution Programs* 3rd edn (Berlin: Springer)

Schwefel H-P and Rudolph G 1995 Contemporary evolution strategies *Advances in Artificial Life (Proc. 3rd Int. Conf. on Artificial Life, Granada, Spain) (Lecture Notes in Artificial Intelligence 929)* ed F Morán *et al* (Berlin: Springer) pp 893–907

Sebald A V and Schlenzig J 1994 Minimax design of neural net controllers for highly uncertain plants *IEEE Trans. Neural Networks* **NN-5** 73–82

28

Generation gap methods

Jayshree Sarma and Kenneth De Jong

28.1 Introduction

The concept of a generation gap is linked to the notion of *nonoverlapping* and *overlapping* populations. In a nonoverlapping model parents and offspring never compete with one another, i.e. the entire parent population is always replaced by the offspring population, while in an overlapping system parents and offspring compete for survival. The term *generation gap* refers to the amount of overlap between parents and offspring. The notion of a generation gap is closely related to selection algorithms and population management issues.

A selection algorithm in an evolutionary algorithm (EA) involves two elements: (i) a selection pool and (ii) a selection distribution over that pool. A selection pool is required for reproduction selection as well as for deletion selection. The key issue in both these cases is 'what does the pool contain when parents are selected and when survivors are selected?'.

In the selection for the reproduction phase, parents are selected to produce offspring and the selection pool consists of the current population. How the parents are selected for reproduction depends on the individual EA paradigm.

In the selection for the deletion phase, a decision has to be made as to which individuals to select for deletion to make room for the new offspring. In nonoverlapping systems the entire selection pool consisting of the current population is selected for deletion: the parent population (μ) is always replaced by the offspring population (λ). In overlapping models, the selection pool for deletion consists of both parents and their offspring. Selection for deletion is performed on this combined set and the actual selection procedure varies in each of the EA paradigms.

Historically, both evolutionary programming and evolution strategies had overlapping populations while the canonical genetic algorithms used nonoverlapping populations.

28.2   Historical perspective

In evolutionary programming (Fogel *et al* 1966), each individual produces one
offspring and the best half from the parent and offspring populations are selected
to form the new population. This is an overlapping system as the parents and
their offspring constantly compete with each other for survival.

In evolution strategies (Schwefel 1981), the $(\mu + \lambda)$ and the (μ, λ) models
correspond to the overlapping and nonoverlapping populations respectively. In
the $(\mu + \lambda)$ system parents and offspring compete for survival and the best μ
are selected. In the (μ, λ) model the number of offspring produced is generally
far greater than the parents. The offspring are then ranked according to fitness
and the best μ are selected to replace the parent population.

Genetic algorithms are based on the two reproductive plans introduced and
analyzed by Holland (1975). In the first plan, R_1, at each time step a single
individual was selected probabilistically using payoff proportional selection to
produce a single offspring. To make room for this new offspring, one individual
from the current population was selected for deletion using a uniform random
distribution.

In the second plan, R_d, at each time step all individuals were
deterministically selected to produce their expected number of offspring. The
selected parents were kept in a temporary storage location. When the process of
recombination was completed, the offspring produced replaced the entire current
population. Thus in R_d, individuals were guaranteed to produce their expected
number of offspring (within probabilistic roundoff).

At that time, from a theoretical point of view, the two plans were viewed
as generally equivalent. However, because of practical considerations relating
to the overhead of recalculating selection probabilities and severe genetic
drift (allele loss) in small populations, most early researchers favored the R_d
approach.

The earliest attempt at evaluating the properties of R_1 and R_d plans was
a set of empirical studies (De Jong 1975) in which a parameter G, called the
generation gap, was defined to introduce the notion of overlapping generations.
The generation gap parameter controls the fraction of the population to be
replaced in each generation. Thus, $G = 1$ (replacing the entire population)
corresponded to R_d and $G = 1/\mu$ (replacing a single individual) represented
R_1.

These early studies (De Jong 1975) suggested that any advantages that
overlapping populations might have were offset by the negative effects of
genetic drift (allele loss). The genetic drift was caused by the high variance
in expected lifetimes and expected number of offspring, mainly because at
that time, generally, modest population sizes were used ($\mu \leq 100$). These
negative effects were shown to increase in severity as G was reduced. These
studies also suggested the advantages of an *implicit* generation overlap. That
is, using the optimal crossover rate of 0.6 and optimal mutation rate of 0.001

(identified empirically for the test suite used) meant that approximately 40% of the offspring were clones of their parents, even for $G = 1$. A later empirical study by Grefenstette (1986) confirmed the earlier results that a larger generation gap value improved performance.

However, early experience with classifier systems (e.g. Holland and Reitman 1978) yielded quite the opposite behavior. In classifier systems only a subset of the population is replaced each time step. Replacing a small number of classifiers was generally more beneficial than replacing a large number or possibly all of them. Here the poor performance observed as the generation gap value increased was attributed to the fact that the population as a whole represented a single solution and thus could not tolerate large changes in its content.

In recent years, computing equipment with increased capacity is easily available and this effectively removes the reason for preferring the R_d approach. The desire to solve more complex problems using genetic algorithms has prompted researchers to develop an alternative to the generational system called the 'steady state' approach, in which typically parents and offspring do coexist (see e.g. Syswerda 1989, Whitley and Kauth 1988).

28.3 Steady state and generational evolutionary algorithms

Steady state EAs are systems in which usually only one or two offspring are produced in each generation. The selection pool for deletion can consist of the parent population only or can be possibly augmented by the offspring produced. The appropriate number of individuals are selected for deletion, based on some distribution, to make room for these new offspring. Generational systems are so named because the entire population is replaced every generation by the offspring population: the lifetime of each individual in the population is only one generation. This is the same as the *nonoverlapping* population systems, while the steady state EA is an *overlapping* population system.

One can conceptually think of a steady state model in evolutionary programming and evolution strategies. For example, from a parent population of μ individuals, a single offspring can be formed by recombination and mutation and can then be inserted into the population. A recent study of the steady state evolutionary programming performed by Fogel and Fogel (1995) concluded that the generational model of evolutionary programming may be more appropriate for practical optimization problems. The first example of the steady state evolutionary strategies is the $(\mu + 1)$ approach introduced by Rechenberg (1973) which had a parent population greater than one ($\mu > 1$). All the parents were then allowed to participate in the reproduction phase to create one offspring. The $(\mu + 1)$ model was not used as it was not feasible to selfadapt the step sizes (Bäck *et al* 1991).

An early example of the steady state model of genetic algorithms is the R_1 model defined by Holland (1975) in which the selection pool for deletion consists only of the parent population and a uniform deletion strategy is used. The R_d

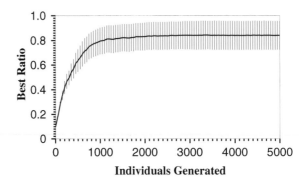

Figure 28.1. The mean and variance of the growth curves of the best in an overlapping system (population size, 50; $G = 1/50$).

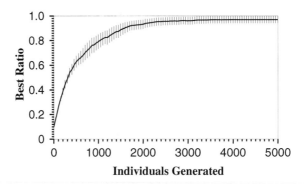

Figure 28.2. The mean and variance of the growth curves of the best in a nonoverlapping system (population size, 50; $G = 1$).

approach is the generational genetic algorithm. Theoretically, the two systems (overlapping systems using uniform deletion and nonoverlapping systems) are considered to be similar in expectation for infinite populations. However, there can be high variance in the expected lifetimes and expected number of offspring when small finite populations are used.

This variance can be highlighted by keeping everything in the two systems constant and changing only one parameter, viz., the number of offspring produced. Figures 28.1 and 28.2 illustrate the average and variance for the growth curve of the best in two systems, producing and replacing only a single individual each generation in one and replacing the entire population each generation in the other. A population size of 50 was used, the best occupied 10% of the initial population, and the curves are averaged over 100 independent runs. Only payoff proportional selection, reproduction, and uniform deletion were used to drive the systems to a state of equilibrium. Notice that in the overlapping system (figure 28.1) the best individuals take over the population only about 80% of the time and the growth curves exhibit much higher variance when compared to the nonoverlapping population (figure 28.2).

This high variance for small generation gap values causes more genetic drift (allele loss). Hence, with smaller population sizes, the higher variance in a steady state system makes it easier for alleles to disappear. Increasing the population size is one way to reduce the the variance (see figure 28.3) and thus offset the allele loss. In summary, the main difference between the generational and steady state systems is higher genetic drift in the latter especially when small population sizes are used with low generation gap values. (See the article by De Jong and Sarma (1993) for more details.)

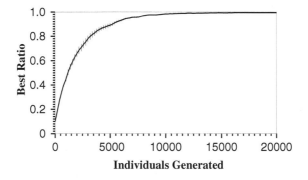

Figure 28.3. The mean and variance of the growth curves of the best in an overlapping system (population size, 200; $G = 1/200$).

So far we have assumed that there is an uniform distribution on the selection pool used for deletion, but most researchers using a steady state genetic algorithm generally use a distribution other than the standard uniform distribution. Syswerda (1991) shows how the growth curves can change when different deletion strategies, such as deleting the least fit, exponential ranking of the members in the selection pool, and reverse fitness, are used. Peck and Dhawan (1995) demonstrate an improvement in the ideal growth behavior of the steady state system when uniform deletion is changed to a first-in–first-out (FIFO) deletion strategy. An early model of a steady state (overlapping) system is GENITOR (Whitley and Kauth 1988, Whitley 1989) which not only uses ranking selection (Chapter 23) instead of proportional selection (Chapter 25) on the selection pool for reproduction but also uses deletion of the worst member as the deletion strategy. The GENITOR approach exhibited significant performance improvement over the standard generational approach.

Using a deletion scheme other than a uniform deletion changes the selection pressure. The selection pressure induced by the different selection schemes can vary considerably. Both these changes can alter the exploration–exploitation balance. Two different studies have shown that improved performance in a steady state system, like GENITOR, is due to higher growth rates and changes in the exploration–exploitation balance caused by using different selection and deletion strategies and is not due to the use of an overlapping model (Goldberg and Deb 1991, De Jong and Sarma 1993).

28.4   Elitist strategies

The cycle of birth and death of individuals is very much linked to the management of the population. Individuals that are born have an associated lifetime. The expected lifetime of an individual is typically one generation, but in some EA systems it can be longer. We now explore this issue in more detail.

Elitist strategies link the lifetimes of individuals to their fitnesses. Elitist strategies are techniques to keep good solutions in the population longer than one generation. Though all individuals in a population can expect to have a lifetime of one generation, individuals with higher fitness can have a longer lifetime when elitist strategies are used.

As stated earlier, the selection pool for deletion is comprised of both the parents and the offspring populations in the overlapping system. This combined population is usually ranked according to fitness and then truncated to form the new population. This method ensures that most of the current individuals with higher fitness survive into the next generation, thus extending their lifetime. In the $(\mu + \lambda)$ evolution strategies, a very strong elitist policy is in effect as the top μ are always kept. In evolutionary programming, a stochastic tournament is used to select the survivors, and hence the elitist policy is not quite as strong as in the evolution strategy case. In the (μ, λ) evolution strategies there is no elitist strategy to preserve the best parents.

Unlike evolution strategies and evolutionary programming, where there is postselection of survivors based on fitness, in generational genetic algorithms there is only preselection of parents for reproduction. Recombination operators are applied to these parents to produce new offspring, which are then subject to mutation. Since all parents are replaced each generation by their offspring, there is no guarantee that the individuals with higher fitness will survive into the next generation. An elitist strategy in generational genetic algorithms is a way of ensuring that the lifetime of the very best individual is extended beyond one generation. Thus, unlike evolutionary programming and evolution strategies, where more than just the best individual survive, in generational genetic algorithms generally only the best individual survives. Steady state genetic algorithms which use deletion schemes other than uniform random deletion have an implicit elitist policy and so automatically extend the lifetime of the higher-fitness individuals in the population.

It should be noted that the elitist strategies were deemed necessary when genetic algorithms are used as function optimizers and the goal is to find a global optimal solution (De Jong 1993). Elitist strategies tend to make the search more exploitative rather than explorative and may not work for problems in which one is required to find multiple optimal solutions.

References

Bäck T, Hoffmeister F and Schwefel H-P 1991 A survey of evolutionary strategies *Proc. 4th Int. Conf. on Genetic Algorithms (San Diego, CA, July 1991)* ed R K Belew and L B Booker (San Mateo, CA: Morgan Kaufmann) pp 2–9

De Jong K A 1975 *An Analysis of the Behavior of a Class of Genetic Adaptive Systems* Phd Dissertation, University of Michigan

——1993 Genetic algorithms are NOT function optimizers *Foundations of Genetic Algorithms 2* ed L D Whitley (San Mateo, CA: Morgan Kaufmann) pp 5–17

De Jong K A and Sarma J 1993 Generation gaps revisited *Foundations of Genetic Algorithms 2* ed L D Whitley (San Mateo, CA: Morgan Kaufmann) pp 19–28

Fogel G B and Fogel D B 1995 Continuous evolutionary programming: analysis and experiments *Cybernet. Syst.* **26** 79–90

Fogel L J, Owens A J and Walsh M J 1996 *Artificial Intelligence through Simulated Evolution* (New York: Wiley)

Goldberg D E and Deb K 1991 A comparative analysis of selection schemes used in genetic algorithms *Foundations of Genetic Algorithms 1* ed G J E Rawlins (San Mateo, CA: Morgan Kaufmann) pp 69–93

Grefenstette J J 1986 Optimization of control parameters for genetic algorithms *IEEE Trans. Syst. Man Cybernet.* **SMC-16** 122–8

Holland J H 1975 *Adaptation in Natural and Artificial Systems* (Ann Arbor, MI: University of Michigan Press)

Holland J H and Reitman J S 1978 Cognitive systems based on adaptive algorithms *Pattern-Directed Inference Systems* ed D A Waterman and F Hayes-Roth (New York: Academic)

Peck C C and Dhawan A P 1995 Genetic algorithms as global random search methods: an alternative perspective *Evolutionary Comput.* **3** 39–80

Rechenberg I 1973 *Evolutionsstrategie: Optimierung technischer Systeme nach Prinzipien der biologischen Evolution* (Stuttgart: Frommann-Holzboog)

Schwefel H-P 1981 *Numerical Optimization of Computer Models* (Chichester: Wiley)

Syswerda G 1989 Uniform crossover in genetic algorithms *Proc. 3rd Int. Conf. on Genetic Algorithms (Fairfax, VA, June 1989)* ed J D Schaffer (San Mateo, CA: Morgan Kaufmann) pp 2–9

——1991 A study of reproduction in generational and steady-state genetic algorithms *Foundations of Genetic Algorithms 1* ed G J E Rawlins (San Mateo, CA: Morgan Kaufmann) pp 94–101

Whitley D 1989 The GENITOR algorithm and selection pressure: why rank-based allocation of reproductive trials is best *Proc. 3rd Int. Conf. on Genetic Algorithms (Fairfax, VA, June 1989)* ed J D Schaffer (San Mateo, CA: Morgan Kaufmann) pp 116–21

Whitley D and Kauth J 1988 *GENITOR: a Different Genetic Algorithm* Colorado State University Technical Report CS-88-101

29

A comparison of selection mechanisms

Peter J B Hancock

29.1 Introduction

Selection provides the driving force behind an evolutionary algorithm. Without it, the search would be no better than random. This section explores the pros and cons of a variety of different methods of performing selection. Selection methods differ in two main ways: the way they aim to distribute reproductive opportunities across members of the population, and the accuracy with which they achieve their aim. The accuracy may differ because of sampling noise inherent in some selection algorithms. There are also other differences that may be significant, such as time complexity and suitability for parallel processing. Crucially for some applications, they also differ in their ability to deal with evaluation noise.

There have been a number of comparisons of different selection methods by a mixture of analysis and simulation, usually on deliberately simplified tasks. Goldberg and Deb (1991) considered a system with just two fitness levels, and studied the time taken for the fitter individuals to take over the population under the action of selection only, verifying their analysis with simulations. Hancock (1994) extended the simulations to a wider range of selection algorithms, and added mutation as a source of variation, to compare effective growth rates. The effects of adding noise to the evaluation function were also considered. Syswerda (1991) compared generational and incremental models on a ten-level takeover problem. Thierens and Goldberg (1994) derived analytical results for rates of growth for a bit counting problem, where the approximately normal distribution of fitness values allowed them to include recombination in their analysis. Bäck (1994) compared takeover times for all the major selection methods analytically and reported an experiment on a 30-dimensional sphere problem. Bäck (1995) compared tournament and (μ, λ) selection more closely. Blickle and Thiele (1995a, b) undertook a detailed analytical comparison of a number of selection methods (note that the second paper corrects an error in the first). Other studies include those of Bäck and Hoffmeister (1991), de la Maza and Tidor (1993) and Pál (1994).

It would be useful to have some objective measure(s) with which to compare selection methods. A general term is selection pressure. The meaning of this is intuitively clear, the higher the selection pressure, the faster the rate of convergence, but it has no strict definition. Analysis of selection methods has concentrated on two measures: takeover time and selection intensity. Takeover time is the number of generations required for one copy of the best string to reproduce so as to fill the population, under the effect only of selection (Goldberg and Deb 1991). Selection intensity is defined in terms of the average fitness before and after selection, $\bar{f}$ and $\bar{f}_{\text{sel}}$, and the fitness variance σ:

$$I = \frac{\bar{f}_{\text{sel}} - \bar{f}}{\sigma}.$$

This captures the notion that it is harder to produce a given step in average fitness between the population and those selected when the fitness variance is low. However, both takeover time and selection intensity depend on the fitness functions, and so theoretical results may not always transfer to a real problem. There is an additional difficulty because the fitness variance itself depends on the selection method, so different methods configured to have the same selection intensity may actually grow at different rates.

Most of the selection schemes have a parameter that controls either the proportion of the population that reproduces or the distribution of reproductive opportunities, or both. One aim in what follows will be to identify some equivalent parameter settings for different selection methods.

29.2 Simulations

A number of graphs from simulations similar to those reported by Hancock (1994) are shown here, along with some analytical and experimental results from elsewhere. The takeover simulation initializes a population of 100 randomly, with rectangular distribution, in the range 0–1, with the exception that one individual is set to 1. The rate of takeover of individuals with the value 1 under the action of selection alone is plotted. Results reported are averaged over 100 different runs. The simulation is thus similar to that used by Goldberg and Deb (1991), but the greater range of fitness values allows investigation of the diversity maintained by the different selection methods. Since some of them produce exponential takeover in such conditions, a second set of simulations makes the problem slightly more realistic by adding mutation as a source of variation to be exploited by the selection procedures. This growth simulation initializes the population in the range 0–0.1. During reproduction, mutation with a Gaussian distribution, mean 0, standard deviation 0.02, is added to produce the offspring, subject to remaining in the range 0–1. Some plots show the value of the best member of the population after various numbers of evaluations, again averaged over 100 different runs. Other plots show the growth of the worst value in the population, which gives an indication of the diversity maintained in the population. Some selection methods are better at preserving such diversity: other things being equal, this seems likely to improve the quality of the overall search (Mühlenbein and Schlierkamp-Voosen 1995, Blickle and Thiele 1995b).

It should be emphasized that fast convergence on these tasks is not necessarily good: they are deliberately simple in an effort to illustrate some of the differences between selection methods and the reasons underlying them. Good selection methods need to balance exploration and exploitation. Before reporting results, we shall consider a number of more theoretical points of similarities and differences.

29.3 Population models

There are two different points in the population cycle at which selection may be implemented. One approach, typical of genetic algorithms (GAs), is to choose individuals from the population for reproduction, usually in some way proportional to their fitness. These are then acted on by the chosen genetic operators to produce the next generation. The other approach, more typical of evolution strategies (ESs) and evolutionary programming (EP), is to allow all the members of the population to reproduce, and then select the better members of the extended population to go through to the next generation. This difference, of allowing all members to reproduce, is sometimes flagged as one of the key differences in approach between ES/EP and GAs. In fact the two approaches may be seen as equivalent once running, differing only in what is called the population. If the extended population typical of the ES and EP approach is labeled simply the population, then it may be seen that, as with the first approach, the best individuals are selected for reproduction and used to generate the new (extended) population. Looked at this way, it is the traditional GA approach that allows all members of the population at least some chance of reproduction, where the methods that use truncation selection restrict the number that are allowed to breed. There remains, however, a difference in philosophy: the traditional GA approach is reproduction according to fitness, while the truncation selection typical of the ES, EP, and breeder GA is more like survival of the fittest. There will also be a difference at startup, with ES/EP initializing μ individuals, while an equivalent GA initializes $\mu + \lambda$.

29.4 Equivalence: expectations and reality

A number of pairs of the common selection algorithms turn out to be, in some respects, equivalent. The equivalence, usually in expected outcome, can hide differences due to sampling errors, or behavior in the presence of noise, that may cause significant differences in practice. This section considers some of these similarities and differences, in order to reduce the number that need be considered in detail in Section 29.5.

29.4.1 Tournament selection and ranking

Goldberg and Deb (1991) showed that simple binary tournament selection (TS) (see Chapter 24) is equivalent to linear ranking (Section 25.2) when set to give two offspring to the top-ranked string ($\beta_{rank} = 2$). However, this is only in expectation: when implemented the obvious way, picking each fresh pair of potential parents from the population with replacement, tournament selection suffers from sampling errors like those produced by roulette wheel sampling, precisely because each tournament is performed separately. A way to reduce this noise is to take a copy of the population and choose pairs for tournament from it *without* replacement. When the copy population is exhausted, another copy is made to select the second half of the new population (Goldberg *et al* 1989). This method ensures that each individual participates in exactly two tournaments, and will not fight itself. It does not eliminate the problem, since, for example, an average individual, that ought to win once, may pick better or worse opponents both times, but it will at least stop several copies of any one being chosen.

The selection pressure generated by tournament selection may be decreased by making the tournaments stochastic. The equivalence, apart from sampling errors, with linear ranking remains. Thus TS with a probability of the better string winning of 0.75 is equivalent to linear ranking with $\beta_{rank} = 1.5$. The selection pressure may be increased by holding tournaments among more than two individuals. For three, the best will expect three offspring, while an average member can expect 0.75 (it should win one quarter of its expected three tournaments). The assignment is therefore nonlinear and Bäck (1994) shows that, to a first approximation, the results are equivalent to exponential nonlinear ranking, where the probability of selection of each rank i, starting at $i = 1$ for the best, is given by $(s - 1)(s^{i-1})/(s^{\mu} - 1)$, where s is typically in the range 0.9–1 (Blickle and Thiele 1995b). (Note that the probabilities as specified by Michalewicz (1992) do not sum to unity (Bäck 1994).) More precisely, they differ in that TS gives the worst members of the population no chance to reproduce. Figure 29.1 compares the expected number of offspring for each rank in a population of 100. The difference results in a somewhat lower population diversity for TS when run at the same growth rate.

Goldberg and Deb (1991) prefer TS to linear ranking on account of its lower time complexity (since ranking requires a sort of the population), and Bäck (1994) argues similarly for TS over nonlinear ranking. However, time complexity is unlikely to be an issue in serious applications, where the evaluation time usually dominates all other parts of the algorithm. The difference is in any case reduced if the noise-reduced version of TS is implemented, since this also requires shuffling the population. For global population models, therefore, ranking, with Baker's sampling procedure (Baker 1987), is usually preferable. TS may be appropriate in incremental models, where only one individual is to be evaluated at a time, and in parallel population models. It may also be

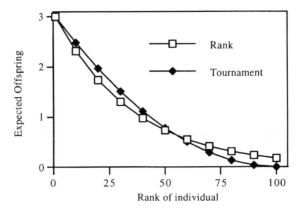

Figure 29.1. Expected number of offspring against rank for tournament selection with tournament size 3 and exponential rank selection with $s = 0.972$.

appropriate in, for instance, game playing applications, where the evaluation itself consists of individuals playing each other.

Freisleben and Härtfelder (1993) compared a number of selection schemes using a meta-level GA, that adjusted the parameters of the GA used to tackle their problem. Tournament selection was chosen in preference to rank selection, which at first sight seems odd, since the only difference is added noise. A possible explanation lies in the nature of their task, which was learning the weights for a neural net simulation. This is plagued with symmetry problems (e.g. Hancock 1992). The GA has to break the symmetries and decide on just one to make progress. It seems possible that the inaccuracies inherent in tournament selection facilitated this symmetry breaking, with one individual having an undue advantage, and thereby taking over the population. Noise is not always undesirable, though there may be more controlled ways to achieve the same result.

29.4.2 *Incremental and generational models*

There is apparently a large division between incremental and generational reproduction models. However, Syswerda (1991) shows that an incremental model where the deletion is at random produces the same expected result as a generational model with the same rank selection for reproduction. Again, however, this analysis overlooks sampling effects. Precisely because incremental models generate only one or two offspring per cycle, they suffer the roulette wheel sampling error. Figure 29.2 shows the growth rate for best and worst in the population for the two models with the same selection pressure (best expecting 1.2 offspring). The incremental model grows more slowly, yet loses diversity more rapidly, an effect characteristic of this kind of sampling error. Incremental models also suffer in the presence of evaluation noise (see Section 29.6).

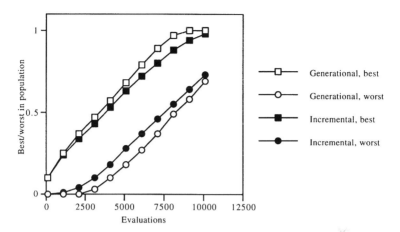

Figure 29.2. The growth rate in the presence of mutation of the best and worst in the population for the incremental model with random deletion and the generational model, both with linear rank selection for reproduction, $\beta_{rank} = 1.2$.

The very highest selection pressure possible from an evolutionary system would arise from an incremental system, where only the best member of the population is able to reproduce, and the worst is removed if the new string is an improvement. Since the rest of the population would thus be redundant, this is equivalent to a $(1+1)$ ES, the dynamics of which are well investigated (Schwefel 1981).

29.4.3 Evolution strategy, evolutionary programming, and truncation selection

Some GA workers allow only the top few members of the population to reproduce (Nolfi *et al* 1990, Mühlenbein and Schlierkamp-Voosen 1993). This is often called truncation selection, and is equivalent to the ES (μ, λ) approach subject only to a difference in what is called the population (see Section 29.3).

EP uses a form of tournament selection where all members of the extended population $\mu + \lambda$ compete with c others, chosen at random with replacement. Those μ that amass the most wins then reproduce by mutation to form the next extended population. This may be seen as a rather softer form of truncation selection, converging to the same result as a $(\mu + \mu)$ ES as the size of c increases. The value of c does not directly affect the selection pressure, only the noise in the selection process.

The EP selection process may be softened further by making the tournaments probabilistic. One approach is to make the probability of the better individual winning dependent on the relative fitness of the pair: $p_i = f_i/(f_i + f_j)$ (Fogel

1988). Although intuitively appealing, this has the effect of reducing selection pressure as the population converges and can produce growth curves remarkably similar to unscaled fitness proportional selection (FPS; Hancock 1994).

29.5 Simulation results

29.5.1 Fitness proportional selection

Simple FPS suffers from sensitivity to the distribution of fitness values in the population, as discussed in Chapter 23. The reduction of selection pressure as the population converges may be countered by moving baseline techniques, such as windowing and sigma scaling. These are still vulnerable to undesirable loss of diversity caused by a particularly fit individual, which may produce many offspring. Rescaling techniques are able to limit the number of offspring given to the best, but may still be affected by the overall spread of fitness values, and particularly by the presence of very poor individuals.

Figure 29.3 compares takeover and growth rates of FPS and some of the baseline adjustment and rescaling methods. The simple takeover rates for the three adjusted methods are rather similar for these scale parameters, with linear scaling just fastest. Simple FPS is so slow it does not really show on the same graph: it reaches only 80% convergence after 40 000 evaluations on this problem. The curves for growth in the presence of mutation are all rather alike: the presence of the mutation maintains the range of fitness values in the population, giving simple FPS something to work on. Note, however, that it still starts off relatively fast and slows down towards the end: probably the opposite of what is desirable. The three scaled versions are still similar, but note that the order has reversed. Windowing and sigma scaling now grow more rapidly precisely because they fail to limit the number of offspring to especially good individuals. A fortuitous mutation is thus better exploited than in the more controlled linear scaling, which leads to the correct result in this simple hill-climbing task, but may not in a more complex real problem.

29.5.2 Ranking

Goldberg and Deb (1991) show that the expected growth rate for linear ranking is proportional to the value of β_{rank}, the number of offspring given to the best individual. For exponential scaling, the selection pressure is proportional to $1 - s$. This makes available a wide range of selection pressures, defined by the value of s, illustrated in figure 29.4. The highest takeover rate available with linear ranking ($\beta_{\text{rank}} = 2$) is also shown. Exponential ranking can go faster with smaller values of s (see table 29.1). Note the logarithmic x-axis on this plot.

With exponential ranking, because of the exponential assignment curve, poor individuals do rather better than with linear ranking, at the expense of those more in the middle of the range. One result of this is that, for parameter settings that give similar takeover times, exponential ranking loses the worse values in the population more slowly, which may help preserve diversity in practice.

29.5.3 Evolution strategies

The selection pressure generated by the ES selection methods have been extensively analyzed, sometimes under the title of truncation selection (see e.g. Bäck 1994). Selection pressure is dependent on the ratio of μ to λ (see

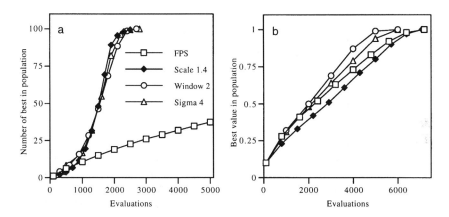

Figure 29.3. (a) The takeover rate for FPS, with windowing, sigma, and linear scaling. (b) Growth rates in the presence of mutation.

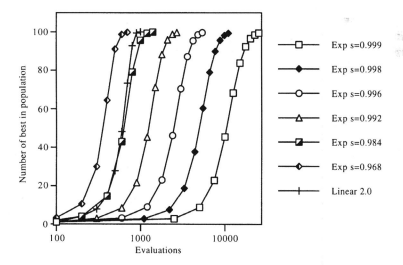

Figure 29.4. The takeover rate for exponential rank selection for a number of values of s, together with that for linear ranking, $\beta_{\text{rank}} = 2$.

table 29.1). One simulation result is shown, in figure 29.5, to make clear the selection pressure achievable by (μ, λ) selection, and indicate its potential susceptibility to evaluation noise, discussed further below.

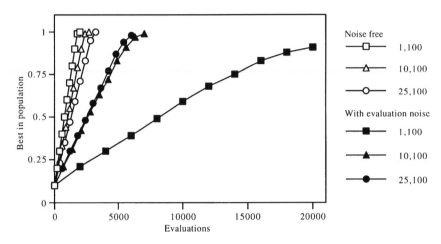

Figure 29.5. The growth rate in the presence of mutation for ES (μ, λ) selection with and without evaluation noise, for $\lambda = 100$ and $\mu = 1, 10,$ and 25.

29.5.4 Incremental models

Goldberg and Deb (1991) show that the Genitor incremental model develops a very high growth rate, compared to that typical of GAs. This is mostly due to the method of deletion, in which the worst member of the population is eliminated (cf the ES truncation approach). As a consequence, the value of β_{rank} used in the linear ranking to select parents has little effect on the takeover rate. Even with $\beta_{\text{rank}} = 1$ (i.e. a random choice of parents), kill-worst converges in around 900 evaluations (cf about 3000 for the scaled FPS variants in figure 29.3(a)). Increasing β_{rank} to its maximum of 2.0 only reduces this to around 600 evaluations.

There are a number of ways to decide which of the population should be removed (Syswerda 1991), such as killing the oldest (also known as FIFO deletion (De Jong and Sarma 1993)); one of the n worst; by inverse rank; or simply at random. The various deletion strategies radically affect the behavior of the algorithm. As discussed above, random deletion resembles a generational model. Kill oldest also produces much softer selection than kill-worst, producing takeover rates similar to generational models with the same selection pressure (see figure 29.6). However, the incremental model starts more quickly and ends more slowly than the generational one.

Syswerda (1991) prefers kill-by-inverse-rank. In his simulations, this produces results similar to kill-worst, but he is using a high inverse selection

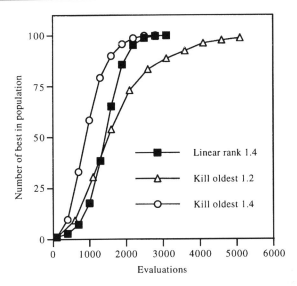

Figure 29.6. The takeover rates for the generational model and the kill-oldest incremental model, both using linear ranking for selection.

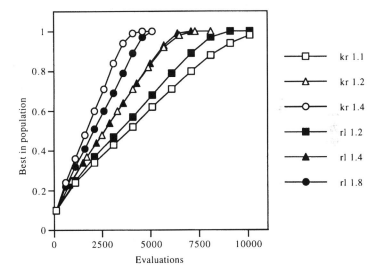

Figure 29.7. Growth rates in the presence of mutation for incremental kill-by-inverse-rank (kr) and generational linear ranking (rl) for various values of β_{rank}.

pressure (exponential ranking with $s = 0.9$). A more controlled result is given by selecting for reproduction from the top and for deletion from the bottom using ranking with the same, more moderate value of β_{rank}. Using linear ranking, the growth rate changes more rapidly than β_{rank}. This is because an increase in β_{rank} has two effects: increasing the probability of picking one of the better members of the population at each step, and increasing the number of steps for which they are likely to remain in the population, by decreasing their probability of deletion. Figure 29.7 compares growth rates in the presence of mutation for kill-by-rank incremental and equivalent generational models. It may be seen that the generational model with $\beta_{\text{rank}} = 1.4$ and the incremental model with $\beta_{\text{rank}} = 1.2$ produce very similar results. Another matched pair at lower growth rates is generational with $\beta_{\text{rank}} = 1.2$ and incremental with $\beta_{\text{rank}} = 1.13$ (not shown).

One of the arguments in favor of incremental models is that they allow good new individuals to be exploited at once, rather than having to wait a generation. It might be thought that any such gain would be rather slight, since although a good new member could be picked at once, it is more likely to have to wait several iterations at normal selection pressures. There is also the inevitable sampling noise to be overcome. De Jong and Sarma (1993) claim that there is actually no net benefit, since adding new fit members has the effect of increasing the average fitness, thus reducing the likelihood of them being selected. However, this argument applies only to takeover problems: when reproduction operators are included the incremental approach can generate higher growth rates. Figure 29.8 compares the growth of an incremental kill-oldest model with a generational model using the same selection scheme. The graph also shows one of the main drawbacks of the incremental models: their sensitivity to evaluation noise, to be discussed in the following section.

29.6   The effects of evaluation noise

Hancock (1994) extended the growth simulations to study the effects of adding evaluation noise. A Gaussian random variable, mean zero, standard deviation 0.2, was added to each underlying true value for use in selection. The true value was used for reproduction. It proved necessary to add this much noise—ten times the standard deviation of the 'signal' mutation—to bring about significant reduction in growth rates for the generational selection models.

The sensitivity of the different selection algorithms to evaluation noise is largely dependent on whether they retain parents for further reproduction. Fully generational models are relatively immune, while most incremental models and those like the $(\mu + \lambda)$ ES that allow parents to compete for retention fare much worse, because individuals receiving a fortuitously good evaluation will be kept. The exception for incremental models is kill-oldest, which maintains the necessary turnover. Figure 29.8 shows the comparison. Kill oldest deteriorates only a little more than the generational model in the presence of noise, while kill-worst, which grows much the fastest in the absence of noise, almost fails completely.

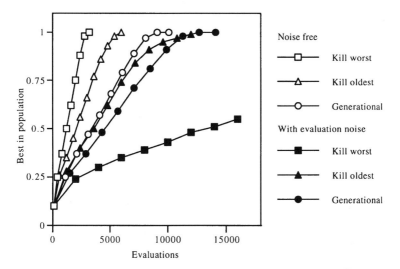

Figure 29.8. Growth in the presence of mutation, with and without evaluation noise, for the generational model with linear ranking and incremental models with kill-worst and kill-oldest, all using $\beta_{rank} = 1.2$ for selection.

Within generational models, there are differences in noise sensitivity. Figure 29.9 compares the growth rates for linear ranking and sigma-scaled FPS, with and without noise. It may be seen that the scaled FPS deteriorates less.

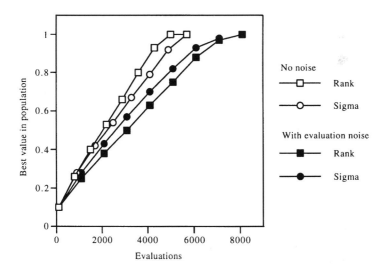

Figure 29.9. Growth in the presence of mutation, with and without evaluation noise, for the generational model with linear ranking, $\beta_{rank} = 1.8$, and sigma-scaled FPS, $s = 4$.

This is caused by sigma scaling's inability to control superfit individuals. A genuinely good individual, that happens to receive a helpful boost from the noise, may be given many offspring by sigma scaling, but will be limited to β_{rank}, in this case 1.8, by ranking. As before, rapid convergence is beneficial in this simple task, but is unlikely to be so in general.

The ES (μ, λ) method can achieve extremely high selection pressures, when it becomes sensitive to evaluation noise in a manner similar to incremental models (figure 29.5). In this case, the problem is that too many reproductions are given to a few strings, whose rating may be overestimated by the noise. Figure 29.5 shows a clear turnaround: as the selection pressure is increased, performance in noise becomes worse.

One approach to countering the effects of noise is to perform two or more evaluations per string, and average the results. Fitzpatrick and Grefenstette (1988) investigated this and concluded that it is better to evaluate only once and proceed with the next generation. A possibly more efficient method is to reevaluate only the apparently fitter individuals. Candidates may be chosen as for reproduction, e.g. by rank. However, experiments with incremental kill-by-rank indicated that the extra evaluations did not pay their way, with convergence taking only a little less than twice as many evaluations in total (Hancock 1994). Hammel and Bäck (1994) compared the effects of reevaluation with an equivalent increase in the population size and showed that reevaluations lead to a better final result. Indeed, on Rastrigan's function, increasing the population size resulted in a deterioration of convergence performance. Taken together, these results suggest a strategy of evaluating only once initially, and keeping the population turning over, but then starting to reevaluate as the population begins to converge. Hammel and Bäck suggest an alternative possibility of incorporating reevaluation as a parameter to be optimized by the evolutionary system itself.

29.7   Analytic comparison

Blickle and Thiele (1995b) perform an extensive analysis of several selection schemes, deriving the dependence of selection intensity on the selection parameters under the assumption of a normal distribution of fitness. Their results, which reassuringly agree with the simulation results here, are shown in an adapted form in table 29.1. They also consider selection variance, confirming that methods such as ES selection that disallow weakest strings from reproduction reduce the population variance more rapidly than those that allow weak strings some chance. Of the methods considered, exponential rank selection gives the highest fitness variance, for the reasons illustrated in figure 29.1. Their conclusion is that exponential rank selection is therefore probably the 'best' of the schemes that they consider.

Table 29.1. Parameter settings that give equivalent selection intensities for ES (μ, λ), TS, and linear and exponential ranking, adapted and extended from Blickle and Thiele (1995b). Under tournament size, p refers to the probability of the better string winning.

I	ES μ/λ	Tournament size	β_{rank}, Lin rank	s, Exp rank, $\lambda = 100$
0.11	0.94	2, $p = 0.6$	1.2	0.996
0.34	0.80	2, $p = 0.8$	1.6	0.988
0.56	0.66	2	2.0	0.979
0.84	0.47	3	—	0.966
1.03	0.36	4	—	0.955
1.16	0.30	5	—	0.945
1.35	0.22	7	—	0.926
1.54	0.15	10	—	0.900
1.87	0.08	20	—	0.809

29.8 Conclusions

The choice of a selection mechanism cannot be made independently of other aspects of the evolutionary algorithm. For instance, Eshelman (1991) deliberately combines a conservative selection mechanism with an explorative recombination operator in his CHC algorithm. Where search is largely driven by mutation, it may be possible to use much higher selection pressures, typical of the ES approach. If the evaluation function is noisy, then most incremental models and others that may retain parents are likely to suffer. Certainly, selection pressures need to be lower in the presence of noise, and, of the incremental models, kill-oldest fares best. Without noise, incremental methods can provide a useful increase in exploitation of good new individuals. Care is needed in the choice of method of deletion: killing the worst provides high growth rates with little means of control. Killing by inverse rank or killing the oldest offers more control. Amongst generational models, the ES (μ, λ) and exponential rank selection methods give the biggest and most controllable range of selection pressures, with the ES method probably most suited to mutation-driven, high-growth-rate systems, and ranking better for slower, more explorative searches, where maintenance of diversity is important.

References

Bäck T 1994 Selective pressure in evolutionary algorithms: a characterization of selection methods *Proc. 1st IEEE Conf. on Evolutionary Computation (Orlando, FL, June 1994)* (Piscataway, NJ: IEEE) pp 57–62

——1995 Generalized convergence models for tournament and (μ, λ)-selection *Proc. 6th Int. Conf. on Genetic Algorithms (Pittsburgh, PA, July 1995)* ed L J Eshelman (San Mateo, CA: Morgan Kaufmann) pp 2–8

Bäck T and Hoffmeister F 1991 Extended selection mechanisms in genetic algorithms *Proc. 4th Int. Conf. on Genetic Algorithms (San Diego, CA, July 1991)* ed R K Belew and L B Booker (San Mateo, CA: Morgan Kaufmann) pp 92–9

Baker J E 1987 Reducing bias and inefficiency in the selection algorithm *Proc. 2nd Int. Conf. on Genetic Algorithms (Cambridge, MA, 1987)* ed J J Grefenstette (Hillsdale, NJ: Erlbaum) pp 14–21

Blickle T and Thiele L 1995a A mathematical analysis of tournament selection *Proc. 6th Int. Conf. on Genetic Algorithms (Pittsburgh, PA, July 1995)* ed L J Eshelman (San Mateo, CA: Morgan Kaufmann) pp 9–16

——1995b *A Comparison of Selection Schemes used in Genetic Algorithms* TIK-Report 11, Computer Engineering and Communication Networks Laboratory (TIK), Swiss Federal Institute of Technology

De Jong K A and Sarma J 1993 Generation gaps revisited *Foundations of Genetic Algorithms 2* ed D Whitley (San Mateo, CA: Morgan Kaufmann) pp 19–28

de la Maza M and Tidor B 1993 An analysis of selection procedures with particular attention paid to proportional and Boltzman selection *Proc. 5th Int. Conf. on Genetic Algorithms (Urbana-Champaign, IL, July 1993)* ed S Forrest (San Mateo, CA: Morgan Kaufmann) pp 124–31

Eshelman L J 1991 The CHC adaptive search algorithm: how to have safe search when engaging in nontraditional genetic recombination *Foundations of Genetic Algorithms* ed G J E Rawlins (San Mateo, CA: Morgan Kaufmann) pp 265–83

Fitzpatrick J M and Grefenstette J J 1988 Genetic algorithms in noisy environments *Machine Learning* **3** 101–20

Fogel D B 1988 An evolutionary approach to the traveling salesman problem *Biol. Cybern.* **60** 139–44

Freisleben B and Härtfelder M 1993 Optimization of genetic algorithms by genetic algorithms *Artificial Neural Nets and Genetic Algorithms* ed R F Albrecht, C R Reeves and N C Steele (Berlin: Springer) pp 392–9

Goldberg D E and Deb K 1991 A comparative analysis of selection schemes used in genetic algorithms *Foundations of Genetic Algorithms* ed G J E Rawlins (San Mateo, CA: Morgan Kaufmann) pp 69–93

Goldberg D E, Korb B and Deb K 1989 Messy genetic algorithms: motivation, analysis and first results *Complex Syst.* **3** 493–530

Hammel U and Bäck T 1994 Evolution strategies on noisy functions: how to improve convergence properties *Parallel Problem solving from Nature—PPSN III (Proc. Int. Conf. on Evolutionary Computation and 3rd Conf. on Parallel Problem Solving from Nature, Jerusalem, October 1994) (Lecture Notes in Computer Science 866)* ed Yu Davidor, H-P Schwefel and R Männer (Berlin: Springer) pp 159–68

Hancock P J B 1992 *Coding strategies for genetic algorithms and neural nets* PhD Thesis, Department of Computer Science, University of Stirling

——1994 An empirical comparison of selection methods in evolutionary algorithms *Evolutionary Computing (AISB Workshop, Leeds, 1994, Selected Papers) (Lecture Notes in Computer Science 865)* ed T C Fogarty (Berlin: Springer) pp 80–94

Michalewicz Z 1992 *Genetic Algorithms + Data Structures = Evolution Programs* (Berlin: Springer)

Mühlenbein H and Schlierkamp-Voosen D 1993 Predictive models for the breeder genetic algorithm *Evolut. Comput.* **1** 25–50

——1995 Analysis of selection, mutation and recombination in genetic algorithms *Evolution as a Computational Process (Lecture Notes in Computer Science 899)* ed W Banzhaf and F H Eckman (Berlin: Springer) pp 188–214

Nolfi S, Elman J L and Parisi D 1990 *Learning and Evolution in Neural Networks* Technical report CRL TR 9019 UCSD

Pál K F 1994 Selection schemes with spatial isolation for genetic optimization *Parallel Problem solving from Nature—PPSN III (Proc. Int. Conf. on Evolutionary Computation and 3rd Conf. on Parallel Problem Solving from Nature, Jerusalem, October 1994) (Lecture Notes in Computer Science 866)* ed Yu Davidor, H-P Schwefel and R Männer (Berlin: Springer) pp 170–9

Schwefel H-P 1981 *Numerical Optimization of Computer Models* (Chichester: Wiley)

Syswerda G 1991 A study of reproduction in generational and steady-state genetic algorithms *Foundations of Genetic Algorithms* ed G J E Rawlins (San Mateo, CA: Morgan Kaufmann) pp 94–101

Thierens D and Goldberg D E 1994 Elitist recombination: an integrated selection recombination GA *Proc. IEEE World Congr. on Computational Intelligence* (Piscataway, NJ: IEEE)

30

Interactive evolution

Wolfgang Banzhaf

30.1 Introduction

The basic idea of interactive evolution (IE) is to involve a human user on-line in the variation–selection loop of the evolutionary algorithm (EA). This is to be seen in contrast to the conventional participation of the user prior to running the EA by defining a suitable representation of the problem (Chapters 14–21), the fitness criterion for evaluation of individual solutions, and corresponding operators (Chapters 31–34) to improve fitness quality. In the latter case, the user's role is restricted to passive observation during the EA run.

The minimum requirement for IE is the definition of a problem representation, together with a determination of population parameters only. Search operators of arbitrary kind as well as selection according to arbitrary criteria might be applied to the representation by the user. The process is much more comparable to the creation of a piece of art, for example, a painting, than to the automatic evolution of an optimized problem solution. In IE, the user assumes an active role in the search process. At the minimum level, the IE system must hold present solutions together with variants presently generated or considered.

Usually, however, automatic means of variation (i.e. evolutionary search operators using random events) are provided with an IE system. In the present context we shall require the existence of automatic means of variation by operators for mutation (Chapter 32) and recombination (Chapter 33) of solutions which are to be defined prior to running the EA.

30.2 History

Dawkins (1986) was the first to consider an elaborate IE system. The evolution of *biomorphs*, as he called them, by IE in a system that he had originally intended to be useful for the design of treelike graphical forms has served as a prototype for many systems developed subsequently. Starting with the contributions of Sims (1991) and the book of Todd and Latham (1992), computer art developed into the present major application area of IE.

IE of grammar-based structures has also been considered (Nguyen and Huang 1994, McCormack 1994). Raw image data have been used more recently for the purpose of evolving forms (Graf and Banzhaf 1995a).

30.3   The problem

The problem IE is trying to address has been encountered in all varieties of EAs that make use of automatic evolution: the existence of nonexplicit conditions, that is, conditions that are not formalizable.

- The absence of a human user in steering and controlling the process of evolution sometimes leads to unnecessary detours from the goal of global optimization. In most of these cases, human intervention into the search and selection process would advance the search rather quickly and allow faster convergence onto the most promising regions of the fitness landscape, or, sometimes, escape from a local optimum. Hence, a mobilization of human knowledge can be achieved by allowing the user to participate in the process.
- Many design processes require subjective judgment relying on human intuition, aesthetical values, or taste. In such cases, the fitness criterion cannot be formulated explicitly, but can only be applied on a comparative case-by-case basis. Direct human participation in IE allows for machine-supported evolution of designs that would otherwise be completely manual.

Thus, IE can be used (i) to accelerate EAs and (ii) in some areas to enable application of EAs altogether.

30.4   The interactive evolution approach

Selection in a standard IE system, as opposed to that in an automatic evolution system, is based on user action. It is typically the selection step that is subjugated to human action, although in less frequent cases the variation process might also be done by hand.

The standard algorithm for IE (following the notation in the introduction) is presented at the top of the next page. As in an automatic evolution system, there are parameters that are required to be fixed *a priori*: μ, λ, Θ_ι, Θ_m, Θ_r, Θ_s. There are, however, also parameters subject to change, Θ'_m, Θ'_r, Θ'_s, depending on the user interaction with the IE system. Both parameter sets together determine the actual effect of mutation, recombination, and selection operators.

A simple variation of the standard algorithm shown overleaf is to allow for population parameters to be also the subject of user interaction with the system. For example, some systems (Graf and Banzhaf 1995a) consider growing populations and a variable number of variants.

A more complicated variant of the standard algorithm would add a sorting process of variants according to a predefined fitness criterion. One step further

Input: $\mu, \lambda, \Theta_\iota, \Theta_m, \Theta_r, \Theta_s$
Output: a^*, the individual last selected during the run, or
                  P^*, the population last selected during the run.
1        $t \leftarrow 0$;
2        $P(t) \leftarrow$ initialize(μ);
3   **while** $(\iota(P(t), \Theta_\iota) \neq$ true) **do**
4   **Input:** Θ_r', Θ_m'
5                $P'(t) \leftarrow$ recombine($P(t), \Theta_r, \Theta_r'$);
6                $P''(t) \leftarrow$ mutate($P'(t), \Theta_m, \Theta_m'$);
7   **Output:** $P''(t)$
8   **Input:** Θ_s'
9                $P(t+1) \leftarrow$ select($P''(t), \mu, \Theta_s, \Theta_s'$);
10               $t \leftarrow t+1$;
        od

is to allow this sorting process to result in a preselection in order to present a smaller number of variants for the interactive selection step. Both methods help the user to concentrate his or her selective action on the most promising variants according to this predefined criterion.

This algorithm is formulated as follows:

Input: $\mu, \lambda, \eta, \Theta_\iota, \Theta_m, \Theta_o, \Theta_r, \Theta_s$
Output: a^*, the individual last selected during the run, or
                  P^*, the population last selected during the run.
1        $t \leftarrow 0$;
2        $P(t) \leftarrow$ initialize(μ);
3   **while** $(\iota(P(t), \Theta_\iota) \neq$ true) **do**
4   **Input:** Θ_r', Θ_m'
5                $P'(t) \leftarrow$ recombine($P(t), \Theta_r, \Theta_r'$);
6                $P''(t) \leftarrow$ mutate($P'(t), \Theta_m, \Theta_m'$);
7                $F(t) \leftarrow$ evaluate($P''(t), \lambda$);
8                $P'''(t) \leftarrow$ sort($P''(t), \Theta_o$);
9                $P''''(t) \leftarrow$ select($P'''(t), F(t), \mu, \eta, \Theta_s$);
10   **Output:** $P''''(t)$
11   **Input:** Θ_s'
12               $P(t+1) \leftarrow$ select($P''''(t), \mu, \Theta_s, \Theta_s'$);
13               $t \leftarrow t+1$;
        od

The newly added parameter Θ_o is used here to specify the predefined order of the result after evaluation according to the predefined criterion. As before, the Θ_x'-parameters are used to specify the user interaction with the system. η is the parameter stating how many of the automatically generated and ordered variants are to be presented to the user. If $\mu + \lambda = \eta$ in a $(\mu + \lambda)$-strategy, or

$\lambda = \eta$ in a (μ, λ)-strategy, all variants will be presented for interactive selection. If, however, $\mu + \lambda > \eta$ and $\lambda > \eta$ respectively, solutions would be preselected and we speak of a hybrid evolution system (having elements of automatic as well as interactive evolution). Other parameters are used in the same way as in the standard algorithm.

30.5   Difficulties

The second, more complicated version of IE requires a predefined fitness criterion, in addition to user action. This trades one advantage of IE systems for another: the absence of any requirement to quantify fitness for a small number of variants to be evaluated interactively by the user.

Interactive systems have one serious difficulty, especially in connection with the automatic means of variation that are usually provided: whereas the generation of variants does not necessarily require human intervention, selection of variants does call the attention of the user. Due to psychological constraints, however, humans can normally select only from a small set of choices. IE systems are thus constrained to present only of the order of ten choices at each point in time from which to choose. Also in sequence, only a limited number of generations can be practically inspected by a user before the user becomes tired.

It is emphasized that this limitation must not mean that the generation of variants has to be restricted to small numbers. Rather the variants have to be properly ordered at least, for a presentation of a subset that can be handled interactively.

30.6   Application areas

An application of IE may be roughly divided into two parts:

(i)   structural evolution by discrete combination of predefined elements and
(ii) parametric evolution of genes coding for quantifiable features of the phenotype.

All application use these parts to various degrees.

In the first part, one has to define the structure elements that might be combined into a correct genotype. Examples are symbolic expressions coding for appearance of points in an image plane (Sims 1991) or elementary geometric figures such as cone and cube (Todd and Latham 1992). In the second part, parameters have to be used to further specify features of these structural elements. Together, this information constitutes the genotype of the future design hopefully to be selected by a user. In a process called *expression* this genotype is then transformed into an image or three-dimensional form that can be displayed as a phenotype for the selection step.

Table 30.1 gives an overview of the presently used IE systems. The reader is advised to consult details with the sources given in the reference list.

Table 30.1. An overview of different IE systems.

Application	Genotypic elements	Phenotype	Source
Lifelike structures	line drawing parameters	biomorphs	Dawkins (1986)
Textures, images	math. functions, image processing operations	(x, y, z) pixel values	Sims (1991)
Animation	math. functions, image processing operations	(x, y, z) pixel values	Sims (1991)
Person tracking	(position of) facial parts	face images	Caldwell and Johnston (1991)
Images, sculptures	geometric forms and visually defined graphical elements	3D rendering of grown objects	Todd and Latham (1992)
Dynamical systems	CA rules, differential equations	system behavior	Sims (1992)
Images, animation	rules, parameters of L-systems	rendered objects	McCormack (1994)
Airplane design	structural elements, e.g. wings, body	airplane drawings	Nguyen and Huang (1994)
Images, design	tiepoints of bitmap images	bitmap images	Graf and Banzhaf (1995a)

Figure 30.1 illustrates some results with runs in different IE systems.

Within the process of genotype–phenotype mapping a (recursive) developmental process is sometimes applied (Dawkins 1986, Todd and Latham 1992) whose results are finally displayed as the image for selection.

30.7 Further developments and perspectives

As of now, the means to generate a small group of variants from which to choose interactively are still not very good. For example, one could imagine a tool for categorizing variants into a number of families of similar design and then present only one representative from each family. In this way, a large population of variants could be used in the background which is invisible to the user but might have beneficial effects in the course of evolution.

Another very interesting area of research is to assign *a posteriori* effective fitness values to members of the population, depending on user action. An individual which is selected more often would be assigned a higher fitness than an individual which is not. This might result in at least a crude measure of the nonquantifiable fitness measures that lie at the heart of IE. One might even adjust the effect the operators have on the population, based on what is observed in the course of evolution directed by the user. In this way, an 'intelligent' system could be created, that is able to learn from actions of the user how to vary the population in order to arrive at good designs.

Another direction of research is to look into involving the user not (only) in the selection process, but in the variation process. Quite often, humans would

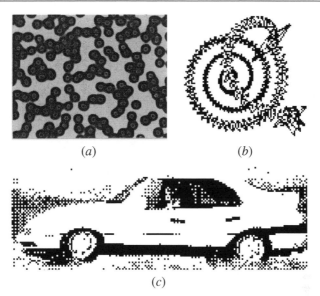

Figure 30.1. Samples of evolved objects: (*a*) dynamical system, cell structure (Sims 1992, © MIT Press); (*b*) artwork by Mutator (Todd and Latham 1992, with permission of the authors); (*c*) hybrid car model (Graf and Banzhaf 1995b, © IEEE Press).

have intuitive ideas for improvement of solutions when observing an automatic evolutionary process taking its steps. These ideas might be used to cut short the search routes an automatic algorithm is following. For this purpose, a user might be allowed to intervene in the process at appropriate interrupt times.

Finally, all sensory inputs could be used for IE. The systematic variation of components of a chemical compound that specifies an odor, for example, could be used to evolve a nice smelling perfume. Taste could as well be subject to interactive evolutionary tools, as could other objects if appropriately mapped to our senses (for instance by virtual reality tools).

With the advent of interactive media in the consumer market, production-on-demand systems might one day include an interactive evolutionary design device that allows the user not only to customize a product design before it goes into production, but also to generate his or her own original design that has never been realized before and usually will never be produced again. This would open up the possibility of evolutionary product design by companies which track their customers' activities and then distribute the best designs they discover.

References

Caldwell C and Johnston V S 1991 Tracking a criminal suspect through 'face-space' with a genetic algorithm *Proc. 4th Int. Conf. on Genetic Algorithms (San Diego, CA, July 1991)* ed R K Belew and L B Booker (San Mateo, CA: Morgan Kaufmann) pp 416–21

Dawkins R 1986 *The Blind Watchmaker* (New York: Norton)

Graf J and Banzhaf W 1995a Interactive evolution of images *Proc. 4th Conf. on Evolutionary Programming (San Diego, CA, March 1995)* ed J R McDonnell, R G Reynolds and D Fogel (Cambridge: MIT Press) pp 53–65

——1995b An expansion operator for interactive evolution *Proc. 2nd IEEE Int. Conf. on Evolutionary Computation (Perth, November–December 1995)* (Piscataway, NJ: IEEE) pp 798–802

McCormack J 1994 Interactive evolution of L-system grammars for computer graphics modelling *Complex Systems* ed T Bossomaier and D Green (Singapore: World Scientific) pp 118–30

Nguyen T C and Huang T S 1994 Evolvable 3D modeling for model-based object recognition systems *Advances in Genetic Programming* ed K Kinnear (Cambridge: MIT Press) pp 459–75

Sims K 1991 Artificial evolution for computer graphics *Comput. Graph.* **25** 319–28

——1992 Interactive evolution of dynamical systems *Toward a Practice of Autonomous Systems* ed F J Varela and P Bourgine (Cambridge, MA: MIT Press) pp 171–8

Todd S and Latham W 1992 *Evolutionary Art and Computers* (London: Academic)

Further reading

This section is intended to give an overview of presently available work in IE and modeling methods which might be interesting to use.

1. Prusinkiewicz P and Lindenmayer A 1991 *The Algorithmic Beauty of Plants* (Berlin: Springer)

   An informative introduction to L-systems and their use in computer graphics.

2. Koza J R 1992 *Genetic Programming* (Cambridge, MA: MIT Press)

   A book describing methods to evolve computer code, mainly in the form of LISP-type S-expressions.

3. Caldwell C and Johnston V 1991 Tracking a criminal suspect through 'face-space' with a genetic algorithm *Proc. Int. Conf. on Genetic Algorithms (San Diego, CA, July 1991)* ed R K Belew and L B Booker (San Mateo, CA: Morgan Kaufmann) pp 416–21

   Very interesting work containing one of the more profane applications of IE.

4. Baker E 1993 Evolving line drawings *Proc. Int. Conf. on Genetic Algorithms (Urbana-Champaign, IL, July 1993)* ed S Forrest (San Mateo, CA: Morgan Kaufmann) p 627

   This contribution discusses new ideas on design using simple style elements for IE.

5. Roston G P 1994 *A Genetic Methodology for Configuration Design* Doctoral Dissertation, Carnegie Mellon University

   Informed discussion of different aspects of using genetic algorithms for design purposes.

31

Introduction to search operators

Zbigniew Michalewicz

Any evolutionary system processes a population of individuals, $P(t) = \{a_1^t, \ldots, a_n^t\}$ (t is the iteration number), where each individual represents a potential solution to the problem at hand. As discussed in Chapters 14–21, many possible representations can be used for coding individuals; these representations may vary from binary strings to complex data structures I.

Each solution a_i^t is evaluated to give some measure of its fitness. Then a new population (iteration $t + 1$) is formed by selecting the more-fit individuals (the selection step of the evolutionary algorithm, see Chapters 22–30). Some members of the new population undergo transformations by means of 'genetic' operators to form new solutions. There are unary transformations m_i (mutation type), which create new individuals by a (usually small) change in a single individual ($m_i : I \rightarrow I$), and higher-order transformations c_j (crossover, or recombination type), which create new individuals by combining parts from several (two or more, up to the population size μ) individuals ($c_j : I^s \rightarrow I$, $2 \leq s \leq \mu$).

It seems that, for any evolutionary computation technique, the representation of an individual in the population and the set of operators used to alter its genetic code constitute probably the two most important components of the system, and often determine the system's success or failure. Thus, a representation of object variables must be chosen along with the consideration of the evolutionary computation operators which are to be used in the simulation. Clearly, the reverse is also true: the operators of any evolutionary system must be chosen carefully in accordance with the selected representation of individuals. Because of this strong relationship between representations and operators, the latter are discussed with respect to some (standard) representations.

In general, Chapters 31–34 provide a discussion on many operators which have been developed since the mid-1960s. Chapter 32 deals with mutation operators. Accordingly, several representations are considered (binary strings, real-valued vectors, permutations, finite-state machines, parse trees, and others) and for each representation one or more possible mutation operators are discussed. Clearly, it is impossible to provide a complete overview of *all* mutation operators, since the number of possible representations is unlimited.

However, Chapter 32 provides a complete description of *standard* mutation operators which have been developed for *standard* data structures.

Chapter 33 deals with recombination operators. Again, as for mutation operators, several representations are considered (binary strings, real-valued vectors, permutations, finite-state machines, parse trees, and others) and for each representation several possible recombination operators are discussed. Recombination operators exchange information between individuals and are considered to be the main 'driving force' behind genetic algorithms, while playing no role in evolutionary programming. There are many important and interesting issues connected with recombination operators; these include properties that recombination operators should have to be useful (these are outlined by Radcliffe (1993)), the number of parents involved in recombination process (Eiben *et al* (1994) described experiments with multiparent recombination operators—so-called *orgies*), or the frequencies of recombination operators.

Chapter 34 discusses some additional variations. These include the Baldwin effect, gene duplication and deletion, and knowledge-augmented operators.

References

Eiben A E, Raue P-E and Ruttkay Zs 1994 Genetic algorithms with multi-parent recombination *Proc. Parallel Problem Solving from Nature* vol 3 (New York: Springer) pp 78–87

Radcliffe N J 1993 Genetic set recombination *Foundations of Genetic Algorithms II (October 1994, Jerusalem)* ed Yu Davidor, H-P Schwefel and R Männer (San Mateo, CA: Morgan Kaufmann) pp 203–19

32

Mutation operators

Thomas Bäck (32.1), *David B Fogel* (32.2, 32.4, 32.6),
Darrell Whitley (32.3) *and Peter J Angeline* (32.5, 32.6)

32.1 Binary strings

Thomas Bäck

The mutation operator currently used in canonical genetic algorithms to manipulate binary vectors (also called binary strings or bitstrings, see Chapter 15) $a = (a_1, \ldots, a_\ell) \in I = \{0, 1\}^\ell$ of fixed length ℓ was originally introduced by Holland (1975, pp 109–11) for general finite individual spaces $I = A_1 \times \ldots \times A_\ell$, where $A_i = \{\alpha_{i_1}, \ldots, \alpha_{i_{k_i}}\}$. According to his definition, the mutation operator proceeds by:

(i) determining the positions $i_1, \ldots, i_h$ ($i_j \in \{1, \ldots, \ell\}$) to undergo mutation by a uniform random choice, where each position has the same small probability p_m of undergoing mutation, independently of what happens at other positions, and

(ii) forming the new vector $a' = (a_1, \ldots, a_{i_1-1}, a'_{i_1}, a_{i_1+1}, \ldots, a_{i_h-1}, a'_{i_h}, a_{i_h+1}, \ldots, a_\ell)$ where $a'_i \in A_i$ is drawn uniformly at random from the set of admissible values at position i.

The original value a_i at a position undergoing mutation is *not* excluded from the random choice of $a'_i \in A_i$; that is, although the position is chosen for mutation, the corresponding value might not change at all. This occurs with probability $1/|A_i|$, such that the effective (realized) mutation probability differs from p_m by a nonneglectible factor of $1/2$ if a binary representation is used.

In order to avoid this problem, it is typically agreed on defining p_m to be the probability of independently inverting each of the variables $a_i \in \{0, 1\}$, such that the mutation operator $m : \{0, 1\}^\ell \to \{0, 1\}^\ell$ produces a new individual $a' = m(a)$ according to

$$a'_i = \begin{cases} a_i & u > p_m \\ 1 - a_i & u \le p_m \end{cases} \tag{32.1}$$

237

where $u \sim U([0, 1))$ denotes a uniform random variable sampled anew for each $i \in \{1, \ldots, \ell\}$.

From a computational point of view, the straightforward implementation of equation (32.1) as a loop calling the random number generator for each position i is extremely inefficient. Since the random variable T describing the distances between two positions to be mutated has a geometrical distribution with $\mathcal{P}\{T = t\} = p_m(1 - p_m)^{t-1}$ and expectation $\mathbf{E}[T] = 1/p_m$, and a geometrical random number can be generated according to

$$t = 1 + \left\lfloor \frac{\ln(1 - u)}{\ln(1 - p_m)} \right\rfloor \tag{32.2}$$

(where $u \sim U([0, 1))$), equation (32.2) provides an efficient method to generate the offset to find the next position for mutation from the current one. If the actual position plus the offset exceeds the vector dimension ℓ, it 'carries over' to the next individual and, if all individuals of the actual population have been processed, to the next generation.

Concerning the importance of mutation for the evolutionary search process, both Holland (1975, p 111) and Goldberg (1989, p 14) emphasize that mutation just serves as a 'background operator', supporting the crossover operator (Section 33.1) by assuring that the full range of allele values is accessible to the search. Consequently, quite small values of $p_m \in [0.001, 0.01]$ were recommended for canonical genetic algorithms (see e.g. De Jong 1975, Grefenstette 1986, Schaffer et al 1989) until recently, when both empirical and theoretical investigations clearly demonstrated the benefits of emphasizing the role of mutation as a search operator in these algorithms. More specifically, some of the important results include:

(i) empirical findings favoring an initially large mutation rate that exponentially decreases over time (Fogarty 1989),

(ii) the theoretical confirmation of the optimality of such an exponentially decreasing mutation rate for simple test functions (Hesser and Männer 1991, 1992, Bäck 1996), and

(iii) the knowledge of a lower bound $p_m = 1/\ell$ for the optimal mutation rate (Bremermann et al 1966, Mühlenbein 1992, Bäck 1993).

It is obvious from these results that not only for evolution strategies and evolutionary programming, but also for canonical genetic algorithms, mutation is an important search operator that cannot be neglected either in practical applications or in theoretical investigations of these algorithms. Moreover, it is also possible to release the user of a genetic algorithm from the problem of finding an appropriate mutation rate control or fine-tuning a fixed value by transferring the strategy parameter self-adaptation principle from evolution strategies and evolutionary programming to genetic algorithms.

32.2 Real-valued vectors

David B Fogel

Mutation generally refers to the creation of a new solution from one and only one parent (otherwise the creation is referred to as a recombination (see Chapter 33). Given a real-valued representation where each element in a population is an n-dimensional vector $x \in \mathbb{R}^n$, there are many methods for creating new elements (offspring) using mutation. These methods have a long history, extending back at least to Bremermann (1962), Bremermann *et al* (1965), and others. A variety of methods will be considered here.

The general form of mutation can be written as

$$x' = m(x) \tag{32.3}$$

where x is the parent vector, m is the mutation function, and x' is the resulting offspring vector. Although there have been some attempts to include mutation operators that do not operate on the specific values of the parents but instead simply choose x' from a fixed probability density function (PDF) (Montana and Davis 1989), such methods lose the inheritance from parent to offspring that can facilitate evolutionary optimization on a variety of response surfaces. The more common form of mutation generates an offspring vector:

$$x' = x + M \tag{32.4}$$

where the mutation M is a random variable. M is often zero mean such that $E(x') = x$; the expected difference between a parent and its offspring is zero.

M can take different forms. For example, M could be the uniform random variable $U(a, b)^n$, where a and b are the lower and upper limits respectively. In this case, a is often set equal to $-b$. The result of applying this operator as M in equation (32.4) yields an offspring within a hyperbox $x + U(-b, b)^n$. Although such a mutation is unbiased with respect to the position of the offspring within the hyperbox, the method suffers from easy entrapment when the parent vector x resides in a locally optimal well that is wider than the available step size. Davis (1989, 1991b) offered a similar operator (known as *creep*) that has a fixed probability of altering each component of x up or down by a bounded small random amount. The only method for alleviating entrapment in such cases relies on probabilistic selection, that is, maintaining a probability for choosing lesser-valued solutions to become parents of the subsequent generations (see Chapter 27). In contrast, unbounded mutation operators do not require such selection methods to guarantee asymptotic global convergence (Fogel 1994, Rudolph 1994).

The primary unbounded mutation PDF for real-valued vectors has been the Gaussian (or 'normal') (Rechenberg 1973, Schwefel 1981, Fogel *et al* 1990, Fogel and Atmar 1990, Bäck and Schwefel 1993, Fogel and Stayton 1994, and many others). The PDF is defined as

$$g(x) = [\sigma(2\pi)^{1/2}]^{-1} \exp[-0.5(x - \mu)^2/\sigma^2].$$

When $\mu = 0$, the parameter σ offers the single control on the scaling of the PDF. It effectively generates a typical step size for a mutation. The use of zero-mean Gaussian mutations generates offspring that are (i) on average no different from their parents and (ii) increasingly less likely to be increasingly different from their parents. Saltations are not completely avoided such that any local optimum can be escaped from in a single iteration, yet they are not so common as to lose all inheritance from parent to offspring.

Other density functions with similar characteristics have also been implemented. Yao and Liu (1996) proposed using Cauchy distributions to aid in escaping from local minima (the Cauchy distribution has a fatter tail than the Gaussian) and demonstrated that Cauchy mutations may offer some advantages across a wide testbed of problems. Montana and Davis (1989) examined the use of Laplace-distributed mutations but there is no evidence that the Laplace distribution is particularly better suited than Gaussian or Cauchy mutations for typical real-valued optimization problems.

In the simplest version of evolution strategies or evolutionary programming, described as a $(1 + 1)$ evolutionary algorithm, a single parent $\boldsymbol{x}$ creates a single offspring $\boldsymbol{x}'$ by imposing a multivariate Gaussian perturbation with mean zero and standard deviation σ on the parent, then selects the better of the two trial solutions as the parent for the next iteration. The same standard deviation is applied to each component of the vector $\boldsymbol{x}$ during mutation. For some problems, the variation of σ (i.e. the step size control parameter in each dimension) can be computed to yield an optimal rate of convergence.

Let the convergence rate be defined as the ratio of the Euclidean distance covered toward the optimum solution to the number of trials required to achieve the improvement. Rechenberg (1973) calculated the convergence rates for two functions:

$$f_1(\boldsymbol{x}) = c_0 + c_1 x_1 \qquad i \in \{2, \ldots, n\} \quad -b/2 \le x_i \le b/2$$
$$f_2(\boldsymbol{x}) = \sum x_i^2$$

where $\boldsymbol{x} = (x_1, \ldots, x_n)^{\mathrm{T}} \in \mathbb{R}^n$. Function f_1 is termed the corridor model and represents a linear function with inequality constraints. Improvement is accomplished by moving along the first axis of the search space inside a corridor of width b. Function f_2 is termed the sphere model and is a simple n-dimensional quadratic bowl.

Rechenberg (1973) showed that the optimum rates of convergence (expected progress toward the optimum) are

$$\sigma = (\pi^{1/2}/2)(b/n)$$

on the corridor model, and

$$\sigma = 1.224\|x\|/n$$

on the sphere model. That is, only a single step size control is needed for optimum convergence. Given these optimum standard deviations for mutation, the optimum probabilities of generating a successful mutation can be calculated as

$$p_1^{\text{opt}} = (2e)^{-1} \approx 0.184$$
$$p_2^{\text{opt}} = 0.270.$$

Noting the similarity of these two values, Rechenberg (1973) proposed the following rule:

The ratio of successful mutations to all mutations should be 1/5. If this ratio is greater than 1/5, increase the variance; if it is less, decrease the variance.

Schwefel (1981) suggested measuring the success probability on-line over $10n$ trials (where there are n dimensions) and adjusting σ at iteration t by

$$\sigma(t) = \begin{cases} \sigma(t-n)\delta & \text{if } p_s < 0.2 \\ \sigma(t-n)\delta & \text{if } p_s > 0.2 \\ \sigma(t-n) & \text{if } p_s = 0.2 \end{cases}$$

with $\delta = 0.85$ and p_s equaling the number of successes in $10n$ trials divided by $10n$, which yields convergence rates of geometric order for both f_1 and f_2 (Bäck et al 1993; see the book by Bäck (1996) for corrections to the update rule offered by Bäck et al (1993)).

The use of a single step size control parameter covering all dimensions simultaneously is of limited robustness. The optimization performance can be improved by using appropriate step sizes in each dimension. This is particularly evident when consideration is given to optimizing a vector of parameters each of different units of dimension (e.g. temperature and pressure). Determining appropriate settings for each of n step sizes poses a significant challenge to the human operator; as such, methods have been proposed for self-adapting the step sizes concurrent to the evolutionary search.

The first efforts in self-adaptation date back at least to the article by Reed et al (1967), but the two most common implementations in use currently derive from the work of Schwefel (1981) and Fogel et al (1991). In each case, the vector of objective variables x is accompanied by a vector strategy parameters σ where σ_i denotes the standard deviation to use when applying a zero-mean Gaussian mutation to that component in the parent vector. The strategy parameters are updated by slightly different methods according to Schwefel (1981) and Fogel et al (1991).

Schwefel (1981) offered the procedure

$$\sigma_i' = \sigma_i \exp(\tau_0 N(0, 1) + \tau N_i(0, 1))$$
$$x_i' = x_i + N(0, \sigma_i')$$

where the constant $\tau \propto 1/[2(n^{1/2})]^{1/2}$, $\tau_0 \propto 1/(2n)^{1/2}$, $N(0, 1)$ is a standard Gaussian random variable sampled once for all n dimensions and $N_i(0, 1)$ is a standard Gaussian random variable sampled anew for each of the n dimensions. The procedure offers a general control for all dimensions and an individualized control for each dimension (Schwefel (1981) also offered a simplified method for self-adapting a single step size parameter σ). The values of σ' are, as shown, log-normal perturbations of their parent's vector σ.

Fogel *et al* (1991) independently offered the procedure

$$x_i' = x_i + N(0, \sigma_i)$$
$$\sigma_i' = \sigma_i + \chi N(0, \sigma_i)$$

where the parents' strategy parameters are used to create the offspring's objective values before being mutated themselves, and the mutation of the strategy parameters is achieved using a Gaussian distribution scaled by χ and the standard deviation for each dimension. This procedure also requires incorporating a rule such that if any component σ_i' becomes negative it is reset to an arbitrary small value ϵ.

Several comparisons have been conducted between these methods. Saravanan and Fogel (1994) and Saravanan *et al* (1995) indicated that the log-normal procedure offered by Schwefel (1981) generated generally superior optimization performance (statistically significant) across a series of standard test functions. Angeline (1996a), in contrast, found that the use of Gaussian mutations on the strategy parameters generated better optimization performance when the objective function was made noisy. Gehlhaar and Fogel (1996) indicated that mutating the strategy parameters before creating the offspring objective values appears to be more generally useful both in optimizing a set of test functions and in molecular docking applications.

Both of the above methods for self-adaptation have been extended to include possible correlation across the dimensions. That is, rather than use n independent Gaussian random perturbations, a multivariate Gaussian mutation with arbitrary covariance can be applied. Schwefel (1981) described a method for incorporating rotation angles α such that new solutions are created by

$$\sigma_i' = \sigma_i \exp(\tau_0 N(0, 1) + \tau N_i(0, 1))$$
$$\alpha_j' = \alpha_j + \beta N_j(0, 1)$$
$$x_i' = x_i + N(0, \sigma_i', \alpha_j')$$

where $\beta \approx 0.0873$ (5°), $i = 1, \ldots, n$ and $j = 1, \ldots, n(n-1)/2$, although it is not necessary to include all possible pairwise correlations in the method. Fogel *et al* (1992) offered a similar method operating directly on the components of the covariance matrix but the method does not guarantee positive definite matrices for $n > 2$, and the conventional method for implementing correlated mutation relies on the use of rotation angles as described above.

Another type of zero-mean mutation found in the literature is the so-called *nonuniform mutation* of Michalewicz (1996, pp 111–2), where

$$x_i'(t) = \begin{cases} x_i(t) + \Delta(t, \text{ub}_i - x_i(t)) & \text{if } u < 0.5 \\ x_i(t) - \Delta(t, x_i(t) - \text{lb}_i) & \text{if } u \geq 0.5 \end{cases}$$

where $x_i(t)$ is the ith parameter of the vector x at generation t, $x_i \in [\text{lb}_i, \text{ub}_i]$, the lower and upper bounds, respectively, u is a random uniform $U(0, 1)$, and the function $\Delta(t, y)$ returns a value in the range $[0, y]$ such that the probability of $\Delta(t, y)$ being close to zero increases as t increases, essentially taking smaller steps on average. Michalewicz *et al* (1994) used the function

$$\Delta(t, y) = yu(1 - t/T)^b$$

where T is a maximal generation number and b is a system parameter chosen by the operator to determine the degree of nonuniformity.

There have been recent attempts to use nonzero-mean mutations on real-valued vectors. Ostermeier (1992) proposed an evolution strategy where the Gaussian mutations applied to the objective vector x are controlled by a vector of expectations μ as well as a vector of standard deviations σ. Ghozeil and Fogel (1996), following earlier work by Bremermann and Rogson (1964), have implemented a polar coordinate mutation in which new offspring are generated by perturbing the parent in a random direction (θ) with a specified step size (r).

32.3   Permutations

Darrell Whitley

32.3.1   Introduction

Mutation operators can be used in a number of ways. Random mutation hillclimbing (Forrest and Mitchell 1993) is a search algorithm which applies a mutation operator to a single string and accepts any improving moves. Some forms of evolutionary algorithms apply mutation operators to a population of strings without using recombination, while other algorithms may combine the use of mutation with recombination.

Any form of mutation which is to be applied to a permutation must yield a string which also represents a permutation. Most mutation operators for permutations are related to operators which have also been used in neighborhood local search strategies. Many of these operators thus can be applied in such as way that they reach a well-defined neighborhood of adjacent states.

32.3.2   2-opt, 3-opt, and k-opt

The most common form of mutation is 2-opt (Lin and Kernighan 1973). Given a sequence of elements

$$A \; B \; C \; D \; E \; F \; G \; H$$

the 2-opt operator selects two points along the string, then reverses the segment between the points. Note that if the permutation is viewed as a circuit as in the traveling salesman problem (TSP), then all shifts of a sequence of N elements are equivalent. It follows that once two cut points have been selected in this circular string, it does not matter which segment is reversed; the effect is the same.

The 2-opt operator can be applied to all pairs of edges in $N(N-1)/2$ steps. This is analogous to one iteration of local search over all variables in a parameter optimization problem. If a full iteration of 2-opt to all pairs of edges fails to find an improving move, then a local optimum has been reached.

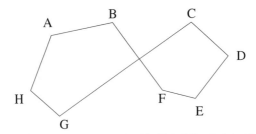

Figure 32.1. A graph.

2-opt is classically associated with the Euclidean TSP. Consider the graph in figure 32.1. If this is interpreted as a Euclidean TSP, then reversing the segment [C D E F] or the segment [G H A B] results in a graph where none of the edges cross and which has lower cost than the graph where the edges cross. Let {A, B, ..., Z} be a set of vertices and (a, b) be the edge between vertices A and B. If vertices {B, C, F, G} in figure 32.1 are connected by the set of edges ((b, c), (b, f), (b, g), (c, f), (c, g) (f, g)), then two triangles are formed when B is connected to F and C is connected to G. To illustrate, create a new graph by placing a new vertex X at the point where the edges (b, f) and (c, g) cross. In the new graph in Euclidean space, the distance represented by edge (b, c) must be less than edges (b, x) + (x, c), assuming B, C, and X are not on a line; likewise, the distance represented by edge (f, g) must be less than edge (f, x) + (x, g). Thus, reversing the segment [C D E F] will always reduce the cost of the tour due to this triangle inequality. For the TSP this leads to the general principle that multiple applications of 2-opt will always yield a tour that has no crossed edges.

One can also look at reversing more than two segments at a time. The 3-opt operator cuts the permutation into three segments and then looks at all possible ways of reordering these segments. There are 3! = 6 ways to order the segments and each segment can be placed in a forward or reverse order. This yields up to $2^3 * 6 = 48$ possible new reorderings of the original permutation. For the symmetric TSP, however, all shifted arrangements of the three segments are equal and all reversed arrangements of the three segments are equal. Thus, the 3! orderings are all equivalent. (By analogy, note that there is only one possible Hamiltonian circuit tour between three cities.) This leaves only $2^3 = 8$ ways of placing each of the segments in a forward or reverse direction, each of which yields a unique tour. Thus, for the symmetric TSP, the cost to test one 3-opt move is eight times greater than the cost of testing one 2-opt move. For other types of scheduling problem, such as resource allocation, reversals and shifts of the complete permutation are not necessarily equivalent and the cost of a 3-opt move may be up to 48 times greater than that of a 2-opt move. Also note that there are $\binom{N}{3}$ ways to break a permutation up into combinations of three segments compared to $\binom{N}{2}$ ways of breaking the permutation into two segments. Thus, the set of all possible 3-opt moves is much larger than the set of possible 2-opt moves. This further increases the cost of performing one pass of 3-opt over all possible ways of partitioning a permutation into three segments compared to a pass of 2-opt over all pairs of possible segments.

One can also use k-opt, where the permutation is broken into k segments, but such an operator will obviously be very costly.

32.3.3   Insert, swap, and scramble operators

The TSP is sensitive to the adjacency of elements in a permutation, so that 2-opt represents a minimal change from one Hamiltonian circuit to another. For resource scheduling applications the permutation represent a priority queue and reversing a segment of a permutation represents a major change in access to available resources. For example, think of the permutation as representing a line of people waiting to buy a limited supply of tickets for different seats on different trains. The relative order of elements in the permutation tends to be important in this case and not the adjacency of the individual elements. In this case, a 2-opt segment reversal impacts many customers and is far from a minor change.

Radcliffe and Surry (1995) argue for representation-independent concepts of mutation and related forms of hillclimbers. Concerning desirable properties of a mutation operator, they state, 'One nearly universal characteristic, however, is that they ensure ... that the entire search space remains accessible from any population, and indeed from any individual. In most case mutation operators can actually move from any point in the search space to any other point directly, but the probability of making "large" moves is very much smaller than that of making "small" moves (at least with small mutation rates)' (p 58). They also suggest that a single mutation should represent a minimal change and look at different types of mutation operator for different representations of the TSP.

For resource allocation problems, a more modest change than 2-opt is to merely select one element and to *insert* it at some other position in the permutation. Syswerda (1991) refers to a variant of this as *position-based mutation* and describes it as selecting two elements and then moving the second element before the first element. Position-based mutation appears to be less general than the *insert* operator, since elements can only be moved forward in position-based mutation.

Similarly, one can select two elements and *swap* the positions of the two elements. Syswerda denotes this as *order-based mutation*. Note that if an element is moved forward or backward one position, this is equivalent to a swap of adjacent elements. One way in which swap can be used as a local search operator is to swap all adjacent elements, or perhaps also all pairs of elements. Finally, Syswerda also defines a *scramble* mutation operator that selects a sublist of permutation elements and randomly reorders (i.e. scrambles) the order of the subset while leaving the other elements in the permutation in the same absolute position. Davis (1991a) also reports on a scramble sublist mutation operator, except that the sublist is explicitly composed of contiguous elements of a permutation. (It is unclear whether Syswerda's scramble operator is also meant to work on contiguous elements or not; an operator that selects a sublist of elements over random positions of the permutation is certainly possible.)

For a problem that involved scheduling a limited number of flight simulators, Syswerda (1991, p 342) reported that when applied individually, the order-based swap mutation operator yielded the best results when compared to position-based mutation and scramble mutation. In this case the swaps were selected randomly rather than being performed over a fixed well-defined neighborhood. Davis (1991, p 81) on the other hand reports that the scramble sublist mutation operator proved to be better than the swap operator on a number of applications.

In conclusion, one cannot make *a priori* statements about the usefulness of a particular mutation operator without knowing something about the type of problem that is to be solved and the representation that is being used for that problem, but in general it is useful to distinguish between permutation problems that are sensitive to adjacency (e.g. the TSP) versus relative order (e.g. resource scheduling) or absolute position, which appears to be the least common.

32.4   Finite-state machines

David B Fogel

Given a finite-state machine representation (Chapter 18) where each element in a population is defined by a 5-tuple

$$M = (Q, T, P, s, o)$$

where Q is a finite set, the set of states, T is a finite set, the set of input symbols, P is a finite set, the set of output symbols, $s : Q \times T \to Q$, the next state function, and $o : Q \times T \to P$, the next output function,
there are various methods for mutating parents to create offspring. Following directly from the definition, five obvious modes of mutation present themselves: (i) change an output symbol, (ii) change a state transition, (iii) add a new state, (iv) delete a state, and (v) change the start state. Each of these will be discussed in turn.

(i)   Changing an output symbol consists of determining a particular state $q \in Q$, and then determining a particular symbol $\tau \in T$. For this pair (q, τ), identify the associated output symbol $\rho \in P$ and change it to a symbol chosen at random over the set P. The probability mass function for selecting a new symbol is typically uniform over the possible symbols in P, but can be chosen to reflect nearness between symbols or other known relationships between the symbols.

(ii) Changing a state transition consists of determining a particular state $q_1 \in Q$, and then determining a particular symbol $\tau \in T$. For this pair (q_1, τ), identify the associated next state q_2 and change it to a state chosen at random over the set Q. The probability mass function for selecting a new symbol is typically uniform over the possible states in Q.

(iii) Adding a state can only be performed when the maximum size of the machine has not been exceeded. The operation is accomplished by increasing the set Q by one element. This new state must be properly defined by generating an associated output symbol ρ_i and next state transition q_i for all input symbols $i = 1, \ldots, |T|$. The generation is typically performed by selecting output symbols and next state transitions with equal probability across their respective sets. Optionally, the new state may also be forced to be connected to the preexisting states by redirecting a randomly selected state transition of a randomly chosen preexisting state to the new state.

(iv) Deleting a state can be performed when the machine has at least two states. The operation is accomplished by decreasing the set Q by one element chosen at random (uniformly). All state transitions from other states that point to the deleted state must be redirected to the remaining states. This is often performed at random, with the new states selected with equal probability.

(v)   Changing the start state can be performed when the machine has at least two states. The operation is accomplished by selecting a state $q \in Q$ to be the new starting state. Again, the selection is typically made uniformly over the available states.

The mutation operation can be implemented with various probabilities assigned to each mode of mutation (Fogel and Fogel 1986), although many of the initial experiments in evolutionary programming used equal probabilities

(Fogel *et al* 1966). Further, multiple mutations can be performed (see e.g. Fogel *et al* 1966), and macromutations can be defined over pairs or higher-order combinations of these primitive operations. Recent efforts by Fogel *et al* (1994, 1995) and Angeline *et al* (1996) have incorporated the use of self-adaptation in mutating finite-state machines.

32.5 Parse trees

Peter J Angeline

Standard genetic programming (Koza 1992), much as with traditional genetic algorithms, discounts mutation's role during evolution, often to an extreme (i.e. a mutation rate of zero). In many genetic programs, no mutation operations are used, which forces population sizes to be quite large in order to ensure access to all the primitives in the primitive language throughout a run.

In order to avoid unnecessarily large population sizes, Angeline (1996b) defines four distinct forms of mutation for parse trees (Chapter 19). The *grow* mutation operator randomly selects a leaf from the tree and replaces it with a randomly generated new subtree (figure 32.2). The *shrink* mutation operator selects an internal node from the tree and replaces the subtree below it with a randomly generated leaf node (figure 32.3). The *switch* mutation operator selects an internal node from the parse tree and reorders its argument subtrees (figure 32.4). Finally, the *cycle* mutation operator selects a random node and replaces it with a new node of the same type (figure 32.5). If a leaf node is selected, then it is replaced by a leaf node. If an internal node is selected, then it is replaced by a function primitive that takes an equivalent number of arguments. Note that the mutation operation defined by Koza (1992) is a combination of a shrink mutation followed by a grow mutation at the same position.

Angeline (1996b) also defines a numerical terminal mutation that manipulates numerical terminals in a parse tree using the Gaussian mutations typically used in evolution strategies and evolutionary programming (see also Bäck 1996, Fogel 1995). In this mutation operation, a single numerical terminal in the parse tree is selected at random and a Gaussian random variable with a user-defined variance is added to its value.

If the application of a mutation operation creates a parse tree that violates the size limitation criteria for the parse tree, typically the operation is revoked and the state of the parse tree prior to the operation is restored. In some cases, when a series of mutations are to be performed, as in Angeline (1996b), the complete set of mutations is executed prior to checking whether the mutated parse tree conforms to the imposed size restrictions.

When evolving typed parse trees as in Montana (1995), mutation must also be sensitive to the return type of the manipulated node. In order to preserve the syntactic constraints, the return type of the node after mutation must be the same. This is accomplished by keeping track of the return types for the various

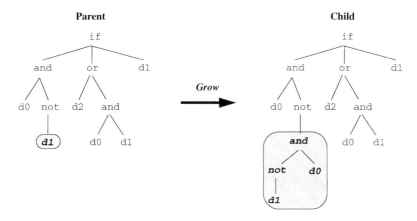

Figure 32.2. An illustration of the grow mutation operator applied to a Boolean parse tree. Given a parent tree to mutate, a terminal node is selected at random (highlighted) and replaced by a randomly generated subtree to produce the child tree.

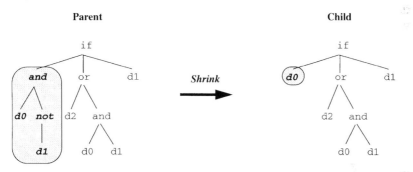

Figure 32.3. An illustration of the shrink mutation operator applied to a Boolean parse tree. Given a parent tree to mutate, an internal function node is selected at random (highlighted) and replaced by a randomly selected terminal to produce the child tree.

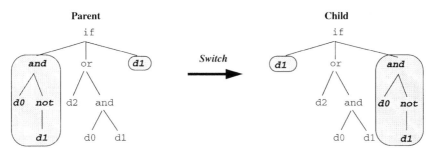

Figure 32.4. An illustration of the switch mutation operator applied to a Boolean parse tree. Given a parent tree to mutate, an internal function node is selected, two of the subtrees below it are selected (highlighted in the figure) and their positions switched to produce the child tree.

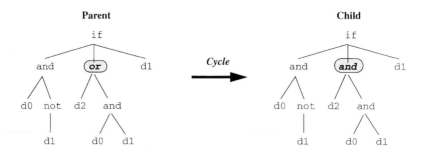

Figure 32.5. An illustration of the cycle mutation operator applied to a Boolean parse tree. Given a parent tree to mutate, a single node, either a terminal or function, is selected at random (highlighted in the parent) and replaced by a randomly selected node with the same number of arguments to produce the child tree.

primitives in the language and restricting mutation to return those primitives with the corresponding type.

32.6   Other representations

David B Fogel and Peter J Angeline

Many real-world applications suggest the use of representations that are hybrids of the canonical representations. One common instance is the simultaneous use of discrete and continuous object variables, with a general formulation of the global optimization problem as follows (Bäck and Schütz 1995):

$$\min\{f(x, d)|x \in M, R^n \supseteq M, d \in N, Z^{n_d} \supseteq N\}.$$

Within evolution strategies and evolutionary programming, the common representation is simply the real-integer vector pair (i.e. no effort is made to encode these vectors into another representation such as binary).

The simple approach to mutating such a representation would be to embed the integers in the real numbers and use the standard methods of mutation (e.g. Gaussian random perturbation) found in evolution strategies and evolutionary programming. The results could be rounded to the integers when dealing with the elements in d. Bäck and Schütz (1995) note, however, that, for a discrete optimization problem, the 'optimum point obtained by rounding the results of the continuous optimization might be different from the true discrete optimum point even for linear objective functions with linear constraints'. Bäck and Schütz (1995) also note the potential problems in optimizing x and d separately (as in the work of Lohmann (1992) and Fogel (1991, 1993) among others) because there may be interdependences between the appropriate mutations to x and d.

Bäck and Schütz (1995) approach the general problem by including a vector of mutation strategy parameters $p_j \in (0, 1)$ and $j = 1, 2, \ldots, d$, where there are d components to the vector d. (Alternatively, fewer strategy parameters could be used.) These strategy parameters are adapted along with the usual step size control strategy parameters for Gaussian mutation of the real-world vector x. The discrete strategy parameters are updated by the formula

$$p'_j = \left(\frac{1 + (1 - p_j)}{p_j \times \exp[-\gamma N_j(0, 1)]} \right)^{-1}$$

where γ is set proportional to $[2(d)^{1/2}]^{-1/2}$. Actual mutation to the parameters in d can be accomplished using an appropriate random variable (e.g. uniform or Poisson).

With regard to mutation in introns, because the introns are not coded into functional behavior (i.e. they do not affect performance in terms of the objective function), the manner in which they are mutated is irrelevant.

In the standard genetic algorithm representation, the semantics of an allele value (how the allele is interpreted) are typically tied to its position in the fixed-length n-ary string. For instance, in a binary string representation, each position signifies the presence or absence of a specific feature in the genome being decoded. The difficulty with such a representation is that with positions in the string representation that are semantically linked, but separated by a large number of intervening positions in the string, crossover has a high probability of disrupting beneficial settings for these two positions. Goldberg *et al* (1989) describe a representation for a genetic algorithm that embodies one approach to addressing this problem. In their messy genetic algorithm (mGA), each allele value is represented as a pair of values, one specifying the actual allele value and one specifying the position the allele occupies. Messy GAs are defined to be of variable length, and Goldberg *et al* (1989) describe appropriate methods for resolving underdetermined or overdetermined genomes. In this representation it is important to note that the semantics are literally carried along with the allele value in the form of the allele's string position.

Diplodic representations, representations that include multiple allele values for each position in the genome, have been offered as mechanisms for modeling cyclic environments. In a diplodic representation, a method for determining which allele value for a gene will be expressed is required to adjudicate when the allele values do not agree. Building on earlier investigations (e.g. Bagley 1967, Hollstein 1971, Brindle 1981), Goldberg and Smith (1987) demonstrate that an evolving dominance map allows quicker adaptation to cyclical environment changes than either a haploid representation or a diploid representation using a fixed dominance mapping. In the article by Goldberg and Smith (1987), a triallelic representation from the dissertation of Hollstein (1971) is used: 1, i, and 0. Both 1 and i map to the allele value of '1', while 0 maps to the allele value of '0' with 1 dominating both i and 0 and 0 dominating i. Thus, the dominance of a 1 over a 0 allele value could be altered via mutation by altering the value to an i. Ng and Wong (1995) extend the multiallele approach to dominance computation by adding a fourth value for a recessive 0. Thus 1 dominates 0 and o while 0 dominates i and o. When both allele values for a gene are dominant or recessive, then one of the two values is chosen randomly to be the dominant value. Ng and Wong (1995) also suggest that the dominance of all of the components in the genome should be reversed when the fitness value of an individual falls by 20% or more between generations.

References

Angeline P J 1996a The effects of noise on self-adaptive evolutionary optimization, *Proc. 5th Ann. Conf. on Evolutionary Programming* ed L J Fogel, P J Angeline and T Bäck (Cambridge, MA: MIT Press) pp 443–50
——1996b Genetic programming's continued evolution *Advances in Genetic Programming* vol 2, ed P J Angeline and K Kinnear (Cambridge, MA: MIT Press) pp 89–110
Angeline P J, Fogel D B and Fogel L J 1996 A comparison of self-adaptation methods for finite state machines in a dynamic environment *Proc. 5th Ann. Conf. on Evolutionary Programming* ed L J Fogel, P J Angeline and T Bäck (Cambridge, MA: MIT Press) pp 441–50
Bäck T 1993 Optimal mutation rates in genetic search *Proc. 5th Int. Conf. on Genetic Algorithms (Urbana-Champaign, IL, July 1993)* ed S Forrest (San Mateo, CA: Morgan Kaufmann) pp 2–8
——1996 *Evolutionary Algorithms in Theory and Practice* (New York: Oxford University Press)
Bäck T, Rudolph G and Schwefel H-P 1993 Evolutionary programming and evolution strategies: similarities and differences *Proc. 2nd Ann. Conf. on Evolutionary Programming (San Diego, CA)* ed D B Fogel and W Atmar (La Jolla, CA: Evolutionary Programming Society) pp 11–22
Bäck T and Schütz M 1995 Evolution strategies for mixed-integer optimization of optical multilayer systems *Proc. 4th Ann. Conf. on Evolutionary Programming (San Diego, CA, March 1995)* ed J R McDonnell, R G Reynolds and D B Fogel (Cambridge, MA: MIT Press) pp 33–51
Bäck T and Schwefel H-P 1993 An overview of evolutionary algorithms for parameter optimization *Evolutionary Comput.* **1** 1–24
Bagley J D 1967 *The Behavior of Adaptive Systems which Employ Genetic and Correlation Algorithms* Doctoral Dissertation, University of Michigan; University Microfilms 68-7556
Bremermann H J 1962 Optimization through evolution and recombination *Self-Organizing Systems* ed M C Yovits, G T Jacobi and G D Goldstine (Washington, DC: Spartan) pp 93–106
Bremermann H J and Rogson M 1964 *An Evolution-type Search Method for Convex Sets* ONR Technical Report, contracts 222(85) and 3656(58)
Bremermann H J, Rogson M and Salaff S 1965 Search by evolution *Biophysics and Cybernetic Systems* ed M Maxfield, A Callahan and L J Fogel (Washington, DC: Spartan) pp 157–67

——1966 Global properties of evolution processes *Natural Automata and Useful Simulations* ed H H Pattec, E A Edelsack, L Fein and A B Callahan (Washington, DC: Spartan) pp 3–41

Brindle A 1981 *Genetic Algorithms for Function Optimization* Doctoral Dissertation, University of Alberta

Davis L 1989 Adapting operator probabilities in genetic algorithms *Proc. 3rd Int. Conf. on Genetic Algorithms (Fairfax, VA, June 1989)* ed J D Schaffer (San Mateo, CA: Morgan Kaufmann) pp 61–9

Davis L 1991a *Handbook of Genetic Algorithms* (New York: Van Nostrand Reinhold)

——1991b A genetic algorithms tutorial *Handbook of Genetic Algorithms* ed L Davis (New York: Van Nostrand Reinhold) pp 1–101

De Jong K A 1975 *An Analysis of the Behaviour of a Class of Genetic Adaptive Systems* PhD Thesis, University of Michigan

Fogarty T C 1989 Varying the probability of mutation in the genetic algorithm *Proc. 3rd Int. Conf. on Genetic Algorithms (Fairfax, VA, 1989)* ed J D Schaffer (San Mateo, CA: Morgan Kaufmann) pp 104–9

Fogel D B 1991 *System Identification through Simulated Evolution* (Needham, MA: Ginn)

——1993 Using evolutionary programming to construct neural networks that are capable of playing tic-tac-toe *Proc. 1993 IEEE Int. Conf. on Neural Networks* (Piscataway, NJ: IEEE) pp 875–80

——1994 Asymptotic convergence properties of genetic algorithms and evolutionary programming: analysis and experiments *Cybern. Syst.* **25** 389–407

——1995 *Evolutionary Computation: Toward a New Philosophy of Machine Intelligence* (New York: IEEE)

Fogel D B and Atmar J W 1990 Comparing genetic operators with Gaussian mutations in simulated evolutionary processes using linear systems *Biol. Cybern.* **63** 111–4

Fogel D B, Fogel L J and Atmar J W 1991 Meta-evolutionary programming *Proc. 25th Asilomar Conf. on Signals, Systems, and Computers* ed R R Chen (Pacific Grove, CA: Maple) pp 540–5

Fogel D B, Fogel L J, Atmar W and Fogel G B 1992 Hierarchic methods of evolutionary programming *Proc. 1st Ann. Conf. on Evolutionary Programming* ed D B Fogel and W Atmar (La Jolla, CA: Evolutionary Programming Society) pp 175–82

Fogel D B and Stayton L C 1994 On the effectiveness of crossover in simulated evolutionary optimization *BioSystems* **32** 171–82

Fogel L J, Angeline P J and Fogel D B 1994 A preliminary investigation on extending evolutionary programming to include self-adaptation on finite state machines *Informatica* **18** 387–98

——1995 An evolutionary programming approach to self-adaptation in finite state machines *Proc. 4th Ann. Conf. on Evolutionary Programming (San Diego, CA, March, 1995)* ed J R McDonnell, R G Reynolds and D B Fogel (Cambridge, MA: MIT Press) pp 355–65

Fogel L J and Fogel D B 1986 *Artificial Intelligence through Evolutionary Programming* Final Report for US Army Research Institute, contract no PO-9-X56-1102C-1

Fogel L J, Owens A J and Walsh M J 1966 *Artificial Intelligence Through Simulated Evolution* (New York: Wiley)

Forrest S and Mitchell M 1993 Relative building-block fitness and the building block hypothesis *Foundations of Genetic Algorithms 2* ed D Whitley (San Mateo, CA: Morgan Kaufmann) pp 109–26

Gehlhaar D K and Fogel D B 1996 Tuning evolutionary programming for conformationally flexible molecular docking *Proc. 5th Ann. Conf. on Evolutionary Programming* ed L J Fogel, P J Angeline and T Bäck (Cambridge, MA: MIT Press) at press

Ghozeil A and Fogel D B 1996 A preliminary investigation into directed mutations in evolutionary algorithms *Parallel Problem Solving from Nature 4* to appear

Goldberg D E 1989 *Genetic Algorithms in Search, Optimization and Machine Learning* (Reading, MA: Addison-Wesley)

Goldberg D E, Korb D E and Deb K 1989 Messy genetic algorithms: motivation, analysis, and first results *Complex Syst.* **3** 493–530

Goldberg D E and Smith R E 1987 Nonstationary function optimization using genetic algorithms with dominance and diploidy *Proc. 2nd Int. Conf. on Genetic Algorithms (Cambridge, MA, 1987)* ed J J Grefenstette (Hillsdale, NJ: Erlbaum) pp 59–68

Grefenstette J J 1986 Optimization of control parameters for genetic algorithms *IEEE Trans. Syst. Man Cybernet.* **SMC-16** 122–8

Hesser J and R Männer 1991 Towards an optimal mutation probability in genetic algorithms *Proc. 1st Conf. on Parallel Problem Solving from Nature (Dortmund, 1990) (Lecture Notes in Computer Science 496)* ed H-P Schwefel and R Männer (Berlin: Springer) pp 23–32

——1992 Investigation of the m-heuristic for optimal mutation probabilities *Parallel Problem Solving from Nature, 2 (Proc. 2nd Int. Conf. on Parallel Problem Solving from Nature, Brussels, 1992)* ed R Männer and B Manderick (Amsterdam: Elsevier) pp 115–24

Holland J H 1975 *Adaptation in Natural and Artificial Systems* (Ann Arbor, MI: University of Michigan Press)

Hollstein R B 1971 *Artificial Genetic Adaptation in Computer Control Systems* Doctoral Dissertation, University of Michigan; University Microfilms 71-23, 773

Koza J R 1992 *Genetic Programming: On the Programming of Computers by Means of Natural Selection* (Cambridge, MA: MIT Press)

Lin S and Kernighan B 1973 An efficient heuristic procedure for the traveling salesman problem *Operations Res.* **21** 498–516

Lohmann, R 1992 Structure evolution in neural systems *Dynamic, Genetic, and Chaotic Programming* ed B Soucek (New York: Wiley) pp 395–411

Michalewicz Z 1996 *Genetic Algorithms + Data Structures = Evolution Programs* 3rd edn (Berlin: Springer)

Michalewicz Z, Logan T and Swaminathan S 1994 Evolutionary operators for continuous convex parameter spaces *Proc. 3rd Ann. Conf. on Evolutionary Programming (San Diego, CA, February 1994)* ed A V Sebald and L J Fogel (Singapore: World Scientific) pp 84–97

Montana D J 1995 Strongly typed genetic programming *Evolutionary Comput.* **3** 199–230

Montana D J and Davis L 1989 Training feedforward neural networks using genetic algorithms *Proc. 11th Int. Joint Conf. on Artificial Intelligence* ed N S Sridharan (San Mateo, CA: Morgan Kaufmann) pp 762–7

Mühlenbein H 1992 How genetic algorithms really work: I mutation and hillclimbing *Parallel Problem Solving from Nature, 2 (Proc. 2nd Int. Conf. on Parallel Problem Solving from Nature, Brussels, 1992)* ed R Männer and B Manderick pp 15–25

Ng K P and Wong K C 1995 A new diploid scheme and dominance change mechanism for non-stationary function optimization *Proc. 6th Int. Conf. on Genetic Algorithms*

(Pittsburgh, PA, July 1995) ed L J Eshelman (San Mateo, CA: Morgan Kaufmann) pp 159–66

Ostermeier A 1992 An evolution strategy with momentum adaptation of the random number distribution *Parallel Problem Solving from Nature, 2 (Proc. 2nd Int. Conf. on Parallel Problem Solving from Nature, Brussels, 1992)* ed R Männer and B Manderick pp 197–206

Radcliffe N and Surry P D 1995 Fitness variance of formae and performance prediction *Foundations of Genetic Algorithms 3* ed D Whitley and M Vose (San Mateo, CA: Morgan Kaufmann) pp 51–72

Rechenberg I 1973 *Evolutionsstrategie: Optimierung technischer Systeme nach Prinzipien der biologischen Evolution* (Stuttgart: Frommann-Holzboog)

Reed J, Toombs R and Barricelli N A 1967 Simulation of biological evolution and machine learning *J. Theor. Biol.* **17** 319–42

Rudolph G 1994 Convergence properties of canonical genetic algorithms *IEEE Trans. Neural Networks* **5** 96–101

Saravanan N and Fogel D B 1994 Learning strategy parameters in evolutionary programming: an empirical study *Proc. 3rd Ann. Conf. on Evolutionary Programming (San Diego, CA, February 1994)* ed A V Sebald and L J Fogel (Singapore: World Scientific) pp 269–80

Saravanan N, Fogel D B and Nelson K M 1995 A comparison of methods for self-adaptation in evolutionary algorithms *BioSystems* **36** 157–66

Schwefel H-P 1981 *Numerical Optimization of Computer Models* (Chichester: Wiley)
——1995 *Evolution and Optimum Seeking* (New York: Wiley)

Schaffer J D, Caruana R A, Eshelman L J and Das R 1989 A study of control parameters affecting online performance of genetic algorithms for function optimization *Proc. 3rd Int. Conf. on Genetic Algorithms (Fairfax, VA, June 1989)* ed J D Schaffer (San Mateo, CA: Morgan Kaufmann) pp 51–60

Syswerda G 1991 Schedule optimization using genetic algorithms *Handbook of Genetic Algorithms* ed L Davis (New York: Van Nostrand Reinhold) pp 332–49

Yao X and Liu Y 1996 Fast evolutionary programming *Proc. 5th Ann. Conf. on Evolutionary Programming* ed L J Fogel, P J Angeline and T Bäck (Cambridge, MA: MIT Press) at press

33

Recombination

Lashon B Booker (33.1), *David B Fogel* (33.2, 33.4, 33.6),
Darrell Whitley (33.3), *Peter J Angeline* (33.5, 33.6)
and A E Eiben (33.7)

33.1 Binary strings

Lashon B Booker

33.1.1 Introduction

In biological systems (see section 5.4), crossing-over is a complex process
that occurs between pairs of chromosomes. Two chromosomes are physically
aligned, breakage occurs at one or more corresponding locations on each
chromosome, and homologous chromosome fragments are exchanged before
the breaks are repaired. This results in a recombination of genetic material
that contributes to variability in the population. In evolutionary algorithms, this
process has been abstracted into syntactic crossing-over (or crossover) operators
that exchange substrings between chromosomes represented as linear strings
of symbols. In this section we describe various approaches to implementing
these computational recombination techniques. Note that, while binary strings
(Chapter 15) are the canonical representation of chromosomes most often
associated with evolutionary algorithms, crossover operators work the same
way on all linear strings regardless of the cardinality of the symbol alphabet.
Accordingly, the discussion in this section applies to both binary and nonbinary
string representations. The obvious caveat is that the syntactic manipulations by
crossover must yield semantically valid results. When this becomes a problem—
for example, when the chromosomes represent permutations (see Chapter 17)—
then other syntactic operations must be used.

33.1.2 Principal mechanisms

The basic crossover operation, introduced by Holland (1975), is a three-step
procedure. First, two individuals are chosen at random from the population

of 'parent' strings generated by the selection operator (see Chapters 22–30). Second, one or more string locations are chosen as breakpoints (or *crossover points*) delineating the string segments to exchange. Finally, parent string segments are exchanged and then combined to produce two resultant 'offspring' individuals. The proportion of parent strings undergoing crossover during a generation is controlled by the *crossover rate*, $p_c \in [0, 1]$, which determines how frequently the crossover operator is invoked. Holland illustrates how to implement this general procedure by describing the simple one-point crossover operator. Given parent strings x and y, a crossover point is selected by randomly choosing an integer $k \sim U(1, \ell - 1)$:

$$
\begin{array}{l}
(x_1 \ldots x_k x_{k+1} \ldots x_\ell) \\
(y_1 \ldots y_k y_{k+1} \ldots y_\ell)
\end{array}
\implies
\begin{array}{l}
(x_1 \ldots x_k y_{k+1} \ldots y_\ell) \\
(y_1 \ldots y_k x_{k+1} \ldots x_\ell).
\end{array}
$$

Two new resultant strings are formed by exchanging the parent substrings to the right of position k. Holland points out that when the overall algorithm is limited to producing only one new individual per generation, one of the resultant strings generated by this crossover operator must be discarded. The discarded string is usually chosen at random.

Holland's general procedure defines a family of operators that can be described more formally as follows. Given a space I of individual strings, a crossover operator is a mapping

$$ r : I \times I \xrightarrow{m} I \times I \qquad r(a, b) = (c, d) $$

where $m \in \mathbf{B}^\ell$ and

$$
c_i = \begin{cases} a_i & \text{if } m_i = 0 \\ b_i & \text{if } m_i = 1 \end{cases}
\qquad
d_i = \begin{cases} b_i & \text{if } m_i = 0 \\ a_i & \text{if } m_i = 1. \end{cases}
$$

Although this formal description characterizes crossover as a binary operator, there are some implementations of crossover involving more than two parents (e.g. the multiparent uniform crossover operator described by Furuya and Haftka (1993) and the scanning crossover and diagonal crossover operators described by Eiben *et al* (1995)).

The binary string m is a *mask* computed for each invocation of the operator from the set of crossover points. This mask identifies which string segments will be exchanged during the crossover operation. Note that the mask m and its complement $1 - m = (1 - m_1 \ldots 1 - m_\ell)$ generate the same (unordered) set of resultant strings. Another way to interpret the mask is as a specification of which parent provided the symbol at each position in a resultant string. A crossover operation can be viewed as the simultaneous occurrence of two *recombination events*, each producing one of the two offspring. The pair $(m, 1 - m)$ can be used to designate these recombination events. Each symbol in a resultant string is either transmitted by the first parent (denoted in the mask by zero) or the second parent (denoted by one). Consequently, the event generating string c above is specified by m and the event generating d is specified by $1 - m$.

A simple pseudocode for implementing one of these crossover operators is:

```
crossover(a, b) :
      sample u ∈ U(0, 1)
      if (u > pc)
      then return(a, b)
      fi
      c := a;
      d := b;
      m := compute_mask();
      for i := 1 to ℓ do
          if (mi = 1)
          then
                ci := bi;
                di := ai;
          fi
      od
      return(c, d);
```

Empirical studies have shown that the best setting for the crossover rate p_c depends on the choices made regarding other aspects of the overall algorithm, such as the settings for other parameters such as population size and mutation rate, and the selection operator used. Some commonly used crossover rates are $p_c = 0.6$ (De Jong 1975), $p_c \in [0.45, 0.95]$ (Grefenstette 1986), and $p_c \in [0.75, 0.95]$ (Schaffer *et al* 1989). Techniques for adaptively modifying the crossover rate have also proven to be useful (Booker 1987, Davis 1989, Srinivas and Patnaik 1994, Julstrom 1995). The pseudocode shown above makes it clear that the differences between crossover operators are most likely to be found in the implementation of the compute_mask() procedure. The following examples of pseudocode characterize the way compute_mask() is implemented for the most commonly cited crossover operators.

One-point crossover. A single crossover point is selected. This operator can only exchange contiguous substrings that begin or end at the endpoints of the chromosome. This is rarely used in practice.

```
      sample u ∈ U(1, ℓ − 1)
      m := 0;
      for i := u + 1 to ℓ do
          mi = 1;
      od
      return m;
```

n-point crossover. This operator, first implemented by De Jong (1975), generalizes one-point crossover by making the number of crossover points a parameter. The value $n = 2$ designating two-point crossover is the choice that minimizes disruptive effects (see the discussion of disruption in Section 33.1.3) and is frequently used in applications. There is no consensus about the advantages and disadvantages of using values $n \geq 3$. Empirical studies on this issue (De Jong 1975, Eshelman *et al* 1989) are inconclusive.

```
sample u₁, . . . , uₙ ∈ U(1, ℓ), u₁ ≤ · · · ≤ uₙ
if ((n mod 2) = 1)
then uₙ₊₁ := ℓ;
fi
m := 0;
for j := 1 to n step 2 do
     for i := uⱼ + 1 to uⱼ₊₁ do
          mᵢ = 1;
     od
od
return m;
```

By convention (De Jong 1975), when n is odd an additional crossover point is assumed to occur at position ℓ. Note that many implementations select the crossover points without replacement—instead of with replacement as indicated here—to guarantee that the crossover points are distinct. Analysis of disruptive effects has shown that there are only small differences in the two approaches (see the discussion of disruption in Section 33.1.3) and no empirical differences in performance have been reported.

Uniform crossover. This is an operator introduced by Ackley (1987a) but most often attributed to Syswerda (1989). (The basic idea can be traced to early work in mathematical population genetics, see Geiringer (1944)). The number of crossover points is not fixed in advance. Instead, the decision to insert a breakpoint is made independently at each string position. This operator is frequently used in applications.

```
m := 0;
for i := 1 to ℓ do
     sample u ∈ U(0, 1)
     if (u ≤ pₓ)
     then mᵢ = 1;
     fi
od
return m
```

The value $p_x = 0.5$ first used by Ackley remains the standard setting for the crossover probability at each position, though it may be advantageous to use smaller values (Spears and De Jong 1991b). When $p_x = 0.5$, every binary string of length ℓ is equally likely to be generated as a mask. In this case, it is often more efficient to implement the operator by using a random integer sampled from $U(0, 2^\ell - 1)$ as the mask instead of constructing the mask one bit at a time.

Punctuated crossover. Rather than computing the crossover mask directly, Schaffer and Morishima (1987) used a binary string of 'punctuation marks' to indicate the location of crossover points for a multipoint crossover operation. The extra information was appended to the chromosome so that the number and location of crossover points could be manipulated by genetic search. The resulting representation used by the punctuated crossover operator is a string of length 2ℓ, $\boldsymbol{x} = (x_1 \ldots x_\ell x_1' \ldots x_\ell')$, where x_i is the symbol at position i and x_i' is a punctuation mark that is 1 if position i is a crossover point and 0 otherwise. The set of crossover points used in a recombination event under punctuated crossover is given by the union of the crossover points specified on each chromosome

```
compute_mask(a, b)
    j := 0;
    for i := 1 to ℓ/2 do
    mᵢ := j;
    mᵢ' := j
        if ((aᵢ' = 1) or (bᵢ' = 1))
        then j = 1 − j;
        fi
    od
    return (m);
```

Note that the symbol and punctuation mark associated with a chromosome position are transmitted together by the punctuated crossover operator. While the idea behind this operator is appealing, empirical tests of punctuated crossover were not conclusive and the operator is not widely used.

In practice, various aspects of these operators are often modified to enhance performance. Consider, for example, the choice of retaining both resultant strings produced by crossover (a common practice) versus discarding one of the offspring. Holland (1975) described an implementation designed to process only one new individual per generation and, consequently, his algorithm discards one of the offspring generated by crossover. Some implementations retain this feature even if they produce more than one new individual per generation. However, empirical studies (Booker 1982) have shown that retaining both offspring can substantially reduce the loss of diversity in the population. Another widespread practice is to restrict the crossover points to those locations where the parent strings have different symbols. This so-called *reduced surrogate* technique (Booker 1987) improves the ability of crossover to produce offspring that are different from their parents.

An implementation technique called *shuffle crossover* was introduced by Eshelman *et al* (1989). The symbols in the parent strings are 'shuffled' by a permutation operator before crossover is invoked. The inverse permutation is applied to the offspring produced by crossover to restore the original symbol ordering. This method can be used to counteract the tendency in n-point crossover ($n \geq 1$) to disrupt sets of symbols that are widely dispersed on the chromosome more than it disrupts symbols which are close together (see the discussion of bias in Section 33.1.4).

The crossover mechanisms described so far are all consistent with the simplest principle of Mendelian inheritance: the requirement that every gene carried by an offspring is a copy of a gene inherited from one of its parents. Radcliffe (1991) points out that this conservation of genetic material during recombination is not a necessary restriction for artificial recombination operators. From the standpoint of conducting a robust exploration of the opportunities represented by the parent strings, it is reasonable to ask whether a crossover operator can generate all possible offspring having some combination of genes found in the parents. Given a binary string representation, the answer for one-point and n-point crossover is no while the answer for shuffle crossover and uniform crossover is yes. (To see this, simply consider the set of possible resultant strings for the parents **0** and **1**.) For nonbinary strings, however, the only way to achieve this capability is to allow the offspring to have genes that are not carried by either parent. Radcliffe used this idea as the basis for designing the *random respectful recombination* operator. This operator generates a resultant string by copying the symbols at positions where the parents are identical, then choosing random values to fill the remaining positions. Note that for binary strings, random respectful recombination is equivalent to uniform crossover with $p_x = 0.5$.

33.1.3 Formal analysis

Mathematical characterizations of crossover. Several characterizations of crossover operators have been formulated to facilitate the formal analysis of recombination and genetic algorithms. Geiringer (1944) characterized recombination in terms of the probability that sets of genes are transmitted from parents to offspring during a recombination event. The behavior of a crossover operator is then completely specified by the probability distribution it induces over the set of all possible recombination events. Geiringer's study of these so-called *recombination distributions* includes a thorough analysis of recombination acting on a population of linear chromosomes in the absence of selection.

In more detail, the recombination distribution associated with a crossover operator is defined as follows. Let $S_\ell = \{1, \ldots, \ell\}$ be the set of ℓ numbers designating the loci in strings of length ℓ. The number of alleles allowed at each locus can be any arbitrary integer. For notational convenience we will identify a crossover mask m with the subset $A \subseteq S_\ell$ which indicates the loci corresponding to the bit positions i where $m_i = 1$. The set A is simply another way to designate the recombination event specified by m. The complementary subset $A' = S_\ell \setminus A$ designates the recombination event specified by $1 - m$. The recombination distribution $\mathcal{R}$ is given by the probabilities $\mathcal{R}(A)$ for each recombination event. Clearly, under Mendelian segregation, $\mathcal{R}(A) = \mathcal{R}(A')$ since all alleles will be transmitted to one offspring or the other. It is also clear that $\sum_{A \subseteq S_\ell} \mathcal{R}(A) = 1$. We can therefore view recombination distributions as probability distributions over the power set 2^{S_ℓ} (Schnell 1961). The marginal recombination distribution $\mathcal{R}_A$, describing the transmission of the loci in A, is given by the probabilities

$$\mathcal{R}_A(B) = \sum_{C \subseteq A'} \mathcal{R}(B \cup C) \qquad B \subseteq A.$$

$\mathcal{R}_A(B)$ is the marginal probability of the recombination event in which one parent transmits the loci in $B \subseteq A$ and the other parent transmits the loci in $A \setminus B$.

Other mathematical characterizations of crossover operators are useful when the chromosomes happen to be binary strings. If the sum $x \oplus y$ denotes component-wise addition in the group of integers modulo 2 and the product xy denotes bitwise multiplication, then the strings produced by a crossover operator with mask m are given by $ma \oplus (1-m)b$ and $mb \oplus (1-m)a$. Liepins and Vose (1992) use this definition to show that a binary operator is a crossover operator if and only if the operator preserves schemata and commutes with addition and bitwise multiplication. Furthermore, they provide two characterizations of the set of chromosomes that can be generated by an operator given an initial pool of parent strings. Algebraically, this set is given by the mathematical closure of the parent strings under the crossover operator. Geometrically, the set is determined by projections defined in terms of the crossover masks associated with the operator. Liepins and Vose prove that these algebraic and geometric characterizations are equivalent.

The dynamics of recombination. Geiringer used recombination distributions to examine how recombination without selection modifies the proportions of individuals in a population over time. Assume that each individual $x \in \{1, 2, \ldots, k\}^\ell$ is a string of length ℓ in a finite alphabet of k characters. We also assume in the following that $B \subseteq A \subseteq S_\ell$. Let $p^{(t)}(x)$ be the frequency of individual x in a population at generation t, and $p_A^{(t)}(x)$ denote the marginal frequency of individuals that are identical to x at the loci in A. That is,

$$p_A^{(t)}(x) = \sum_y p^{(t)}(y) \qquad \text{for each } y \text{ satisfying } y_i = x_i \; \forall i \in A.$$

Geiringer derives the following important recurrence relations:

$$p^{(t+1)}(z) = \sum_{A,x,y} \mathcal{R}(A) p^{(t)}(x) p^{(t)}(y) \qquad (33.1)$$

$$\text{where} \quad \begin{cases} A \subseteq S_\ell \text{ is arbitrary} \\ y_i = z_i \; \forall i \in A \\ x_i = z_i \; \forall i \in A' \end{cases}$$

$$p^{(t+1)}(z) = \sum_{A,B,x,y} \mathcal{R}_A(B) p^{(t)}(x) p^{(t)}(y) \qquad (33.2)$$

$$\text{where} \quad \begin{cases} B \subseteq A \subseteq S_\ell \text{ are arbitrary subsets} \\ x_i \neq z_i, y_i = z_i \; \forall i \in B \\ x_j = z_j, y_j \neq z_j \; \forall j \in A \setminus B \\ x_k = y_k = z_k \; \forall k \in A' \end{cases}$$

$$p^{(t+1)}(z) = \sum_{A \subseteq S_\ell} \mathcal{R}(A) p_A^{(t)}(z) p_{A'}^{(t)}(z). \qquad (33.3)$$

These recurrence relations are equivalent, complete characterizations of how recombination changes the proportion of individuals from one generation to the next. Equation (33.1) has the straightforward interpretation that alleles appear in offspring if and only if they appear in the parents and are transmitted by a recombination event. Each term on the right-hand side of (33.1) is the probability of a recombination event between parents having the desired alleles at the loci that are transmitted together. A string z is the result of a recombination event A whenever the alleles of z at loci A come from one parent and the alleles at loci A' come from the other parent. The change in frequency of an individual string is therefore given by the total probability of all these favorable occurrences. Equation (33.2) is derived from (33.1) by collecting terms based on marginal recombination probabilities. Equation (33.3) is derived from (33.1) by collecting terms based on marginal frequencies of individuals.

The last equation is perhaps the most significant, since it leads directly to a theorem characterizing the expected distribution of individuals in the limit.

Theorem (Geiringer's theorem II). If ℓ loci are arbitrarily linked, with the one exception of 'complete linkage', the distribution of transmitted alleles 'converges toward independence'. The limit distribution is given by

$$\lim_{t \to \infty} p^{(t)}(z) = \prod_{i=1}^{\ell} p^{(0)}(z_i)$$

which is the product of the ℓ marginal distributions of alleles from the initial population.

This theorem tells us that, in the limit, random mating and recombination without selection lead to chromosome frequencies corresponding to the simple product of initial allele frequencies. A population in this state is said to be in *linkage equilibrium* or Robbins' equilibrium (Robbins 1918). This result holds for all recombination operators that allow any two loci to be separated by recombination.

Note that Holland (1975) sketched a proof of a similar result for schema frequencies and one-point crossover. Geiringer's theorem applied to schemata gives us a much more general result. Together with the recurrence equations, this work paints a picture of 'search pressure' from recombination acting to reduce departures from linkage equilibrium for all schemata.

Subsequent work has carefully analyzed the dynamics of this convergence to linkage equilibrium (Christiansen 1989). It has been proven, for example, that the convergence rate for any particular schema is given by the probability of the recombination event specified by the schema's defining loci. In this view, an important difference between crossover operators is the rate at which, undisturbed by selective pressures, they drive schemata to their equilibrium proportions. These results from mathematical population genetics have only recently been applied to evolutionary algorithms (Booker 1993, Altenberg 1995).

Disruption analysis. Many formal studies of crossover operators focus specifically on the way recombination disrupts and constructs schemata. Holland's (1975) original analysis of genetic algorithms derived a bound for the disruptive effects of one-point crossover. This bound was based on the probability $\ell(\xi)/(\ell - 1)$ that a single crossover point will fall within the defining length $\ell(\xi)$ of a schema ξ. Bridges and Goldberg (1987) subsequently provided an exact expression for the probability of disruption for one-point crossover. Spears and De Jong (1991a) generalized these results to provide exact expressions for the disruptive effects of n-point and uniform crossover.

Recombination distributions provide a convenient framework for analyzing these disruptive effects (Booker 1993). The first step in this analysis is to derive the marginal distributions for one-point, n-point, and uniform crossover. Analyses using recombination distributions can be simplified for binary strings if we represent individual strings using index sets (Christiansen 1989). Each binary string x can be represented uniquely by the subset $A \subseteq S_\ell$ using the convention that A designates the loci where $x_i = 1$ and A' designates the loci where $x_i = 0$. In this notation S_ℓ represents the string $\mathbf{1}$, $\emptyset$ represents the string $\mathbf{0}$, and A' represents the binary complement of the string represented by A. Index sets can greatly simplify expressions involving individual strings. Consider, for example, the marginal frequency $p_A(x)$ of individuals that are identical to x at the loci in A. The index set expression

$$p_A(B) = \sum_{C \subseteq A'} p(B \cup C) \qquad B \subseteq A$$

makes it clear that $p_A(B)$ involves strings having the allele values given by B at the loci designated by A. Note that $p_\emptyset(B) = 1$ and $p_{S_\ell}(B) = p(B)$.

With this notation we can also succinctly relate recombination distributions and schemata. If A designates the defining loci of a schema ξ and $B \subseteq A$ specifies the alleles at those loci, then the frequency of ξ is given by $p_A(B)$ and the marginal distribution $\mathcal{R}_A$ describes the transmission of the defining loci of ξ. In what follows we will assume, without loss of generality, that the elements of the index set A for a schema ξ are in increasing order so that the kth element $A_{(k)}$ is the locus of the kth defining position of ξ. This means, in particular, that the outermost defining loci of ξ are given by the elements $A_{(1)}$ and $A_{(O(\xi))}$ where $O(\xi)$ is the order of ξ. It will be convenient to define the following property relating the order of a schema to its defining length $\delta(\xi)$.

Definition. The *kth component of defining length* for schema ξ, $\delta_k(\xi)$, is the distance between the kth and $k + 1$st defining loci, $1 \le k < O(\xi)$, with the convention that $\delta_0(\xi) \equiv \ell - \delta(\xi)$.

Note that the defining length of a schema is equal to the sum of its defining length components:

$$\delta(\xi) = \sum_{k=1}^{O(\xi)-1} \delta_k(\xi) = A_{(O(\xi))} - A_{(1)}.$$

Given these preliminaries, we can proceed to describe the recombination distributions for specific crossover operators.

One-point crossover. Assume exactly one crossover point in a string of length ℓ, chosen between loci i and $i + 1$ with probability $1/(\ell - 1)$ for $i = 1, 2, \ldots, \ell - 1$. The only recombination events with nonzero probability are $S_x = [1, x]$ and $S'_x = [x + 1, \ell - 1]$ for $x = 1, 2, \ldots, \ell - 1$. The probability of each event is

$$\mathcal{R}^1(S_x) = \mathcal{R}^1(S'_x) = \frac{1}{2(\ell - 1)}$$

since each parent is equally likely to transmit the indicated loci. The marginal distribution $\mathcal{R}^1_A$ for an arbitrary index set A can be expressed solely in terms of these recombination events. We will refer to these events as the primary recombination events.

Now for any arbitrary event $B \subseteq A$ there are two cases to consider:

(i) $B = \emptyset$. This corresponds to the primary recombination events S_x, $x < A_{(1)}$ and S'_x, $x \ge A_{(O(\xi))}$. There are $\ell - 1 - \delta(\xi)$ such events.

(ii) $B \neq \emptyset$. These situations involve the primary events S_x, $A_{(1)} \leq x < A_{(O(\xi))}$. The events B having nonzero probability are given by $B_i = \{A_{(1)}, \ldots, A_{(i)}\}$, $1 \leq i < O(\xi)$. For each i, there are $\delta_i(\xi)$ corresponding primary events.

The complete marginal distribution is therefore given by

$$
\mathcal{R}_A^1(B) = \begin{cases}
\dfrac{\ell - 1 - \delta(\xi)}{2(\ell - 1)} & \text{if } B = \emptyset \text{ or } B = A \\[2ex]
\dfrac{\delta_i(\xi)}{2(\ell - 1)} & \text{if } B = B_i,\ 1 \leq i < O(\xi) \\[2ex]
0 & \text{otherwise.}
\end{cases}
$$

Note that if we restrict our attention to disruptive events, we obtain the familiar result

$$
1 - (\mathcal{R}_A^1(\emptyset) + \mathcal{R}_A^1(A)) = 1 - 2\left(\frac{\ell - 1 - \delta(\xi)}{2(\ell - 1)}\right) = 1 - \left(1 - \frac{\delta(\xi)}{\ell - 1}\right) = \frac{\delta(\xi)}{\ell - 1}.
$$

n-point crossover. The generalization to n crossover points in a string of length ℓ uses the standard convention (De Jong 1975) that when the number of crossover points is odd, a final crossover point is defined at position zero. We also assume that all the crossover points are distinct, which corresponds to the way multipoint crossover is often implemented. Given these assumptions, there are $2\binom{\ell}{n}$ nonzero recombination events if n is even or $n = \ell$, and $2\binom{\ell-1}{n}$ such events if n is odd. Since the n points are randomly selected, these events are equally likely to occur.

We derive an expression for the marginal distributions in the same way as we proceeded for one-point crossover. First we identify the relevant recombination events, then we count them up and multiply by the probability of a single event. Identification of the appropriate recombination events begins with the observation (De Jong 1975) that crossover does not disrupt a schema whenever an even number of crossover points (including zero) fall between successive defining positions. We can use this to identify the configurations of crossover points that transmit all the loci in $B \subseteq A$ and none of the loci in $A \setminus B$. Given any two consecutive elements of A, there should be an even number of crossover points between them if they both belong to B or $A \setminus B$. Otherwise there should be an odd number of crossover points between them. This can be formalized as a predicate $\mathcal{X}_A$ that tests these conditions for a marginal distribution $\mathcal{R}_A$

$$
\mathcal{X}_A(B, n, i) = \begin{cases}
1 & \text{if } n \text{ is even and } \{A_{(i)}, A_{(i-1)}\} \cap B = \emptyset \text{ or } \{A_{(i)}, A_{(i-1)}\} \\
1 & \text{if } n \text{ is odd and } \{A_{(i)}, A_{(i-1)}\} \cap B \neq \emptyset \text{ or } \{A_{(i)}, A_{(i-1)}\} \\
 & \quad\quad \text{where } 2 \leq i \leq O(\xi) \\
0 & \text{otherwise.}
\end{cases}
$$

The recombination events can be counted by simply enumerating all possible configurations of crossover points and discarding those not associated with the marginal distribution. The following function $\mathcal{N}_A$ computes this count recursively (as suggested by the disruption analysis of Spears and De Jong (1991a)):

$$\mathcal{N}_A(B,n,i) = \begin{cases} \sum_{j=0}^{n} \binom{\delta_{i-1}(\xi)}{j} \mathcal{X}_A(B,j,i)\mathcal{N}_A(B,n-j,i-1) \\ \hspace{5cm} 2 < i \leq \mathrm{O}(\xi) \\ \\ \binom{\delta_1(\xi)}{n} \mathcal{X}_A(B,n,2) \hspace{2cm} i = 2. \end{cases}$$

Putting all the pieces together, we can now give an expression for the complete marginal distribution.

$$\mathcal{R}_A^n(B) = \begin{cases} \dfrac{\sum_{j=0}^{n} \binom{\delta_0(\xi)}{j} \mathcal{N}_A(B,n-j,\mathrm{O}(\xi))}{2\binom{\ell}{n}} & \text{if } n \text{ is even or } n = \ell \\ \\ \dfrac{\sum_{j=0}^{n} \binom{\delta_0(\xi)-1}{j} \mathcal{N}_A(B,n-j,\mathrm{O}(\xi))}{2\binom{\ell-1}{n}} & \text{otherwise.} \end{cases}$$

Uniform crossover. The marginal distribution $\mathcal{R}_A^{u(p)}$ for parametrized uniform crossover with parameter p is easily derived from previous analyses (Spears and De Jong 1991b). It is given by

$$\mathcal{R}_A^{u(p)}(B) = p^{|B|}(1-p)^{|A\setminus B|}.$$

Figure 33.1 shows how the marginal probability of transmission for second-order schemata—$2\mathcal{R}_A^n(A)$ and $2\mathcal{R}_A^{u(0.5)}$, $|A| = 2$—varies as a function of defining length. The shape of the curves depends on whether n is odd or even. Since the curves indicate the probability of transmitting schemata, the area above each curve can be interpreted as a measure of potential schema disruption. This interpretation makes it clear that two-point crossover is the best choice for minimizing disruption. Spears and De Jong (1991a) have shown that this property of two-point crossover remains valid for higher-order schemata.

Note that these curves are not identical to the family of curves for nondisruptive crossovers given by Spears and De Jong. The difference is that Spears and De Jong assume crossover points are selected randomly with

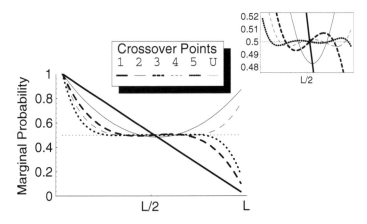

Defining Length for 2nd Order Hyperplanes

Figure 33.1. Transmission probabilities for second-order schemata. The inset shows the behavior of these curves in the vicinity of the point $L/2$.

replacement. This means that their measure $P_{2,\text{even}}$ is a polynomial function of the defining length having degree n, with n identical solutions to the equation $P_{2,\text{even}} = 1/2$ at the point $\ell/2$. The function $\mathcal{R}_A^n(A)$, on the other hand, has n distinct solutions to the equation $2\mathcal{R}_A^n(A) = 1/2$ as shown in the upper right-hand corner of figure 33.1. This property stems from our assumption that crossover points are distinct and hence selected without replacement.

Finally, regarding the construction of schema, Holland (1989) has analyzed the expected waiting time to construct a new schema that falls in the intersection of two schemas already established in a population. He gives examples showing that the waiting time for one-point crossover to construct the new schema can be several orders of magnitude shorter than the waiting time for mutation. Thierens and Goldberg (1993) also examine this property of recombination by analyzing so-called *mixing events*—recombination events in which building blocks from the parents are juxtaposed or 'mixed' to produce an offspring having more building blocks than either parent. Using the techniques of dimensional analysis they show that, given only simple selection and uniform crossover, effective mixing requires a population size that grows exponentially with the number and length of the building blocks involved. This indicates that additional mechanisms may be needed to achieve effective mixing in genetic algorithms.

33.1.4 Crossover bias

In order to effectively use any inductive search operator, it is important to understand whatever tendencies the operator may have to prefer one search outcome over another. Any such tendency is called an inductive *bias*. Random search is the only search technique that has no bias. It has long been recognized that an appropriate inductive bias is *necessary* in order for inductive search to proceed efficiently and effectively (Mitchell 1980). Two types of bias have been attributed to crossover operators in genetic search: *distributional bias* and *positional bias* (Eshelman *et al* 1989).

Distributional bias refers to the number of symbols transmitted during a recombination event and the extent to which some quantities might be more likely to occur than others. This bias is significant because it is correlated with the potential number of schemata from each parent that can be recombined by the crossover operator. An operator has distributional bias if the probability distribution for the number of symbols transmitted from a parent is not uniform. Both one-point and two-point crossover are free of distributional bias. The n-point ($n > 2$) crossover operators have a distributional bias that is well approximated by a binomial distribution with mean $\ell/2$ for large n. Uniform crossover has a strong distributional bias, with the expected number of symbols transmitted given by a binomial distribution with expected value $p_x\ell$. More recently, Eshelman and Schaffer (1993) have emphasized the expected value of the number of symbols transmitted rather than the distribution of those numbers. The bias defined by this criterion, though clearly similar to distributional bias, is referred to as *recombinative bias*.

Positional bias characterizes how much the probability that a set of symbols will be transmitted intact during a recombination event depends on the relative

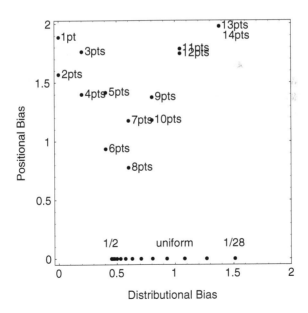

Figure 33.2. One view of the crossover bias 'landscape' generated using quantitative measures derived from recombination distributions.

positions of those symbols on the chromosome. This bias is important because it indicates which schemata are likely to be inherited by offspring from their parents. It is also indicative of the extent to which these schemata will appear in new contexts that can help distinguish the genuine instances of co-adaptation from spurious linkage effects. Holland's (1975) analysis of one-point crossover pointed out that the shorter the defining length of a schema, the more likely it is to be transmitted intact during the crossover operation. Consequently, one-point crossover has a strong positional bias. Analyses of n-point crossover (Spears and De Jong 1991a) lead to a similar conclusion for those operators, though the amount of positional bias varies with n (Booker 1993). Uniform crossover has no positional bias, which is one of the primary reasons it is widely used. Note that shuffle crossover was designed to remove the positional bias from one-point and n-point crossover. Eshelman and Schaffer (1993) have revised their view of positional bias, generalizing the notion to something they now call *schema bias*. An operator has no schema bias if schemata of the same order are equally likely to be disrupted regardless of their defining length.

Recombination distributions can be used to derive quantitative measures of crossover bias (Booker 1993). The overall bias 'landscape' for various crossover operators based on these measures is summarized in figure 33.2.

33.2 Real-valued vectors

David B Fogel

Recombination acts on two or more elements in a population to generate at least one offspring. When the elements are real-valued vectors (Chapter 16), recombination can be implemented in a variety of forms. Many of these forms derive from efforts within the evolution strategies community because of their long involvement with continuous optimization problems. The simpler versions, however, have been popularized within research in genetic algorithms.

For two parent real-valued vectors x_1 and x_2, each of dimension n, one-point crossover is performed by selecting a random crossover point k and exchanging the elements occurring after point k in x_1 with those that occur after point k in x_2 (see figures 33.3 and 33.4). This operator can be extended to a two-point crossover in which two crossover points k_1 and k_2 are selected at random and the segment in between these points is exchanged between parents (see figure 33.5). Extensions to greater multiple-point crossover operators follow naturally.

The one-point and two-point operators attempt to recombine vector segments. Alternatively, individual elements can be recombined without regard to longer segments in which they reside by using a uniform recombination operator. Given two parents x_1 and x_2, one or more offspring are created by randomly selecting each next element from either parent (see figure 33.6). Typically, each parent has an equal chance of contributing the next element. This procedure was offered early on by Reed *et al* (1967) and was reintroduced within

Parents

$$x_1 = x_{1,1}x_{1,2} \ldots x_{1,k}x_{1,k+1} \ldots x_{1,d}$$
$$x_2 = x_{2,1}x_{2,2} \ldots x_{2,k}x_{2,k+1} \ldots x_{2,d}$$

Offspring

$$x_1' = x_{1,1}x_{1,2} \ldots x_{1,k}x_{2,k+1} \ldots x_{2,d}$$
$$x_2' = x_{2,1}x_{2,2} \ldots x_{2,k}x_{1,k+1} \ldots x_{1,d}$$

Figure 33.3. For one-point crossover, two parents are chosen and a crossover point k is selected, typically uniformly across the components. Two offspring are created by interchanging the segments of the parents that occur from the crossover point to the ends of the string.

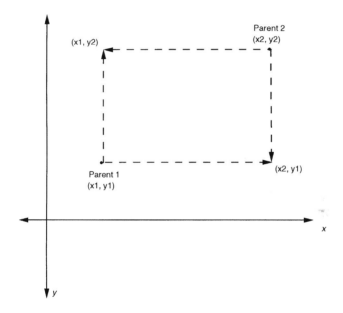

Figure 33.4. A two-dimensional illustration of the potential offspring under a one-point crossover operator applied to real-valued parents.

the genetic algorithm community by Syswerda (1989). A similar procedure is also used within evolution strategies and termed 'discrete recombination' (see below, and also see the *uniform scan* operator of Eiben *et al* (1994), which is applied to multiple parents).

In contrast to the crossover type recombination operators that exchange information between parents, intermediate recombination operators attempt to average or blend components across multiple parents. A canonical version acts

Parents

$$\boldsymbol{x}_1 = x_{1,1}x_{1,2} \ldots x_{1,k_1}x_{1,k_1+1} \cdots x_{1,k_2}x_{1,k_2+1} \cdots x_{1,d}$$
$$\boldsymbol{x}_2 = x_{2,1}x_{2,2} \ldots x_{2,k_1}x_{2,k_1+1} \cdots x_{2,k_2}x_{2,k_2+1} \cdots x_{2,d}$$

Offspring

$$\boldsymbol{x}'_1 = x_{1,1}x_{1,2} \ldots x_{2,k_1}x_{2,k_1+1} \cdots x_{2,k_2}x_{1,k_2+1} \cdots x_{2,d}$$
$$\boldsymbol{x}'_2 = x_{2,1}x_{2,2} \ldots x_{1,k_1}x_{1,k_1+1} \cdots x_{1,k_2}x_{2,k_2+1} \cdots x_{2,d}$$

Figure 33.5. For two-point crossover, two parents are chosen and two crossover points, k_1 and k_2, are selected, typically uniformly across the components. Two offspring are created by interchanging the segments defined by the points k_1 and k_2.

Parents

$$\boldsymbol{x}_1 = x_{1,1}x_{1,2} \ldots x_{1,d}$$
$$\boldsymbol{x}_2 = x_{2,1}x_{2,2} \ldots x_{2,d}$$

Offspring

$$\boldsymbol{x}'_1 = x_{1,1}x_{2,2} \ldots x_{1,d}$$
$$\boldsymbol{x}'_2 = x_{1,1}x_{1,2} \ldots x_{2,d}$$

Figure 33.6. For uniform crossover, each element of an offspring (here two offspring are depicted) is selected from either parent. The example shows that the first element in both offspring were selected from the first parent. In some applications such duplication is not allowed. Typically each parent has an equal chance of contributing each element to an offspring.

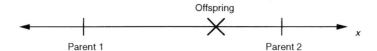

Figure 33.7. A geometrical interpretation of intermediate recombination applied to two parents in a single dimension.

on two parents $\boldsymbol{x}_1$ and $\boldsymbol{x}_2$, and creates an offspring $\boldsymbol{x}'$ as the weighted average:

$$x'_i = \alpha x_{1i} + (1 - \alpha)x_{2i}$$

where α is a number in $[0, 1]$ and $i = 1, \ldots, n$ (figure 33.7). If $\alpha = 0.5$, then the operation is a simple average at each component. Note that this operator can be extended to act on more than two parents (i.e. a multirecombination) by the operation

$$x'_i = \alpha_1 x_{1i} + \alpha_2 x_{2i} + \ldots + \alpha_k x_{ki}$$

subject to

$$\sum \alpha_i = 1 \qquad i = 1, \ldots, k$$

where there are k individuals involved in the multirecombination. This general procedure is also known as *arithmetic crossover* (Michalewicz 1996, p 112) and has been described in various other terms in the literature.

In a more generalized manner, recombination operators can take the following forms (Bäck *et al* 1993, Fogel 1995, pp 146–7):

$$x_i' = \begin{cases} x_{S,i} & (33.4) \\ x_{S,i} \quad \text{or} \quad x_{T,i} & (33.5) \\ x_{S,i} + u(x_{T,i} - x_{S,i}) & (33.6) \\ x_{S_j,i} \quad \text{or} \quad x_{T_j,i} & (33.7) \\ x_{S_j,i} + u_i(x_{T_j,i} - x_{S_j,i}) & (33.8) \end{cases}$$

where S and T denote two arbitrary parents, u is a uniform random variable over $[0, 1]$, and i and j index the components of a vector and the vector itself, respectively. The versions are no recombination (33.4), discrete recombination (or uniform crossover) (33.5), intermediate recombination (33.6), and (33.7) and (33.8) are the global versions of (33.5) and (33.6), respectively, extended to include more than two parents (up to as many as the entire population size).

There are several other variations of crossover operators that have been applied to real-valued vectors.

(i)   The *heuristic crossover* of Wright (1994) takes the form

$$x' = u(x_2 - x_1) + x_2$$

where u is a uniform random variable over $[0, 1]$ and x_1 and x_2 are the two parent vectors subject to the condition that x_2 is not worse than x_1. Michalewicz (1996, p 129) noted that this operator uses values of the objective function to determine a direction to search.

(ii) The *simplex crossover* of Renders and Bersini (1994) selects $k > 2$ parents (say the set J), determines the best and worst individuals within the selected group (say x_1 and x_2, respectively), computes the centroid of the group without x_2 (say c) and computes the reflected vector x' (the offspring) obtained from the vector x_2 as

$$x' = c + (c - x_2).$$

(iii) The *geometrical crossover* of Michalewicz *et al* (1996) takes two parents x_1 and x_2 and produces a single offspring x' as

$$x' = [(x_{11}x_{21})^{0.5}, \ldots, (x_{1n}x_{2n})^{0.5}].$$

This operator can be generalized to a multiparent version:

$$x' = [(x_{11}^{\alpha^1} x_{21}^{\alpha^2} \ldots x_{k1}^{\alpha^k}), \ldots, (x_{1n}^{\alpha^1} x_{2n}^{\alpha^2} \ldots x_{kn}^{\alpha^k})].$$

(iv) The *fitness-based scan* of Eiben *et al* (1994) takes multiple parents and generates an offspring where each component is selected from one of the parents with a probability corresponding to the parent's relative fitness. If a parent has fitness $f(i)$, then the likelihood of selecting each individual component from that parent is $f(i)/\sum f(j)$, where $j = 1, \ldots, k$ and there are k parents involved in the operator.

(v) The diagonal multiparent crossover of Eiben *et al* (1994) operates much like n-point crossover, except that in creating k offspring from k parents, $c \geq 1$ crossover points are chosen and the first offspring is constructed to contain the first segment from parent 1, the second segment from parent 2, and so forth. Subsequent offspring are similarly constructed from a rotation of segments from the parents.

33.3 Permutations

Darrell Whitley

33.3.1 Introduction

An obvious attribute of permutation problems (see Chapter 17) is that simple crossover operators fail to generate offspring that are permutations. Consider the following example of simple one-point crossover, when one parent is denoted with capital letters and the other with lower-case letters:

```
String 1: A B C D E F G H I
          \/
          /\
String 2: h d a e i c f b g

Offspring 1: A B C e i c f b g
Offspring 2: h d a D E F G H I.
```

Neither of the two offspring represents a legal permutation. Offspring 1 duplicates elements *B* and *C* while omitting elements *H* and *D*. Offspring 2 has just the opposite problem: it duplicates *H* and *D* while omitting *B* and *C*.

Davis (1985) and Goldberg and Lingle (1985) defined some of the first operators for permutation problems. One variant of Davis's order crossover operator can be described as follows.

Davis's order crossover. Pick two permutations for recombination. Denote the first parent as the *cut string* and the other the *filler string*. Select two *crossover* points. Copy the sublist of permutation elements between the crossover points from the cut string directly to the offspring, placing them in the same absolute position. This will be referred to as the crossover section. Next, *starting at the second crossover point*, find the next element in the filler string that does not appear in the offspring. Starting at the second crossover point, place the element from the filler string into the next available slot in the offspring. Continue moving the next unused element from the filler string to the offspring. When the end of the filler string (or the offspring) is reached, wrap around to the beginning of the string. When done in this way, Davis's order crossover has the property that Radcliffe (1994) describes as pure recombination: when two identical parents are recombined the offspring will also be identical with the parents. If one does not start copying elements from the filler string starting at the second crossover point, the recombination may not be pure.

The following is an example of Davis's order crossover, where dots represent the crossover points. The underscore symbol in the crossover section corresponds to empty slots in the offspring.

```
Parent 1: A B . C D E F . G H I
Crossover-section: _ _ C D E F _ _ _

Parent 2: h d . a e i c . f b g
Available elements in order: b g h a i

Offspring: a i C D E F b g h.
```

Note that the elements in the crossover section preserve relative order, absolute position, and adjacency from parent 1. The elements that are copied from the filler string preserve only the relative order information from the second parent.

Partially mapped crossover (PMX).   Goldberg and Lingle (1985) introduced the partially mapped crossover operator (PMX). PMX shares the following attributes with Davis's order crossover. One parent string is designated as parent 1, the other as parent 2. Two crossover sites are selected and all of the elements in parent 1 between the crossover sites are directly copied to the offspring. This means that PMX also defines a crossover section in the same manner as order crossover.

```
Parent 1: A B . C D E . F G
Crossover-section: _ _ C D E _ _

Parent 2: c f . e b a . d g.
```

The difference between the two operators is in how PMX copies elements from parent 2 into the open slots in the offspring after a crossover section has been defined. Denote the parents as P1 and P2 and the offspring as OS; let $P1_i$ denote the ith element of permutation P1. The following description of selecting elements from P2 to place in the offspring is based on the article by Whitley and Yoo (1995).

For those elements between the crossover points in parent 2, if element $P2_i$ has already been copied to the offspring, take no action. In the example given here, element e in parent 2 requires no processing. We will consider the rest of the elements by considering the positions in which they appear in the crossover section. If the next element at position $P2_i$ in parent 2 has not already been copied to the offspring, then find $P1_i = P2_j$; if position j has not been filled in the offspring then assign $OS_j = P2_i$. In the example given here, the next element in the crossover section of parent 2 is b which is in the same position as D in parent 1. Element D is located in parent 2 with index 6 and the offspring at OS_6 has not been filled. Copy b to the offspring in the corresponding position. This yields

```
Offspring: _ _ C D E b _.
```

A problem occurs when we try to place element A in the offspring. Element A in parent 2 maps to element E in parent 1; E falls in position 3 in parent 2, but position 3 has already been filled in the offspring. The position in the offspring is filled by C, so we now find element C in parent 2. The position is unoccupied in the offspring, so element A is placed in the offspring at the position occupied by C in parent 2. This yields

```
Offspring: a _ C D E b _.
```

All of the elements in parent 1 and parent 2 that fall within the crossover section have now been placed in the offspring. The remaining elements can be placed by directly copying their positions from parent 2. This yields

```
Offspring: a f C D E b g.
```

33.3.2   Order and position crossover

Syswerda's (1991) order crossover-2 and position crossover are different from either PMX or Davis's order crossover in that there is no contiguous block which is directly passed to the offspring. Instead several elements are randomly selected by absolute position.

Order crossover-2.   This operator starts by selecting K random positions in parent 2, where the parents are of length L. The corresponding elements from parent 2 are then located in parent 1 and reordered so that they appear in the same relative order as they appear in parent 2. Elements in parent 1 that do not correspond to selected elements in parent 2 are passed directly to the offspring. For example,

```
Parent 1: A B C D E F G
Parent 2: C F E B A D G
Selected Elements: * * *.
```

The selected elements in parent 2 are F, B, and A. Thus, the relevant elements are reordered in parent 1.

```
Reorder A B _ _ _ F _ from parent 1 which yields f b _ _ _ a _.
```

All other elements are copied directly from parent 1.

```
(f b _ _ _ a _) combined with (_ _ C D E _ G) yields f b C D E a G.
```

Position crossover. Syswerda defines a second operator called *position* crossover. Using the same example that was used to illustrate Syswerda's order crossover-2, first pick $L - K$ elements from parent 1 which are to be directly copied to the offspring. These elements are copied by position. This yields

```
_ _ C D E _ G.
```

Next scan parent 2 from left to right and place place each element which does not yet appear in the offspring in the next available position. This yields the following progression:

```
# # C D E # G => f # C D E # G
              => f b C D E # G
              => f b C D E a G.
```

Obviously, in this case the two operators generate exactly the same offspring. Jim Van Zant first pointed out the similarity of these two operators in the electronic newsgroup *The Genetic Algorithm Digest.* Whitley and Yoo (1995) show the two operators to be identical using the following argument.

Assume there is one way to produce a target string S by recombining two parents. Given a pair of strings which can be recombined to produce string S, the probability of selecting the K key positions using order crossover-2 required to generate a specific string S is $\binom{L}{K}^{-1}$, while for position crossover the probability of picking the $L - K$ key elements that will produce exactly the same effect is $\binom{L}{L-K}^{-1}$. Since $\binom{L}{K} = \binom{L}{L-K}$ the probabilities are identical.

Now assume there are R unique ways to recombine two strings to generate a target string S. The probabilities for each unique recombination event are equal as shown by the argument in the preceding paragraph. Thus the sum of the probabilities for the various ways of ways of generating S are equivalent for order crossover-2 and position crossover. Since the probabilities of generating any string S are identical, the operators are identical in expectation.

This also means that in practice there is no difference between using order crossover-2 and position crossover as long as the parameters of the operators are adjusted to reflect their complementary nature. If position crossover is used so that $\mathcal{X}\%$ of the positions are initially copied to the offspring, then order crossover is identical if $(100 - \mathcal{X})\%$ positions are selected as relative order positions.

33.3.3   Uniform crossover

Davis' uniform crossover (Davis 1991, p 80) is identical to position crossover and order crossover-2, except that two offspring are generated. A bitstring is used to denote the selection of positions. Offspring 1 copies the elements directly from parent 1 in those positions in the bitstring marked by a '1' bit. Offspring 2 copies the elements from parent 2 in those positions marked by '0' bits. Both offspring then copy the remaining elements from the other parent in relative order.

33.3.4   Edge recombination

Edge recombination was introduced as a specialized operator for the traveling salesman problem (TSP). The motivation behind this operator is that it should preserve the adjacency between permutation elements, since the cost of a tour in a TSP is directly related to the set of adjacency relationships (i.e. the distances between cities) that exists between permutation elements. The original edge recombination operator has gone through three revisions and enhancements over the years. First, the basic idea behind edge recombination is introduced.

Since adjacency information directly translates into cost, the adjacency information from two parent strings is extracted and stored in an adjacency list called the edge table. The edge table really just combines the two tours into a single graph. Recombination occurs by building an offspring using the adjacency information stored in the edge table; in other words, it tries to find a new Hamiltonian circuit in the graph created by merging the two parent strings. Finding a Hamiltonian circuit in an arbitrary graph is itself a nondeterministic-polynomial-time (NP) complete problem and edge recombination must sometimes add edges not contained in the edge table in order to generate a legal tour. The various enhancements to edge recombination attempt to reduce the number of 'foreign edges' (edges not found in the edge table) that must be introduced into an offspring during recombination in order to maintain a feasible tour.

In the original edge recombination operator, no information was maintained about common edges that were shared by both parents. As a result the operator sometimes failed to place an edge in the offspring that appeared in both parents, resulting in a kind of 'mutation by omission' (Whitley *et al* 1991). To solve this problem, information about shared edges was added to the edge table. Edges shared by the two parents are marked with a + symbol. The algorithm can be described as follows.

Consider the following tours as parents to be recombined:

> Parent 1: g d m h b j f i a k e c
> Parent 2: c e k a g b h i j f m d.

An edge list is constructed for each city in the tour. The edge list for some city a is composed of all of the cities in the two parents that are adjacent to city a. If some city is adjacent to a in both parents, this entry is flagged (using a plus sign). Figure 33.8 shows the edge table which is the collective set of edge lists for all cities.

The algorithm for edge recombination is as follows.

(i)   Pick a random city as the initial *current city*. Remove all references to this city from the edge table.

(ii) Look at the adjacency list of the current city. If there is a common edge (flagged by +), go to that city next. (Unless the initial city is the current city, there can be only one common edge; if two common edges existed,

city	edge list	city	edge list
a	+k, g, i	g	a, b, c, d
b	+h, g, j	h	+b, i, m
c	+e, d, g	i	h, j, a, f
d	+m, g, c	j	+f, i, b
e	+k, -c	k	+e, +a
f	+j, m, i	m	+d, f, h

Figure 33.8. An edge table.

one was used to reach the current city.) Otherwise from the cities on the current adjacency list pick the next city to be the one whose own adjacency list is shortest. Ties are broken randomly. Once a city is visited, references to the city are removed from the adjacency list of other cities and it is no longer reachable from other cities.

(iii) Repeat step 2 until the tour is complete or a city has been reached that has no entries in its adjacency list. If not all cities have been visited, randomly pick a new city to start a new partial tour.

Using the edge table in figure 33.8, city a is randomly chosen as the first city in the tour. City k is chosen as the second city in the tour since the edge (a, k) occurs in both parent tours. City e is chosen from the edge list of city k as the next city in the tour since this is the only city remaining in k's edge list. This procedure is repeated until the partial tour contains the sequence [a k e c].

At this point there is no deterministic choice for the fifth city in the tour. City c has edges to cities d and g, which both have two unused edges remaining. Therefore city d is randomly chosen to continue the tour. The normal deterministic construction of the tour then continues until position 7. At position 7 another random choice is made between cities f and h. City h is selected and the normal deterministic construction continues until we arrive at the following partial tour: [a k e c d m h b g].

In this situation, a failure occurs since there are no edges remaining in the edge list for city g. When a potential failure occurs during *edge-3* recombination, we attempt to continue construction at a previously unexplored terminal point in the tour.

A *terminal* is a city which occurs at either end of a partial tour, where all edges in the partial tour are inherited from the parents. The terminal is said to be *live* if that city still has entries in its edge list; otherwise it is said to be a *dead* terminal. Because city a was randomly chosen to start the tour in the previous example, it serves as a new terminal in the event of a failure. Conceptually this is the same as inverting the partial tour to build from the other end.

When a failure occurs, there is at most one *live* terminal in reserve at the opposite end of the current partial tour. In fact, it is not guaranteed to be live, since the construction of the partial tour could isolate this terminal city. Once both terminals of the current partial tour are found to be dead, a new partial tour must be initiated. Note that no local information is employed.

We now continue construction of the partial tour [a k e c d m h b g]. The tour segment is reversed (i.e. [g b h m d c e k a]). Then city i is added to the tour after city a. The tour is then constructed in the normal fashion. In this case, there are no further failures. The final offspring tour is [g b h m d c e k a i f j]. The offspring produced has a single foreign edge ([j–g].)

When a failure occurs at both ends of the subtour, edge-3 recombination starts a new partial tour. However, there is one other possibility, which has been described as part of the edge-4 operator (Dzubera and Whitley 1994) but which has not been widely tested.

Assume that the first partial tour has been constructed such that both ends of the construction lack a *live terminal* by which to continue. Since only one partial tour has been constructed and since initially every city has at least two edges in the edge table, there must be edges internal to the current partial tour that represent possible edges to the *terminal cities* of the partial tour. The edge-4 operator attempts to exploit this fact by inverting part of the partial tour so that a terminal city is reconnected to an edge which is both internal to the partial tour and which appeared in the original edge list of the terminal city. This will cause a previously visited city in the partial tour to move to a terminal position. If this newly created terminal has cities remaining in its (old) edge list, the offspring construction can continue. If it does not, one can look for other internal edges that will allow an inversion. Details on the edge-4 recombination operator are given by Dzubera and Whitley (1994).

If one is using just a recombination operator and a mutation operator, then edge recombination works very well as an operator for the TSP, at least compared to other recombination operators, but if one is hybridizing such that tours are being produced by recombination, then improved using 2-opt, then both the empirical and the theoretical evidence suggests that Mühlenbein's MPX operator may be more effective (Dzubera and Whitley 1994).

33.3.5 Maximal preservative crossover

Mühlenbein (1991, p 331) offers the following pseudocode for the maximal preservative crossover (MPX) operator. (Numbering of the pseudocode has been added for clarity.)

PROC crossover(receiver, donor, offspring)

(i)   Choose position $0 <= i <$ nodes and length $b_{low} <= k <= b_{up}$ randomly.
(ii) Extract the string of edges from position i to position $j = (i + k)$ MOD nodes from the mate (donor). This is the crossover string.
(iii) Copy the crossover string to the offspring.
(iv) Add successively further edges until the offspring represents a valid tour. This is done in the following way.

(a) IF an edge from the receiver parent starting at the last city in the offspring is possible (does not violate a valid tour)

(b) THEN add this edge from the receiver

(c) ELSE IF an edge from the donor starting at the last city in the offspring is possible

(d) THEN add this edge from the donor

(e) ELSE add that city from the receiver which comes next in the string; this adds a new edge, which we will mark as an implicit mutation.

The following example illustrates the MPX operator.

```
Receiver:          G D M H B J F I A K E C
Donor:             c e k a g b h i j f m d
Initial segment:   _ _ k a g _ _ _ _ _ _ _.
```

Note that G is connected to D in the receiver, and that element D through element I can be taken from the receiver without duplicating any of the elements already in the offspring. This produces the partial tour

```
_ _ k a g D M H B J F I.
```

At this point, there is no edge in either parent that is connected to I and has that not been used. Here MPX skips cities in the receiver until it finds one which has not been used. In this case, it reaches E. This causes E and C to be added to the tour to yield

```
E C k a g D M H B J F I.
```

Note that MPX does not transmit adjacency information from parents to offspring as effectively as the various edge recombination operators, since it uses less lookahead to avoid a break in the tour construction. At the same time, when it must introduce a new edge that does not appear in either parent, it skips to a nearby city in the tour rather than picking a random edge. Assuming that the tour is partially optimized (for example, if the tour has been improved via 2-opt) then a city nearby in the tour should also be a city nearby in Euclidean space. This, coupled with the fact that an initial segment is copied from one of the parents, appears to give MPX an advantage when when combined with an operator such as 2-opt. Gorges-Schleuter (1989) implemented a variant of MPX that has some notable features that are somewhat like Davis's order crossover operator. A full description of Gorges-Schleuter's MPX is given by Dzubera and Whitley (1994).

33.3.6 Cycle crossover

The operators discussed so far are aimed at preserving adjacency information (such as edge recombination) or relative order information (such as Davis's uniform order-based crossover). Operators may also emphasize position. Cycle crossover partitions two parents into a set of cycles: a cycle is a subset of elements which is located on a corresponding subset of positions on both the two parent strings. Consider the following example from Oliver *et al* (1987) where the permutation elements correspond to the alphabetic characters with numbers to indicate position:

```
Parent 1:  h k c e f d b l a i  g  j
Parent 2:  a b c d e f g h i j  k  l
Positions: 1 2 3 4 5 6 7 8 9 10 11 12.
```

To find a cycle, pick a position from either parent. Starting with position 1, elements (h, a) belong to cycle 1. The elements (h, a) also appear in positions 8 and 9. Thus the cycle is expanded to include positions (1, 8, 9) and the new elements i and l are added to the corresponding subset. Elements i and l appear in positions 10 and 12, which also causes j to be added to the subset of elements in the cycle. Note that adding j adds no new elements, so the cycle terminates. Cycle 1 includes elements (h, a, i, l, j) in positions (1, 8, 9, 10, 12).

Note that element (c) in position 3 forms a unary cycle of one element. Aside from the unary cycle at element c (denoted U), Oliver *et al* note that there are three cycles between this set of parents:

```
Parent 1:    h k c e f d b l a i g j
Parent 2:    a b c d e f g h i j k l
Cycle Label: 1 2 U 3 3 3 2 1 1 1 2 1.
```

Recombination can occur by picking some cycles from one parent and the remaining cycles from the alternate parent. Note that all elements in the offspring occupy the same positions as in one of the two parents. However, few applications seem to be position sensitive and cycle crossover is less effective at preserving adjacency information (as in the TSP) or relative order information (as in resource scheduling) compared to other operators.

33.3.7 Merge crossover

Blanton and Wainwright (1993) construct permutation recombination operators for multiple vehicle routing with time and capacity constraints. The following example of the merge crossover operator MX1 uses a global precedence vector. Given any two elements in the permutation, the global precedence vector indicates which element has higher priority for processing. Elements which appear earlier in the vector have higher precedence. In vehicle routing each customer has a time window in which they must be served, which can be translated into a global precedence vector: for example, customer X should be served before customer Y because the time window for X closes before the time window for Y. The following example illustrates the operator:

```
Parent 1:    C F G B A H D I E J
Parent 2:    E B G J D I C A F H
Precedence:  A B C D E F G H I J.
```

A single offspring is constructed. In this case, starting at position 1, we compare C and E from the two parents; since C has higher precedence, it is placed in the offspring. Because C has already been allocated a position in the offspring, the C which appears later in parent 2 is exchanged with the E in the initial position of parent 2. This yields

```
Parent 1:   C F G B A H  D  I E J
Parent 2:   C B G J D I <E> A F H
Precedence: A B C D E F  G  H I J
```

where the moved E element is bracketed: <E>. Going to position 2, B has higher precedence than F, so B is kept in position 2. Also, elements F and B are exchanged in parent 1, which yields

```
Parent 1:   C B G <F> A H  D  I E J
Parent 2:   C B G  J  D I <E> A F H
Precedence: A B C  D  E F  G  H I J.
```

Note that one need not actually build a separate offspring, since both parents are in effect transformed into copies of the same offspring. The resulting offspring in the above example is

```
Offspring: C B G F A H D E I J.
```

The MX-2 operator is similar, except that when an element is added to the offspring it is deleted from both parents instead of being swapped. Thus, the process works as follows:

```
Parent 1:   C F G B A H D I E J
Parent 2:   E B G J D I C A F H
Precedence: A B C D E F G H I J.
```

C is added to the offspring and deleted from both parents

```
Parent 1:   _ F G B A H D I E J
Parent 2:   E B G J D I _ A F H
Offspring:  C.
```

Instead of now moving to the second element of each permutation, the first remaining elements in the parents are compared: in this case, E and F are the first elements and E is chosen and deleted. The parents are now represented as follows:

```
Parent 1:   _ F G B A H D I _ J
Parent 2:   _ B G J D I _ A F H
Offspring:  C E.
```

Element B is chosen to fill position 3 in the offspring, and the construction continues to produce the offspring

```
Offspring: C E B F G A H D I J.
```

Note that, over time, this class of operator will produce offspring that are closer to the precedence vector—even if no selection is applied.

33.3.8 Some other operators

Other interesting operators have been introduced over the years for permutation problems. Fox and McMahon (1991) introduced an intersection operator that extracts features common to both parents. Eshelman (1991) used a similar strategy to build a recombination operator that extracts all common subtours for the TSP, and assigns all other elements using local search (2-opt) over an otherwise random assignment. Fox and McMahon also constructed a union operator. In this case, each permutation is converted into a binary matrix representation and the offspring is the logical-or of the matrices representing the parents.

Radcliffe and Surry (1995) have also introduced new operators for the TSP, largely by looking at different representations and then defining appropriate operators with respect to the representations. These representations include the permutation representation, the undirected edge representation, the directed edge representation, and the corner representation.

33.4 Finite-state machines

David B Fogel

Recombination can be applied to logical structures such as finite-state machines. There have been a variety of proposals to accomplish this in the literature. Recall that a finite-state machine (Chapter 18) is a 5-tuple:

$$M = (Q, T, P, s, o)$$

where Q is a finite set, the set of states, T is a finite set, the set of input symbols, P is a finite set, the set of output symbols, $s : Q \times T \to Q$, the next state function, and $o : Q \times T \to P$, the next output function. Perhaps the earliest proposal to recombine finite-state machines in simulated evolution can be found in the work of Fogel (1964) and Fogel *et al* (1966, pp 21–3). The following extended quotation (Fogel *et al* 1966, p 21) may be insightful:

> The recombination of individuals of opposite sex appears to benefit natural evolution. By analogy, why not retain worthwhile traits that have survived separate evolution by combining the best surviving machines through some genetic rule; mutating the product to yield offspring? Note that there is no need to restrict this mating to the best two surviving 'individuals'. In fact, the most obvious genetic rule, majority logic, only becomes meaningful with the combination of more than two machines.

Fogel *et al* (1966) suggested drawing a single state diagram which expresses the majority logic of an array of finite-state machines. Each state of the majority logic machine is the composite of a state from each of the original machines.

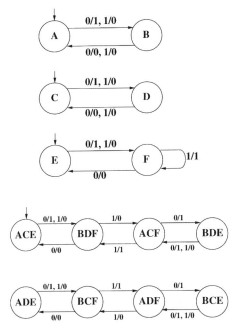

Figure 33.9. Three parent machines (top) are joined by a majority logic operator to form another machine (bottom). The initial state of each machine is indicated by a short arrow pointing to that state. Each state in the majority logic machine is a combination of the states of the three parent machines with the output symbol being chosen as the majority decision of two of the three parent machines. For example, the state BDF in the majority logic machine is determined by examining the states B, D, and F in each of the individual machines. For an input symbol of 0, all three states respond with a 0, therefore this symbol is chosen for the output to an input of 0 in state BDF. For an input symbol of 1, two of the three states respond with a 0, thus, this being the majority decision, this symbol is chosen for the output to an input of 1 in state BDF. Note that several states of the majority logic machine are isolated from the start state and could never be expressed.

Thus the majority machine may have a number of states as great as the product of the number of states in the original machines. Each transition of the majority machine is described by that input symbol which caused the respective transition in the original machines, and by that output symbol which results from the majority element logic being applied to the output symbols from each of the original machines (figure 33.9). If there are only two parents to recombine in this manner, the majority logic machine reduces to the better of the two parents.

Zhou and Grefenstette (1986) used recombination on finite-state automata applied to binary sequence induction problems. The finite-state automata were defined in terms of a 5-tuple:

$$(Q, S, \delta, q_0, F)$$

where Q is a finite set of states, S is a finite input alphabet, $q_0 \in Q$ is the initial state, δ is the transition function mapping the Cartesian product of Q and S into Q, and F is the set of final states, a subset of Q. The chosen representation was

$$(X_1, Y_1, F_1), (X_2, Y_2, F_2), \ldots, (X_8, Y_8, F_8)$$

where each (X_i, Y_i, F_i) represented the state i, X_i and Y_i corresponded to the destination state of the zero and one arrows from state i, respectively, and F_i was a three-bit code where the first two bits were used to indicate whether or not there existed an arrow from state i, and the third bit showed whether the state i was a final state. The maximum number of states was set to eight. The details of how recombination was implemented on this representation are not obvious from the article by Zhou and Grefenstette (1986) but it is reasonable to infer that a simple one-point crossover operator was applied.

Fogel and Fogel (1986) used recombination in a similar manner on finite-state machines by exchanging single states between machines (i.e. output symbol and next-state transitions for each input symbol for a particular state). Birgmeier (1996) also used a similar method implemented as uniform crossover between two machines by state. One offspring was produced from two parents by choosing each row in the transition table from either parent (with specific procedures for handling parents with differing numbers of states). Birgmeier (1996) also offered a new *joining* operator where the offspring's size is the sum of the two parents' number of states. Both the output and transition matrices from the two parents are juxtaposed in the offspring and some of the entries are randomly reset to point to a state in the other half, thus joining the new machines into one.

33.5 Crossover: parse trees

Peter J Angeline

From an evolutionary computation view, crossover, in its most basic form, is an operator that exchanges representational material between two parent structures to produce offspring. Occasionally, it is important to introduce additional constraints on the crossover operation to ensure that the created children observe certain necessary constraints of the representation or problem environment.

Parse tree representations (Chapter 19), as typically used in genetic programming (Koza 1992), require that the crossover operation produce offspring that are also valid parse trees. In order to remain a valid parse tree, the structure must have only terminals at the leaf positions of the tree and only function nodes at each of its internal positions. In addition, each function node of the parse tree must have the correct number of subtrees below it, one for each argument that the function requires.

Often in genetic programming, a simplification is made so that all functions and terminals in the primitive language return the same data type. This is referred to as the *closure principle* (Koza 1992). The effect is to reduce the number of syntactic constraints on the programs so that the complexity of the crossover operation is minimized.

The recursive structure of parse tree representations makes the definition of crossover for tree representations that adhere to the above caveats surprisingly simple. Cramer (1985) initially defined the now standard subtree crossover for parse trees shown in figure 33.10. First, a random subtree is selected and removed from one of the parents. Note that this leaves a hole in the parent such that there exists a function that has a null value for one of its parameters. Next, a random subtree is extracted from the second parent and inserted at the point in the first parent where its subtree was removed. Now the hole in the first parent is again filled. The process is completed by inserting the subtree extracted from the first parent into the position in the second parent where its subtree was removed. As long as only complete subtrees are swapped between parents and the closure principle holds, this simple crossover operation is guaranteed to produce syntactically valid offspring every execution.

Typically, when evolving parse tree representations, a user-defined limit on the maximum size of any tree in the population is provided. Subtree crossover will often increase the size of a given parent such that, over a number of generations, individuals in an unrestricted population may grow to swamp the available computational resources. Given a user-defined restriction on subtree size, expressed as a limit according to either the depth of a tree or the number of nodes it contains, crossover must enforce this limit. When a crossover operation is executed that creates one or more offspring that violate the size limitation, the crossover operation is invalidated and the offspring are restored to their original forms. What happens next is a matter of choice. Some systems will reject both children and revert back to selecting two new parents. Other systems attempt crossover repeatedly either until both offspring fall within the size limit or until a specified number of attempts is reached. Given the nature of the crossover operation, the likelihood of performing a valid crossover operation in a small number of attempts, say five, is fairly good.

Koza (1992) popularized the use of subtree crossover for manipulating parse tree representations in genetic programming. The subtree swapping crossover of Koza (1992) shares much with the subtree crossover defined by Cramer (1985) with a few minor differences. The foremost difference is a bias introduced by Koza (1992) to limit the probability that a leaf node is selected as the subtree from a parent during crossover. The reasoning for this bias according to Koza (1992) is that, in most trees, the number of leaf nodes will be roughly equivalent to the number of nonleaf nodes.

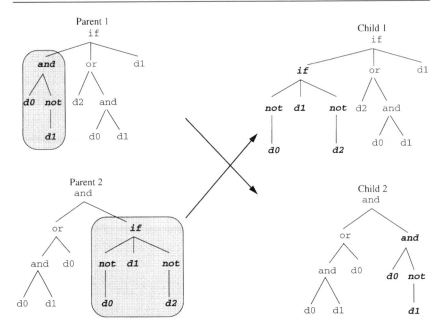

Figure 33.10. An illustration of the crossover operator for parse trees. A subtree is selected at random from each parent, extracted, and exchanged to create two offspring trees.

Consequently, the number of subtrees of depth one will be approximately the number of subtrees of depth greater than one. Merely swapping a leaf between parents to produce children half of the time will not tend to greatly advance the evolutionary process, so, during crossover in a genetic program, the probability that a leaf node is selected is controlled by a bias term called the leaf frequency. Typically, the leaf frequency is set at about 10%, meaning that 10% of the time when a subtree is selected a leaf node will be chosen in a parent while the rest of the time only nonleaf nodes will be chosen. Koza (1992) offers no empirical validation of this bias term or its assumed value.

Often it is important to violate the closure principle and allow multiple types in the parse tree representation in order to more effectively solve a given problem. This implies that there are some functions such that they cannot be used as arguments to certain other functions. Crossover in such typed parse trees, as described by Montana (1995), proceeds much as in subtree crossover with one caveat to compensate for the additional constraint of multiple return types. First, a random node is selected in the first parent's parse tree. The return type of the root of the subtree is determined and the selection of crossover points in the second parent is restricted to only those subtrees that have identical return types. This ensures that the syntactic constraints in both parents are upheld.

When evolving genetic programs using *automatically defined functions* (ADFs), Koza (1994) uses a slightly modified version of subtree crossover. When crossing two genetic programs with ADFs, if the crossover position in the first tree is selected to be within a particular subroutine then only crossover points in the corresponding subroutine in the second parent are considered. This is similar to the typed crossover of Montana (1995) except that, rather than restricting the crossover positions in the second parent based on the type of subtree extracted from the first, it restricts the selection using the functional origin of the initially selected subtree.

33.6 Other representations

Peter J Angeline and David B Fogel

The use of recombination on the alternative mixed-integer representations, and those using introns, does not generally vary from the standard usage. All of the available options of discrete and intermediate recombination apply to the mixed-integer format offered by Bäck and Schütz (1995). Introns are used with the belief that they will enhance the chances for crossover to recombine building blocks. Moreover, Wu and Lindsay (1995) suggest that the addition of introns can have an equivalent effect of varying crossover probabilities across a chromosome, and state 'the advantages of the noncoding segment method including the fact that the genetic algorithm does not need to be modified to handle variable crossover probabilities and that crossover location calculations are much simpler'.

33.7 Multiparent recombination

A E Eiben

33.7.1 Introduction

To make the following survey unambiguous we have to start with setting some conventions on terminology. The term *population* will be used for a multiset of individuals that undergoes selection and reproduction. This terminology is maintained in genetic algorithms, evolutionary programming, and genetic programming, but in evolution strategies all μ individuals in a (μ, λ) or $(\mu + \lambda)$ strategy are called parents. We, however, use the term *parents* only for those individuals that are selected to undergo recombination. In other words, parents are those individuals that are actually used as inputs for a recombination operator; the *arity of a recombination operator* is the number of parents it uses. The next notion is that of a *donor*, being a parent that actually contributes to (at least one of) the alleles of the child(ren) created by the recombination operator. This contribution can be for instance the delivery of an allele, as in uniform crossover in canonical genetic algorithms (GAs), or the participation in an averaging operation, as in intermediate recombination in evolutionary strategies (ESs). As an illustration consider a steady-state GA where 100 individuals form the population and two of them are chosen as parents to undergo uniform crossover to create one single offspring. If, by pure chance, the offspring only inherits alleles from parent 1, then parent 1 is a donor, and parent 2 is not.

33.7.2 Miscellaneous operators

We begin this survey with papers where the multiparent aspect has an incidental character. By an incidental character we mean that the operator is defined and used in a specific application and has, for instance, a certain fixed arity, or, even if the definition is general and would allow comparison between different number of parents, this aspect is not given attention.

The recombination mechanism of Kaufman (1967) is applied for evolving models for a given process, where a model is an array of a number of blocks, and models may differ in the numbers of blocks they contain. Recombination of four models to create one new model is defined as follows. The size of the child (the number of blocks) is equal to the size of each of its parents with probability 0.25. The ith block of the child is chosen with equal probability from those parents that have at least i blocks. Let us note that there is an exception to this latter rule of choosing one of the parents' blocks, but that exception has a very problem-specific reason; therefore we rather present the general idea here.

In an extensive study on bit vector function optimization, stochastic iterated genetic hill climbing (SIGH) is studied and compared with other techniques, such as GAs, iterated hill climbing, and simulated annealing ?(Ackley 1987b). SIGH applies a sophisticated probabilistic voting mechanism with time-dependent probability distributions (cooling), where m 'voters' (m being the size of the population) determine the values of a new bitstring. SIGH is shown to be better than a GA with one-point and uniform crossover on four out of the six test functions and the overall conclusion is that it is 'competitive in speed with a variety of existing algorithms'.

In the introductory paper on the parallel GA ASPARAGOS (Mühlenbein 1989), *p-sexual voting recombination* is applied for the quadratic assignment problem. Let us remark that the name p-sexual is somewhat misleading, as there are no different genders and no restriction on having one representative of each gender for recombination. The voting recombination produces one child of p parents based on a threshold value v. It determines the ith allele of the child by comparing the ith alleles of the selected parent individuals. If the same allele is found more often than the threshold v, this allele is included in the child; other bits are filled in randomly. In the experiments the values $p = 7$ and $v = 5$ are used and it 'worked surprisingly well', but comparison between this scheme and ordinary two-parent recombination was not performed.

An interesting attempt to combine GAs with the simplex method resulted in the ternary *simplex crossover* (Bersini and Seront 1992). If x^1, x^2, x^3 are the three parents sorted in decreasing order of fitness, then the simplex crossover generates one child x by the following two rules.

(i)   If $x_i^1 = x_i^2$ then $x_i = x_i^1$;

(ii) if $x_i^1 \neq x_i^2$ then $x_i = x_i^3$ with probability p and $x_i = 1 - x_i^3$ with probability $1 - p$.

Using the value $p = 0.8$, the simplex GA performed better than the standard GA on the DeJong functions. The authors remark that applying a modified crossover on more than three parents 'is worth to try'.

The problem of placing actuators on space structures is addressed by Furuya and Haftka (1993). The authors compare different crossovers: among others they use uniform crossover with two as well as with three parents in a GA using integer representation. Based on the experimental results they conclude that the use of three parents did not improve the performance. This might be related to another conclusion, indicating that for this problem mutation is an efficient operator and crossover might not be important. Uniform crossover with an arbitrary number of parents is also used by Aizawa (1994) as part of a special schema sampling procedure in a GA, but the multiparent feature is only a side-effect and is not investigated.

A so-called *triadic crossover* is introduced and tested by Pál (1994) for a multimodal spin–lattice problem. The triadic crossover is defined in terms of two parents and one extra individual, chosen randomly from the population. The operator creates one child; it takes the bits in positions where the first parent and the third individual have identical bits from this parent and the rest of the bits from the other parent. Clearly, the result is identical to the outcome of a voting crossover on these three individuals as parents. Although the paper is primarily concerned with different selection schemes, a comparison between triadic, one-point, and uniform crossover is made, where triadic crossover turned out to deliver the best results.

33.7.3 Operators with undefined arity

In the introduction to this section we defined the arity of a recombination operator as the number of parents it uses. In some cases this number depends on the outcomes of random drawings; the operator is called without knowing in advance how many parents will be applied. In this section we treat three mechanisms of this kind.

Global recombination in ESs allows the use of more than two recombinants (Bäck 1996, Schwefel 1995). In ES there are two basic types of recombination, intermediate and discrete recombination, both having a standard two-parent variant and a global variant. Given a population of μ individuals global recombination creates one offspring x by the following mechanism.

$$x_i = \begin{cases} x_i^{S_i} \text{ or } x_i^{T_i} & \textit{global discrete recombination} \\ x_i^{S_i} + \chi_i(x_i^{T_i} - x_i^{S_i}) & \textit{global intermediate recombination} \end{cases}$$

where the two parents x^{S_i}, x^{T_i} $(S_i, T_i \in \{1, \ldots, \mu\})$ are redrawn for each i anew and so is the contraction factor χ_i. The above definition applies to the object variables as well as the strategy parameter part; that is, for the mutation stepsizes (σ) and the rotation angles (α). Observe that the multiparent character of global recombination is the consequence of redrawing the parents x^{S_i}, x^{T_i} for each coordinate i. Therefore, probably more than two individuals contribute to the offspring x, but their number is not defined in advance. It is clear that investigations on the effects of different numbers of parents on algorithm performance could not be performed in the traditional ES framework. The option of using multiple parents can be turned on or off, that is, global recombination can be used or not, but the arity of the recombination operator is not tunable. Experimental studies on global versus two-parent recombination are possible, but so far there are almost no experimental results available on this subject. Schwefel (1995) notes that 'appreciable acceleration' is obtained by changing to a bisexual from an asexual scheme (i.e., adding recombination using two parents to the mutation-only algorithm), but only a 'slight further increase' is obtained when changing from bisexual to multisexual recombination (i.e., using global recombination instead of the two-parent variant). Recall the remark on the name p-sexual voting. The terms bisexual and multisexual are not appropriate either for the same reason: individuals have no gender or sex, and recombination can be applied to any combination of individuals.

Gene-pool recombination (GPR) was introduced by Mühlenbein and Voigt (1996) as a multiparent recombination mechanism for discrete domains. It is defined as a generalization of two-parent recombination (TPR). Applying GPR is preceded by selecting a gene pool consisting of would-be parents. Applying GPR the two parent alleles of an offspring are randomly chosen for each locus with replacement from the gene pool and the offspring allele is computed 'using any of the standard recombination schemes for TPR'. Theoretical analysis on infinite populations shows that GPR is mathematically more tractable than TPR. If n stands for the number of variables (loci), then the evolution with proportional selection and GPR is fully described by n equations, while TPR needs 2^n equations for the genotypic frequencies. In practice GPR converges about 25% faster than TPR for Onemax. The authors conclude that GPR separates the identification and the search of promising areas of the search space better; besides it searches more reasonably than does TPR. Voigt and Mühlenbein (1995) extend GPR to continuous domains by combining it with uniform fuzzy two-parent recombination (UFTPR) from Voigt *et al* (1995). The resulting uniform fuzzy gene-pool recombination (UFGPR) outperforms UFTPR on the spherical function in terms of realized heritability, giving it a higher convergence speed. The convergence of UFGPR is shown to be about 25% faster than that of UFTPR.

A very particular mechanism is the *linkage evolving genetic operator* (LEGO) as defined by Smith and Fogarty (1996). The mechanism is designed to detect and propagate blocks of corresponding genes of potentially varying length during the evolution. Punctuation marks in the chromosomes denote the beginning and the end of each block and more chromosomes with the appropriately positioned punctuation marks are considered as donors of a whole block during the creation of a child. Although the multiparent feature is only a side-effect, LEGO is a mechanism where more than two parents can contribute to an offspring.

33.7.4 Operators with tunable arity

Unary reproduction operators, such as mutation, are often called asexual, based on the biological analogies. Sexual reproduction traditionally amounts to two-parent recombination in evolutionary computation (EC), but the operators discussed in the previous section show that the sexual character of recombination can be intensified, in the sense that more than two parents can be recombined. Nevertheless, this intensification is not graded: the multiparent option can be turned on or off, but the extent of sexuality (the number of parents) cannot be tuned. In this section we consider recombination operators that make sexuality a graded, rather than a Boolean, feature by having an arity that can vary. In other words, the operators we survey here are called with a certain number of parents as input, and this number can be modified by the user.

An early paper mentioning multiparent recombination is that of Bremermann *et al* (1966) on solving linear equations. It presents the definition of three different multiparent recombination mechanisms, called m-tuple mating. Given m binary parent vectors $x^1, \ldots, x^m$, the *majority mating* mechanism creates one offspring vector x by choosing

$$x_i = \begin{cases} 0 & \text{if half or more of the parents have } x_i^j = 0 \\ 1 & \text{otherwise.} \end{cases}$$

Another mating mechanism for m binary parent vectors is called *mating by crossing over*. Describing it in contemporary terms, the mechanism works by selecting $m - 1$ crossover points (identical in each parent) and then composing one child by selecting exactly one segment from each parent. The third operator is called *mating by averaging* and it is defined for vectors of continuous variables. Quite naturally, the child x of parents $x^1, \ldots, x^m$ is defined by

$$x_i = \sum_{j=1}^{m} \lambda_j x_i^j$$

where $\sum_{j=1}^{m} \lambda_j = 1$. Unfortunately, only very little is reported on the performance of these operators. It is remarked that using majority mating and mating by crossing over the results were somewhat inconclusive; no definite benefit was obtained. Using mating by averaging, however, led to 'spectacular effects' within a linear programming scheme, but these effects are not specified.

Scanning crossover has been introduced as a generalization and extension of uniform crossover in GAs creating one child from r parents (Eiben 1991, Eiben *et al* 1994). The name is based on the following general procedure scanning parents and thus building the child from left to right. Let $x^1, \ldots, x^r$ be the selected parents of length L and let x denote the child.

> **procedure** scanning:
>
> **begin**
>
> > INITIALIZE position markers as $i_1 = \ldots = i_r := 1$;
> >
> > *% mark 1st position in each parent*
> >
> > **for** $i = 1$ **to** $i = L$
> >
> > > CHOOSE $j \in \{1, \ldots, r\}$;
> > >
> > > $x_i := x_{i_j}^{j}$;   *% ith allele of x is the i_jth allele of x^j*
> > >
> > > UPDATE position markers $i_1, \ldots, i_r$;
>
> **end**

The above procedure provides a general framework for a certain style of multiparent recombination, where the precise execution, hence the exact definition of the operator, depends on the mechanisms of CHOOSE and UPDATE. In the simplest case the UPDATE operation can shift the markers one position to the right; that is, $i_j := i_j + 1$, $j \in \{1, \ldots, r\}$, can be used. This is appropriate for bitstrings, integer, and floating-point representation. Scanning can also be easily adapted to order-based representation, where each individual is a permutation, if the UPDATE operation shifts to the first allele which is not in the child yet:

$$i_j := \min\{k \mid k \geq i_j, \ x_k^j \notin \{x_1, \ldots, x_{i_j}\}\} \qquad j \in \{1, \ldots, r\}.$$

Observe that, because of the term $k \geq i_j$ above, a marker can remain at the same position after an UPDATE, and will only be shifted if the allele standing at that position is included in the child. This guarantees that each offspring will be a permutation.

Depending on the mechanism of choosing a parent (and thereby an allele) there are three different versions of scanning. The choice can be deterministic, choosing a parent containing the allele with the highest number of occurrences and breaking ties randomly (*occurrence-based scanning*). Alternatively it can be random, either unbiased, following a uniform distribution thus giving each parent an equal chance to deliver its allele (*uniform scanning*), or biased by the fitness of the parents, where the chance of being chosen is fitness proportional (*fitness-based scanning*). Uniform scanning for $r = 2$ is the same as uniform crossover, although creating only one child, and it also coincides with discrete recombination in evolution strategies. The occurrence-based version is very much like the voting or majority mating mechanism discussed before, but without the threshold v or with $v = \lfloor m/2 \rfloor$ respectively. The effect of the number of parents in scanning crossover has been studied in several papers. An overview of these studies is given in the next subsection.

Diagonal crossover has been introduced as a generalization of one-point crossover in GAs (Eiben *et al* 1994). In its original form diagonal crossover creates r children from r parents by selecting $r-1$ crossover points in the parents and composing the children by taking the resulting r chromosome segments from the parents 'along the diagonals'. Later on, a one-child version was introduced (van Kemenade *et al* 1995). Figure 33.11 illustrates both variants. It is easy to see that for $r = 2$ diagonal crossover coincides with one-point crossover, and in some sense it also generalizes traditional two-parent n-point crossover. To be precise, if we define (r, s) *segmentation crossover* as working on r parents with s crossover points, diagonal crossover becomes its $(r, r - 1)$ version, its $(2, n)$ variant coincides with n-point crossover, and one-point crossover is an instance of both schemes for $(r, s) = (2, 1)$ as parameters. The effect of operator arity for diagonal crossovers will be also discussed in the next subsection.

A recombination mechanism with tunable arity in ES is proposed by Schwefel and Rudolph (1995). The $(\mu, \kappa, \lambda, \rho)$ ES provides the possibility of freely adjusting the number of parents (called ancestors by the authors). The parameter ρ stands for the number of parents and global recombination is redefined for any given set $\{x^1, \ldots, x^\rho\}$ of parents as

$$x_i = \begin{cases} x_i^j & \rho\text{ary discrete recombination} \\ (1/\rho) \sum_{k=1}^{\rho} x_i^k & \rho/\rho \text{ intermediate recombination} \end{cases}$$

where $j \in \{1, \ldots, \rho\}$ is uniform randomly chosen for each i independently. Let us note that, in the original paper, the above operators are called uniform crossover and global intermediate recombination respectively. We introduce the names ρary discrete recombination and ρ/ρ intermediate recombination respectively here for the sake of a consequent terminology. (A reason for using the term ρ/ρ intermediate recombination instead of ρary intermediate recombination is given below, in the paragraph discussing a paper by Eiben and Bäck (1997).) Observe that ρary discrete recombination coincides with uniform scanning crossover, while ρ/ρ intermediate recombination is a special case of mating by averaging. At this time there are no experimental results available on the effect of ρ within this framework.

Related work in ESs also uses ρ as the number of parents as an independent parameter for recombination (Beyer 1995). For purposes of a theoretical analysis it is assumed that all parents are different, uniform randomly chosen from the population of μ individuals. Beyer defines the ρ/ρ intermediate recombination

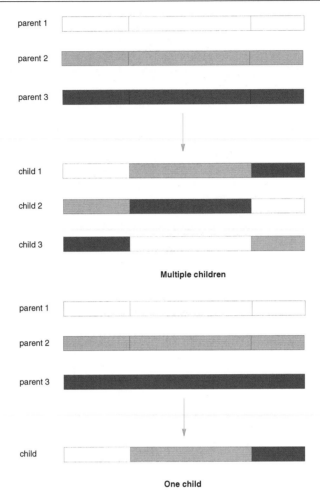

Figure 33.11. Diagonal crossover (top) and its one-child version (bottom) for three parents.

and ρary discrete recombinations similarly to Schwefel and Rudolph (1995) and denotes them as intermediate (μ/ρ_I) recombination and dominant (μ/ρ_D) recombination, respectively. The $(\mu/\rho, \lambda)$ evolution strategy is studied on the spherical function for the special case of $\rho = \mu$. By this latter assumption it is not possible to draw conclusions on the effect of ρ, but the analysis shows that the optimal progress rate $\hat{\varphi}^*$ of the $(\mu/\mu, \lambda)$ ES is a factor of μ higher than that of the (μ, λ) ES, for both recombination mechanisms. Beyer hypothesizes that recombination has a statistical error correction effect, called genetic repair, and this effect can be improved by using more than two parents for creating offspring (Beyer 1996).

Another generalization of global intermediate recombination in evolution strategies is proposed by Eiben and Bäck (1997). The new operator is applied after selecting ρ parent individuals from the population of μ, and the resampling of two donors x^{S_i} and x^{T_i} for each i takes only these ρ individuals into consideration. Note that this operator is also ρary, just like the ρ/ρ intermediate recombination as defined above, but utilizes only two donors for each allele of the offspring. To express this difference, this operator is called $\rho/2$ *intermediate recombination* and the operator of Beyer (1995) and Schwefel and Rudolph (1995) is called ρ/ρ intermediate recombination. Observe, that the $\rho/2$ intermediate recombination is a true generalization of the original intermediate recombination: the case of $\rho = 2$ coincides with local intermediate recombination, while for $\rho = \mu$, it is equal to global intermediate recombination.

While intermediate recombination is based on taking the arithmetical average of the real-valued alleles of the parents, the geometrical average is computed by the *geometrical crossover*. Michalewicz *et al* (1996) present the definition for any ($k \geq 2$) number of parents, where the offspring of the parents $\{x^1, \ldots, x^k\}$ is defined as

$$x^{k+1} = \langle (x_1^1)^{\alpha_1} (x_1^2)^{\alpha_2} \ldots (x_1^k)^{\alpha_k}, \ldots, (x_n^1)^{\alpha_1} (x_n^2)^{\alpha_2} \ldots (x_n^k)^{\alpha_k} \rangle$$

where n is the chromosome length and $\alpha_1 + \ldots + \alpha_k = 1$. The experimental part of the paper is, however, based on the two-parent version, hence there are no results on the effect of using more than two parents with this operator.

The same holds for the so-called *sphere crossover* (Schoenauer and Michalewicz 1997); the authors give the general definition for k parents, but the experiments are restricted to the two-parent version. In the general case the offspring of parents $\{x^1, \ldots, x^k\}$ is defined as

$$x^{k+1} = \left\langle \left[\alpha_1 (x_1^1)^2 + \ldots + \alpha_k (x_1^k)^2 \right]^{1/2}, \ldots, \left[\alpha_1 (x_n^1)^2 + \ldots + \alpha_k (x_n^k)^2 \right]^{1/2} \right\rangle.$$

33.7.5 *The effects of higher operator arities*

In recent years quite a few papers have studied the effect of operator arity on EA performance, some even in combination with varying selective pressure. Here we give a brief summary of these results, sorted by articles.

The performance of scanning crossover for different numbers of parents is studied by Eiben *et al* (1994) in a generational GA with proportional selection. Bit-coded GAs for function optimization (DeJong functions F1–4 and a function from Michalewicz) as well as order-based GAs for graph coloring and the TSP are tested with different mechanisms to CHOOSE. In the bit-coded case more parents perform better than two; for the TSP and graph coloring two parents are advisable. Comparing different biases in choosing the child allele, on four out of the five numerical problems fitness-based scanning outperforms the other two and occurrence-based scanning is the worst operator.

Eiben *et al* (1995) investigate diagonal crossover, compared to the classical two-parent n-point crossover and uniform scanning in a steady-state GA with linear-ranked biased selection ($b = 1.2$) and worst-fitness deletion. The test suite consists of two two-dimensional problems (F2 and a function from Michalewicz) and four scalable functions (after Ackley, Griewangk, Rastrigin, and Schwefel). The performance of diagonal crossover and n-point crossover shows a significant correspondence with r and n, respectively. The best performance is always obtained with high values, between 10 and 15, where 15 was the maximum tested. Besides, diagonal crossover is always better than n-point crossover using the same number of crossover points ($r = n - 1$), thus representing the same level of disruptiveness. For scanning the relation between r and performance is less clear, although the best performance is achieved for more than two parents on five out of the six test functions.

The interaction between selection pressure and the parameters r for diagonal crossover and n for n-point crossover is investigated by van Kemenade *et al* (1995). A steady-state GA with tournament selection (tournament size between one and six) combined with random deletion and worst-fitness deletion was applied to the Griewangk and the Schwefel functions. The disruptiveness of both operators increases in parallel as the values for r and n are raised, but the experiments show that diagonal crossover consistently outperforms n-point crossover. The best option proves to be low selection pressure and high r in diagonal crossover combined with worst-fitness deletion.

Motivated by the difficulties of characterizing the shapes of numerical objective functions, the effects of operator arity are studied on fitness landscapes with controllable ruggedness by Eiben and Schippers (1996). The NK landscapes of Kauffman (1993), where the level of epistasis, hence the ruggedness of the landscape, can be tuned by the parameter K, are used for this purpose. The multiple-child and the one-child version of diagonal crossover and uniform scanning are tested within a steady-state GA with linear-ranked biased selection ($b = 1.2$) and worst-fitness deletion for $N = 100$ and different values of K. Two kinds of epistatic interaction, nearest-neighbor interaction (NNI) and random-neighbor interaction (RNI), are considered. Similarly to earlier findings (Eiben *et al* 1995), the tests show that the performance of uniform scanning cannot be related to the number of parents. The two versions of diagonal crossover behave identically, and for both operators there is a consequent improvement when increasing r. However, as the epistasis (ruggedness of the landscape) grows from $K = 1$ to $K = 5$ the advantage of more parents becomes smaller. On landscapes with significantly high epistasis ($K = 25$) the relationship between operator arity and algorithm performance seems to diminish. We illustrate these observations with a figure showing the error (deviation of the best individual from the optimum) at termination for the case of NNI in figure 33.12. The final conclusions of this investigation can be very well related to works of Schaffer and Eshelman (1991), Eshelman and Schaffer (1993) and Hordijk and Manderick (1995) on the usefulness of (two-parent)

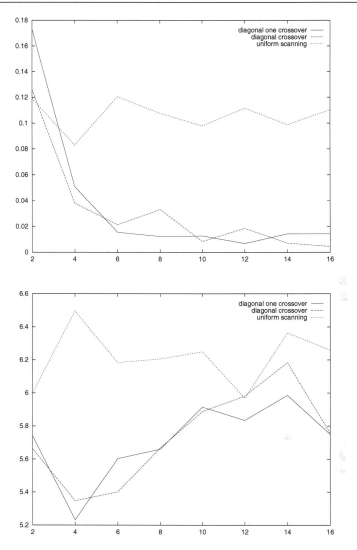

Figure 33.12. Illustration of the effect of the number of parents (horizontal axis) on the error at termination (vertical axis) on NK landscapes with NNI, $N = 100$, $K = 1$ (top), $K = 25$ (bottom).

recombination. It seems that if and when crossover is useful, that is, on mildly epistatic problems, then multiparent crossover can be more useful than the two-parent variants.

The results of an extensive study of diagonal crossover for numerical optimization in GAs are reported by Eiben and van Kemenade (1997). Diagonal crossover is compared to its one-offspring version and n-point crossover on a test suite consisting of eight functions, monitoring the speed, that is, the total number of evaluations, the accuracy, that is, the median of the best objective function value found (all functions have an optimum of zero), and the success rate, that is, the percentage of runs where the global optimum is found. In most cases an increase of performance can be achieved by increasing the disruptivity of the crossover operator (using higher values of n for n-point crossover), and even more improvement is achieved if the disruptivity of the crossover operator *and* the number of parents is increased (using more parents for diagonal crossover). This study gives a strong indication that for diagonal crossover an advantageous multiparent effect does exist, that is, (i) using this operator with more than two parents increases GA performance and (ii) this improvement is not only the consequence of the increased number of crossover points.

A recent investigation of Eiben and Bäck (1997) addresses the working of multiparent recombination operators in continuous search spaces, in particular within ESs. This study compares $\rho/2$ intermediate recombination, ρary discrete recombination, which is identical to uniform scanning crossover, and diagonal crossover with one child. Experiments are performed on unimodal landscapes (sphere model and Schwefel's double sum), multimodal functions with regularly arranged optima and a superimposed unimodal topology (Ackley, Griewangk, and Rastrigin functions) and on the Fletcher–Powell and the Langermann functions that have an irregular, random arrangement of local optima. On the Fletcher–Powell function multiparent recombination does not increase evolutionary algorithm (EA) performance; besides for the unimodal double sum increasing operator arity decreases performance. Other conclusions seem to depend on the operator in question; the greatest consequent improvement on raising the number of parents is obtained for diagonal crossover.

33.7.6 Conclusions

The idea of applying more than two parents for recombination in an evolutionary problem solver occurred as early as the 1960s (Bremermann *et al* 1966). Several authors have designed and applied recombination operators with higher arities for a specific task, or used an existing operator with an arity higher than two (Kaufman 1967, Mühlenbein 1989, Bersini and Seront 1992, Furuya and Haftka 1993, Aizawa 1994, Pál 1994). Nevertheless, investigations explicitly devoted to the effect of operator arity on EA performance are still scarce; the study of the phenomenon of multiparent recombination has just begun. What would such a study mean? Similarly to the question of whether binary reproduction operators (crossover with two parents) have advantages over unary ones (using mutation only), it can be investigated whether or not using more than two parents is advantageous. In the case of operators with tunable arity this question can be refined and the relationship between operator arity and algorithm performance can be studied. It is, of course, questionable whether multiparent recombination can be considered as one single phenomenon showing one behavioral pattern.

The survey presented here discloses that there are (at least) three different types of multiparent mechanism with tunable arity:

(i)   operators based on allele frequencies among the parents, such as majority mating, voting recombination, ρary discrete recombination, or scanning crossover;

(ii) operators based on segmenting and recombining the parents, such as mating by crossing over, diagonal crossover, or (r, s) segmentation crossover;

(iii) operators based on numerical operations, in particular averaging of (real-valued) alleles, such as mating by averaging, ρ/ρ intermediate recombination, $\rho/2$ intermediate recombination, and geometrical and spherical crossover.

A *priori* it cannot be expected that these different schemes show the same response to raising operator arities. There are also experimental results supporting differentiation among various multiparent mechanisms. For instance, there seems to be no clear relationship between the number of parents and the performance of uniform scanning crossover, while the opposite is true for diagonal crossover (Eiben and Schippers 1996).

Another aspect multiparent studies have to take into consideration is the expected different behavior on different types of fitness landscape. As no single technique would work on every problem, multiparent mechanisms will have their limitations too. Some studies indicate that on irregular landscapes, such as NK landscapes with relatively high K values (Eiben and Schippers 1996), or the Fletcher–Powell function (Eiben and Bäck 1997), they do not work. On the other hand, on the same Fletcher–Powell function Eiben and van Kemenade (1997) observed an advantage of increasing the number of parents for diagonal crossover in a GA framework using bit coding of variables, although they also found indications that this can be an artifact, caused simply by the increased disruptiveness of the operator for higher arities. Investigations on multiparent effects related to fitness landscape characteristics smoothly fit into the tradition of studying the (dis)advantages of two-parent crossovers under different circumstances (Schaffer and Eshelman 1991, Eshelman and Schaffer 1993, Spears 1993, Hordijk and Manderick 1995).

Let us also touch on the issue of practical difficulties when using multiparent recombination operators. Introducing operator arity as a new parameter implies an obligation of setting its value. Since so far there are no reliable heuristics for setting this parameter, finding good values may require numerous tests, prior to 'real' application of the EA. A solution may be based on previous work on adapting (Davis 1989) or self-adapting (Spears 1995) the frequency of applying different operators. Alternatively, a number of competing subpopulations could be used in the spirit of Schlierkamp-Voosen and Mühlenbein (1996). According to the latter approach each different arity is used within one subpopulation and subpopulations with greater progress, that is, with more powerful operators, become larger. A first assessment of this technique can be found in an article by

Eiben *et al* (1998a). Another recent result indicates the advantage of using more parents in the context of constraint satisfaction problems (Eiben *et al* 1998b).

Concluding this survey we can note the following. Even though there are no biological analogies of recombination mechanisms where more than two parent genotypes are mixed in one single recombination act, formally there is no necessity to restrict the arity of reproduction mechanisms to one (mutation) or two (crossover) in computer simulations. Studying the phenomenon of multiparent recombination has just begun, but there is already substantial evidence that applying more than two parents can increase the performance of EAs. Considering multiparent recombination mechanisms is thus a sound design heuristic for practitioners and a challenge for theoretical analysis.

References

Ackley D H 1987a *A Connectionist Machine for Genetic Hillclimbing* (Boston, MA: Kluwer)
——1987b An empirical study of bit vector function optimization *Genetic Algorithms and Simulated Annealing* ed L Davis (San Mateo, CA: Morgan Kaufmann) pp 170–215
Aizawa A N 1994 Evolving SSE: a stochastic schemata exploiter *Proc. 1st IEEE Conf. on Evolutionary Computation (Orlando, FL, 1990)* (Piscataway, NJ: IEEE) pp 525–9
Altenberg L 1995 The schema theorem and Price's theorem *Foundations of Genetic Algorithms 3* ed L Whitley and M Vose (San Mateo, CA: Morgan Kaufmann)
Bäck T 1996 *Evolutionary Algorithms in Theory and Practice* (New York: Oxford University Press)
Bäck T, Rudolph G and Schwefel H-P 1993 Evolutionary programming and evolution strategies: similarities and differences *Proc. 2nd Ann. Conf. on Evolutionary Programming (San Diego, CA)* ed D B Fogel and W Atmar (La Jolla, CA: Evolutionary Programming Society) pp 11–22
Bäck T and Schütz M 1995 Evolution strategies for mixed-integer optimization of optical multilayer systems *Proc. 4th Ann. Conf. on Evolutionary Programming (San Diego, CA, March 1995)* ed J R McDonnell, R G Reynolds and D B Fogel (Cambridge, MA: MIT Press) pp 33–51
Bersini H and Seront G 1992 In search of a good evolution–optimization crossover *Parallel Problem Solving from Nature, 2 (Proc. 2nd Int. Conf. on Parallel Problem Solving from Nature, Brussels, 1992)* ed R Männer and B Manderick (Amsterdam: Elsevier–North-Holland) pp 479–88
Beyer H-G 1995 Toward a theory of evolution strategies: on the benefits of sex—the $(\mu/\mu, \lambda)$ theory *Evolutionary Comput.* **3** 81–111
——1996 Basic principles for a unified EA-theory *Evolutionary Algorithms and their Applications Workshop (Dagstuhl, 1996)*
Birgmeier M 1996 Evolutionary programming for the optimization of trellis-coded modulation schemes *Proc. 5th Ann. Conf. on Evolutionary Programming* ed L J Fogel, P J Angeline and T Bäck (Cambridge, MA: MIT Press)
Blanton J and Wainwright R 1993 Multiple vehicle routing with time and capacity constraints using genetic algorithms *Proc. 5th Int. Conf. on Genetic Algorithms (Urbana-Champaign, IL, July 1993)* ed S Forrest (San Mateo, CA: Morgan Kaufmann) pp 452–9

Booker L B 1982 *Intelligent Behavior as an Adaptation to the Task Environment* Doctoral Dissertation, Department of Computer and Communication Sciences, University of Michigan

——1987 Improving search in genetic algorithms *Genetic Algorithms and Simulated Annealing* ed L Davis (San Mateo, CA: Morgan Kaufmann)

——1993 Recombination distributions for genetic algorithms *Foundations of Genetic Algorithms 2* ed L Whitley (San Mateo, CA: Morgan Kaufmann)

Bremermann H, Rogson M and Salaff S 1966 Global properties of evolution processes *Natural Automata and Useful Simulations* ed H Pattee, E Edlsack, L Fein and A Callahan (Washington, DC: Spartan) pp 3–41

Bridges C L and Goldberg D E 1987 An analysis of reproduction and crossover in a binary-coded genetic algorithm *Proc. 2nd Int. Conf. on Genetic Algorithms (Cambridge, MA, 1987)* ed J J Grefenstette (Cambridge, MA: Erlbaum) pp 9–13

Christiansen F B 1989 The effect of population subdivision on multiple loci without selection *Mathematical Evolutionary Theory* ed M W Feldman (Princeton, NJ: Princeton University Press)

Cramer N L 1985 A representation for the adaptive generation of simple sequential programs *Proc. 1st Int. Conf. on Genetic Algorithms and Their Applications* ed J J Grefenstette (Hillsdale, NJ: Erlbaum) pp 183–7

Davis L 1985 Applying adaptive algorithms to epistatic domains *Proc. Int. Joint Conf. on Artificial Intelligence*

——1989 Adapting operator probabilities in genetic algorithms *Proc. 3rd Int. Conf. on Genetic Algorithms (Fairfax, VA, June 1989)* ed J D Schaffer (San Mateo, CA: Morgan Kaufmann) pp 61–9

——(ed) 1991 *Handbook of Genetic Algorithms* (New York: Van Nostrand Reinhold)

De Jong K A 1975 *An Analysis of the Behavior of a Class of Genetic Adaptive Systems* Doctoral Dissertation, Department of Computer and Communication Sciences, University of Michigan

Dzubera J and Whitley D 1994 Advanced correlation analysis of operators for the traveling salesman problem *Parallel Problem Solving from Nature—PPSN III (Proc. Int. Conf. on Evolutionary Computation and 3rd Conf. on Parallel Problem Solving from Nature, Jerusalem, October 1994) (Lecture Notes in Computer Science 866)* ed Yu Davidor, H-P Schwefel and R Männer (Berlin: Springer) pp 68–77

Eiben A 1991 A method for designing decision support systems for operational planning *PhD Thesis* Eindhoven University of Technology

Eiben A and Bäck T 1997 An empirical investigation of multi-parent recombination operators in evolution strategies *Evolutionary Comput.* **5** 347–65

Eiben A E, Raué P-E and Ruttkay Zs 1994 Genetic algorithms with multiparent recombination *Parallel Problem Solving from Nature—PPSN III (Proc. Int. Conf. on Evolutionary Computation and 3rd Conf. on Parallel Problem Solving from Nature, Jerusalem, October 1994) (Lecture Notes in Computer Science 866)* ed Yu Davidor, H-P Schwefel and R Männer (Berlin: Springer) pp 77–87

Eiben A and Schippers C 1996 Multi-parent's niche: *n*-ary crossovers on NK-landscapes *Proc. 4th Int. Conf. on Parallel Problem Solving from Nature (Berlin, 1996) (Lecture Notes in Computer Science 1141)* ed H-M Voigt, W Ebeling, I Rechenberg and H-P Schwefel (Berlin: Springer) pp 319–28

Eiben A, Sprinkhuizen-Kuyper I and Thijssen B 1998a Competing crossovers in an adaptive GA framework *Proc. 5th IEEE Conf. on Evolutionary Computation (Anchorage, AK, May 1998)* (Piscataway, NJ: IEEE) at press

Eiben A, van der Hauw J and van Hemert J 1998b Graph coloring with adaptive evolutionary algorithms *J. Heuristics* **4** 25–46

Eiben A and van Kemenade C 1997 Diagonal crossover in genetic algorithms for numerical optimization *J. Control Cybernet.* **26** 447–65

Eiben A E, van Kemenade C H M and Kok J N 1995 Orgy in the computer: multi-parent reproduction in genetic algorithms *Advances in Artificial Life (Proc. 3rd. Eur. Conf. on Artificial Life, Granada) (Lecture Notes in Artificial Intelligence 929)* ed F Morán, A Moreno, J J Merelo and P Chacón (Berlin: Springer) pp 934–45

Eshelman L J 1991 The CHC adaptive search algorithm: how to have safe search when engaging in nontraditional genetic recombination *Foundations of Genetic Algorithms* ed G Rawlins (San Mateo, CA: Morgan Kaufmann)

Eshelman L J, Caruana R A and Schaffer J D 1989 Biases in the crossover landscape *Proc. 3rd Int. Conf. on Genetic Algorithms (Fairfax, VA, June 1989)* ed J D Schaffer (San Mateo, CA: Morgan Kaufmann) pp 10–19

Eshelman L and Schaffer J 1993 Crossover's niche *Proc. 5th Int. Conf. on Genetic Algorithms (Urbana-Champaign, IL, 1993)* ed S Forrest (San Mateo, CA: Morgan Kaufmann) pp 9–14

Fogel D B 1995 *Evolutionary Computation: Toward a New Philosophy of Machine Intelligence* (Piscataway, NJ: IEEE)

Fogel L J 1964 *On the Organization of Intellect* Doctoral Dissertation, UCLA

Fogel L J and Fogel D B 1986 Artificial intelligence through evolutionary programming *Final Report for US Army Research Institute* contract no PO-9-X56-1102C-1

Fogel L J, Owens A J and Walsh M J 1966 *Artificial Intelligence Through Simulated Evolution* (New York: Wiley)

Fox B R and McMahon M B 1991 Genetic operators for sequencing problems *Foundations of Genetic Algorithms* ed G J E Rawlins (San Mateo, CA: Morgan Kaufmann) pp 284–300

Furuya H and Haftka R T 1993 Genetic algorithms for placing actuators on space structures *Proc. 5th Int. Conf. on Genetic Algorithms (Urbana-Champaign, IL, July 1993)* ed S Forrest (San Mateo, CA: Morgan Kaufmann) pp 536–42

Geiringer H 1944 On the probability theory of linkage in Mendelian heredity *Ann. Math. Stat.* **15** 25–57

Goldberg D and Lingle R Jr 1985 Alleles, loci, and the traveling salesman problem *Proc. 1st Int. Conf. on Genetic Algorithms (Pittsburgh, PA, July 1995)* ed J J Grefenstette (Hillsdale, NJ: Erlbaum)

Gorges-Schleuter M 1989 ASPARAGOS: an asynchronous parallel genetic optimization strategy *Proc. 3rd Int. Conf. on Genetic Algorithms (Fairfax, VA, June 1989)* ed J D Schaffer (San Mateo, CA: Morgan Kaufmann)

Grefenstette J 1986 Optimization of control parameters for genetic algorithms *IEEE Trans. Syst. Man Cybern.* **SMC-16** 122–8

Holland J H 1975 *Adaptation in Natural and Artificial Systems* (Ann Arbor, MI: University of Michigan Press)

——1989 Searching nonlinear functions for high values *Appl. Math. Comput.* **32** 255–74

Hordijk W and Manderick B 1995 The usefulness of recombination *Advances in Artificial Life (Proc. 3rd Int. Conf. on Artificial Life, Granada) (Lecture Notes in Artificial*

Intelligence 929) ed F Morán, A Moreno, J J Merelo and P Chacón (Berlin: Springer) pp 908–19

Julstrom B A 1995 What have you done for me lately? Adapting operator probabilities in a steady-state genetic algorithm *Proc. 6th Int. Conf. on Genetic Algorithms (Pittburgh, PA, July 1995)* ed L J Eshelman (San Francisco, CA: Morgan Kaufmann) pp 81–7

Kauffman S 1993 *Origins of Order: Self-Organization and Selection in Evolution* (New York: Oxford University Press)

Kaufman H 1967 An experimental investigation of process identification by competitive evolution *IEEE Trans. Syst. Sci. Cybernet.* **SSC-3** 11–6

Koza J R 1992 *Genetic Programming: on the Programming of Computers by Means of Natural Selection* (Cambridge, MA: MIT Press)

——1994 *Genetic Programming II: Automatic Discovery of Reusable Programs* (Cambridge, MA: MIT Press)

Liepins G and Vose M 1992 Characterizing crossover in genetic algorithms *Ann. Math. Artificial Intell.* **5** 27–34

Michalewicz Z 1996 *Genetic Algorithms + Data Structures = Evolution Programs* 3rd edn (Berlin: Springer)

Michalewicz Z, Nazhiyath G and Michalewicz M 1996 A note on the usefulness of geometrical crossover for numerical optimization problems *Proc. 5th Ann. Conf. on Evolutionary Programming* ed L J Fogel, P J Angeline and T Bäck (Cambridge, MA: MIT Press) pp 305–12

Mitchell T M 1980 The need for biases in learning generalizations *Technical Report* CBM-TR-117, Department of Computer Science, Rutgers University

Montana D J 1995 Strongly typed genetic programming *Evolutionary Comput.* **3** 199–230

Mühlenbein H 1989 Parallel genetic algorithms, population genetics and combinatorial optimization *Proc. 3rd Int. Conf. on Genetic Algorithms (Fairfax, VA, 1989)* ed J D Schaffer (San Mateo, CA: Morgan Kaufmann) pp 416–21

——1991 Evolution in time and space—the parallel genetic algorithm *Foundations of Genetic Algorithms* ed G J E Rawlins (San Mateo, CA: Morgan Kaufmann)

Mühlenbein H and Voigt H-M 1996 Gene pool recombination in genetic algorithms *Meta-Heuristics: Theory and Applications* ed I Osman and J Kelly (Dordrecht: Kluwer) pp 53–62

Oliver I, Smith D and Holland J 1987 A study of permutation crossover operators on the traveling salesman problem *Proc. 2nd Int. Conf. on Genetic Algorithms (Cambridge, MA, 1987)* ed J J Grefenstette (Hillsdale, NJ: Erlbaum) pp 224–30

Pál K 1994 Selection schemes with spatial isolation for genetic optimization *Parallel Problem Solving from Nature—PPSN III (Proc. Int. Conf. on Evolutionary Computation and 3rd Conf. on Parallel Problem Solving from Nature, Jerusalem, 1994) (Lecture Notes in Computer Science 866)* ed Yu Davidor, H-P Schwefel and R Männer (Berlin: Springer) pp 170–9

Radcliffe N J 1991 Forma analysis and random respectful recombination *Proc. 4th Int. Conf. on Genetic Algorithms (San Diego, CA, July 1991)* ed R K Belew and L B Booker (San Mateo, CA: Morgan Kaufmann) pp 222–9

——1994 The algebra of genetic algorithms *Ann. Math. Artificial Intell.* **10** 339–84

Radcliffe N and Surry P D 1995 Fitness variance of formae and performance prediction *Foundations of Genetic Algorithms 3* ed D Whitley and M Vose (San Mateo, CA: Morgan Kaufmann) pp 51–72

Reed J, Toombs R and Barricelli N A 1967 Simulation of biological evolution and machine learning *J. Theor. Biol.* **17** 319–42

Renders J-M and Bersini H 1994 Hybridizing genetic algorithms with hill-climbing methods for global optimization: two possible ways *Proc. 1st IEEE Conf. on Evolutionary Computation (Orlando, FL, June 1994)* (Piscataway, NJ: IEEE) pp 312–7

Robbins R B 1918 Some applications of mathematics to breeding problems, III *Genetics* **3** 375–89

Schaffer J D, Caruana R A, Eshelman L J and Das R 1989 A study of control parameters affecting online performance of genetic algorithms for function optimization *Proc. 3rd Int. Conf. on Genetic Algorithms (Fairfax, VA, June 1989)* ed J D Schaffer (San Mateo, CA: Morgan Kaufmann) pp 51–60

Schaffer J D and Eshelman L 1991 On crossover as an evolutionary viable strategy *Proc. 4th Int. Conf. on Genetic Algorithms (San Diego, CA, 1991)* ed R Belew and L Booker (San Mateo, CA: Morgan Kaufmann) pp 61–8

Schaffer J D and Morishima A 1987 An adaptive crossover distribution mechanism for genetic algorithms *Proc. 2nd Int. Conf. on Genetic Algorithms (Cambridge, MA, 1987)* ed J J Grefenstette (Hillsdale, NJ: Erlbaum) pp 36–40

Schlierkamp-Voosen D and Mühlenbein H 1996 Adaptation of population sizes by competing subpopulations *Proc. 3rd IEEE Conf. on Evolutionary Computation* (Piscataway, NJ: IEEE) pp 330–5

Schnell F W 1961 Some general formulations of linkage effects in inbreeding *Genetics* **46** 947–57

Schoenauer M and Michalewicz Z 1997 Boundary operators for constrained parameter optimization problems *Proc. 7th Int. Conf. on Genetic Algorithms (East Lansing, MI, July 1997)* ed T Bäck (San Mateo, CA: Morgan Kaufmann) pp 320–9

Schwefel H-P 1995 *Evolution and Optimum Seeking* (New York: Wiley)

Schwefel H-P and Rudolph G 1995 Contemporary evolution strategies *Advances in Artificial Life. (Proc. 3rd Int. Conf. on Artificial Life, Granada) (Lecture Notes in Artificial Intelligence 929)* ed F Morán, A Moreno, J J Merelo and P Chacón (Berlin: Springer) pp 893–907

Smith J and Fogarty T 1996 Recombination strategy adaptation via evolution of gene linkage *Proc. 3rd IEEE Conf. on Evolutionary Computation* (Piscataway, NJ: IEEE) pp 826–31

Spears W 1993 Crossover or mutation? *Foundations of Genetic Algorithms—2* ed L Whitley (San Mateo, CA: Morgan Kaufmann) pp 221–38

——1995 Adapting crossover in evolutionary algorithms *Proc. 4th Ann. Conf. on Evolutionary Programming (San Diego, CA, 1995)* ed J R McDonnell, R G Reynolds and D B Fogel (Cambridge, MA: MIT Press) pp 367–84

Spears W M and De Jong K A 1991a An analysis of multi-point crossover ed G Rawlins *Foundations of Genetic Algorithms* (San Mateo, CA: Morgan Kaufmann)

——1991b On the virtues of parameterized uniform crossover *Proc. 4th Int. Conf. on Genetic Algorithms (San Diego, CA, July 1991)* ed R K Belew and L B Booker (San Mateo, CA: Morgan Kaufmann) pp 230–6

Srinivas M and Patnaik L M 1994 Adaptive probabilities of crossover and mutation in genetic algorithms *IEEE Trans. Syst. Man Cybernet.* **SMC-24** 656–67

Syswerda G 1989 Uniform crossover in genetic algorithms *Proc. 3rd Int. Conf. on Genetic Algorithms (Fairfax, VA, June 1989)* ed J D Schaffer (San Mateo, CA: Morgan Kaufmann) pp 2–9

——1991 Schedule optimization using genetic algorithms *Handbook of Genetic Algorithms* ed L Davis (New York: Van Nostrand Reinhold) pp 332–49

Thierens D and Goldberg D E 1993 Mixing in genetic algorithms *Proc. 5th Int. Conf. on Genetic Algorithms (Urbana-Champaign, IL, July 1993)* ed S Forrest (San Mateo, CA: Morgan Kaufmann) pp 38–45

van Kemenade C, Kok J and Eiben A 1995 Raising GA performance by simultaneous tuning of selective pressure and recombination disruptiveness *Proc. 2nd IEEE Conf. on Evolutionary Computation (Perth, 1995)* (Piscataway, NJ: IEEE) pp 346–51

Voigt H-M and Mühlenbein H 1995 Gene pool recombination and utilization of covariances for the breeder genetic algorithm *Proc. 2nd IEEE Int. Conf. on Evolutionary Computation (Perth, 1995)* (Piscataway, NJ: IEEE) pp 172–7

Voigt H-M, Mühlenbein H and Cvetković D 1995 Fuzzy recombination for the breeder genetic algorithm *Proc. 6th Int. Conf. on Genetic Algorithms (Pittsburgh, PA, 1995)* ed L J Eshelman (San Mateo, CA: Morgan Kaufmann) pp 104–11

Whitley D, Starkweather T and Shaner D 1991 Traveling salesman and sequence scheduling: quality solutions using genetic edge recombination *Handbook of Genetic Algorithms* ed L Davis (New York: Van Nostrand Reinhold) pp 350–72

Whitley D and Yoo N-W 1995 Modeling simple genetic algorithms for permutation problems *Foundations of Genetic Algorithms 3* ed D Whitley and M Vose (San Mateo, CA: Morgan Kaufmann) pp 163–84

Wright A H 1994 Genetic algorithms for real parameter optimization *Foundations of Genetic Algorithms* ed G Rawlins (San Mateo, CA: Morgan Kaufmann) pp 205–18

Wu A S and Lindsay R K 1995 Empirical studies of the genetic algorithm with noncoding segments *Evolutionary Comput.* **3** 121–48

Zhou H and Grefenstette J J 1986 Induction of finite automata by genetic algorithms *Proc. 1986 IEEE Int. Conf. on Systems, Man, and Cybernetics (Atlanta, GA)* pp 170–4

34

Other operators

Russell W Anderson (34.1), *David B Fogel* (34.2) *and*
Martin Schütz (34.3)

34.1 The Baldwin effect

Russell W Anderson†

34.1.1 Interactions between learning and evolution

In the course of an evolutionary optimization, solutions are often generated with low phenotypic fitness even though the corresponding genotype may be close to an optimum. Without additional information about the local fitness landscape, such genetic *near misses* would be overlooked under strong selection. Presumably, one could rank near misses by performing a local search and scoring them according to distance from the nearest optimum. Such evaluations are essentially the goal of hybrid algorithms (Chapters 11–13, Balakrishnan and Honavar 1995), which combine global search using evolutionary algorithms and local search using individual learning algorithms. Hybrid algorithms can exploit learning either actively (via Lamarckian inheritance) or passively (via the Baldwin effect).

Under Lamarckian algorithms, performance gains from individual learning are mapped back into the genotype used for the production of the next generation. This is analogous to Lamarckian inheritance in evolutionary theory—whereby characters acquired during a parent's lifetime are passed on to their offspring. Lamarckian inheritance is rejected as a biological mechanism under the modern synthesis, since it is difficult to envision a process by which acquired information can be transferred into the gametes. Nevertheless, the practical utility of Lamarckian algorithms has been demonstrated in some evolutionary optimization applications (Ackley and Littman 1994, Paechter *et al* 1995). Of course, these algorithms are limited to problems where a reverse mapping from the learned phenotype to genotype is possible.

† This work was supported by the Public Health Foundation and the Kett Foundation. The author wishes to thank David Fogel and Peter Turney for encouragement and comments.

However, even under purely Darwinian selection, individual learning influences evolutionary processes, but the underlying mechanisms are subtle. The 'Baldwin effect' is one such mechanism, whereby learning facilitates the assimilation of new genetic innovations (Baldwin 1896, Morgan 1896, Osborn 1896, Waddington 1942, Hinton and Nowlan 1987, Maynard Smith 1987, Anderson 1995a, Turney *et al* 1996). Learning allows an individual to complete and exploit partial genetic programs and thereby survive. In other words, learning guides evolution by assigning 'partial credit' for genetic near misses. Individuals with useful genetic variations are thus maintained by learning, and the corresponding genes increase in frequency in the subsequent generation. As genetic components necessary for a complex structure accumulate in the gene pool, functions that previously required supplemental learning are replaced by genetically determined systems.

Empirical studies can quantify the benefits of incorporating individual learning into evolutionary algorithms (Belew 1989, French and Messinger 1994, Nolfi *et al* 1994, Whitley *et al* 1994, Cecconi *et al* 1995). However, a theoretical treatment of the effects of learning on evolution can strengthen our intuition for *when* and *how* to implement such approaches. This section presents an overview of the principles underlying the Baldwin effect, beginning with a brief history of the elucidation and development in evolutionary biology. Computational models of the Baldwin effect are reviewed and critiqued. The Baldwin effect is then analyzed using standard quantitative genetics. Given reasonable assumptions of the effects of learning on fitness and its associated costs, this theoretical approach builds and strengthens conventional intuition about the effects of individual learning on evolution. Finally, issues concerning problem formulation, learning algorithms, and algorithmic design are discussed.

34.1.2 The Baldwin effect in evolutionary biology

Complex biological structures require the coordinated expression of several genes in order to function properly. Determining how such structures arise through evolution is problematic because it is often difficult to envision the evolutionary advantage offered by intermediate forms. Without additional developmental mechanisms, individuals with incomplete genetic programs would gain no evolutionary advantage over those devoid of any genetic components.

Baldwin (1896), Osborn (1896), and Morgan (1896) proposed how individual learning can facilitate the evolution of complex genetic structures by protecting partial genetic innovations, or 'ontogenetic variations': '[learning] supplements such partial co-ordinations, makes them functional, and so keeps the creature alive' (Baldwin 1896). Baldwin further proposed how this individual advantage of learning guides the process of evolution: 'the variations which were utilized for ontogenetic adaptation in the earlier generation, being thus kept in existence, are utilized more widely in the subsequent generation'

(Baldwin 1896). Over evolutionary time, abilities that were previously maintained by adaptive systems can be replaced by genetically determined systems (i.e. instincts). Waddington proposed an analogous interaction between developmental processes and evolution, whereby developmental adaptations 'guide' or 'canalize' evolutionary change (Waddington 1942, Hinton and Nowlan 1987). Formal mathematical or analytical models quantifying the Baldwin effect did not appear in the literature until fairly recently.

The model of Hinton and Nowlan. The first quantitative model demonstrating the Baldwin effect was constructed by Hinton and Nowlan (1987). They used a computer simulation to study the effects of individual learning on the evolution of a population of neural networks. They considered an extremely difficult problem, where a network conferred a fitness advantage only if it was fully functioning (all connections wired correctly). Each network was given 20 possible connections, specified by 20 genes.

Briefly consider the difficulty of finding this solution using a pure genetic algorithm. Under a binary genetic coding scheme (allelic values of either 'correct' or 'incorrect'), the probability of randomly generating a functional net is 2^{20}. Note that a net with 19 out of 20 correct connections is no better off than one with no correct connections. The corresponding fitness landscape has a singularity at the correct solution with no useful gradient information, analogous to a putting green (figure 34.1). Finding this solution by a pure genetic algorithm, then, is the evolutionary equivalent of a 'hole in one'. Of course, given a large enough random population, an evolutionary algorithm could theoretically find this solution in one generation.

Hinton and Nowlan modeled a modified version of this problem, where genes were allowed *three* alternative forms (alleles): present (1), absent (0), or 'plastic' (?). Connections specified by plastic alleles could be varied by random trials during the individual's life span. This allowed an individual to complete and exploit a partially hard-wired network. Hence, genetic near misses (e.g. 19 out of 20 correct genes) could quickly learn the remaining connection(s) and differentially survive. The presence of plastic alleles, therefore, softened the fitness landscape (figure 34.1). Hinton and Nowlan described the effect of learning ability in their simulation as follows: '[learning] alters the shape of the search space in which evolution operates and thereby provides good evolutionary paths towards sets of co-adapted alleles'. The second aspect of the Baldwin effect (genetic assimilation) was manifested in the mutation of plastic alleles into genetically fixed alleles.

Issues raised with computational models. Hinton and Nowlan's paper is regarded as a landmark contribution to understanding the interactions between learning and evolution (Mitchell and Belew 1995) and has inspired a proliferation of modeling studies (Fontanari and Meir 1990, Ackley and Littman

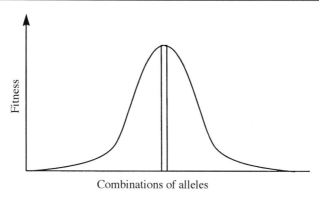

Figure 34.1. Schematic representation of the fitness landscape in the model of Hinton and Nowlan. A two-dimensional representation of genome space in the problem considered by Hinton and Nowlan (1987). The horizontal axis represents all possible gene combinations, and the vertical axis represents relative fitness. Without learning, only one combination of alleles correctly completes the network; hence only one genotype has higher fitness, and no gradient exists. The presence of plastic alleles radically alters this fitness landscape. Assume a correct mutation occurs in one of the 20 genes. The advent of a new correct gene only partially solves the problem. Learning allows individuals close (in Hamming space) to complete the solution. Thus, these individuals will be slightly more fit than individuals with no correct genes. Useful genes will thereby be increased in subsequent generations. Over time, a large number of correct genes will accumulate in the gene pool, leading to a completely genetically determined structure.

1991, 1994, Whitley and Gruau 1993, Whitley *et al* 1994, Balakrishnan and Honavar 1995, Turney 1995, 1996, Turney *et al* 1996). Considering the rather specific assumptions of their model, it is useful to contemplate which aspects of their results are general properties. Among the issues raised by this and subsequent studies are the degree of biological realism, the nature of the fitness landscape, the computational cost of learning, and the role of learning in static fitness landscapes.

First, the model's assumption of plastic alleles that can mutate into permanent alleles seems biologically spurious. However, the Baldwin effect can be manifested in the evolution of a biological structure regardless of the genetic basis of that structure or the mechanisms underlying the learning process (Anderson 1995a). The Baldwin effect is simply a consequence of individual learning on genetic evolution. Subsequent studies have demonstrated the Baldwin effect using a variety of learning algorithms. Turney (1995, 1996) has observed a Baldwin effect in a class of hybrid algorithms, combining a genetic algorithm (GENESIS) and an inductive learning algorithm, where the Baldwin effect was manifested in shifting biases in the inductive learner. French and Messinger (1994) investigated the Baldwin effect under various forms of phenotypic plasticity. Cecconi *et al* (1995) observed the Baldwin effect in a

GA+NN hybrid (a hybrid of a genetic algorithm and a neural network), as did Nolfi *et al* (1994) and Whitley and Gruau (1993). Unemi *et al* (1994) demonstrated the Baldwin effect in a GA+RL hybrid (GA and reinforcement learning; in particular, they studied Q-learning). Whitley *et al* (1994) studied the Baldwin effect with a hybrid of a GA and a simple hill climbing algorithm. Finally, it is interesting to note that genetic mechanisms closely analogous to the plastic alleles of Hinton and Nowlan may be in effect in evolutionary interactions between natural and adaptive antibodies (Anderson 1995b, 1996a). Nevertheless, it is difficult to see how this particular model could be generalized to learning in neural systems.

Second, the model of Hinton and Nowlan assumed an extremely rugged fitness landscape. The assumption of an 'all-or-nothing' fitness landscape has apparently led some to assert that a nonlinear selection function is *necessary* for a Baldwin effect to occur (Hightower *et al* 1996). This claim is not supported by rigorous analysis. Learning can alter the shape of *any* fitness landscape and therefore can affect evolutionary trajectories. For example, consider linear directional selection. If learning only serves to change the *slope* of the selection function, it will by definition affect its severity.

Third, the observation that 'learning *facilitates* evolution,' has often been interpreted as 'learning *accelerates* evolution'. Although several empirical studies have demonstrated increased convergence rates for hybrid algorithms (Parisi *et al* 1991, Turney 1995, Ackley and Littman 1991, 1994, Balakrishnan and Honavar 1995), this more general claim is untenable under many conditions. Intuitively, learning can *slow* genetic change by protecting otherwise less optimal genotypes from selection. Furthermore, individual adaptive abilities can represent an enormous investment of resources (consider the cerebral cortex in man!). Since individual learning accrues a computational or biological cost, the *costs and benefits* of learning must be weighed before drawing such conclusions.

Fourth, most current hybrid algorithm applications operate on a fixed problem, or static fitness landscape. An exception is a study by Unemi *et al* (1994), which involves a simulated robot in a maze. They show that the ability to learn is initially beneficial, but it will eventually be selected out of the gene pool, unless the maze changes dynamically with each new individual trial. Ultimately, learning has no selective advantage in fixed environments, since, presumably, once the optimal genotype is found, exploration away from this optimum only reduces fitness (Stephens 1993, Via 1993, Anderson 1995a). The studies by Hinton and Nowlan (1987) and Fontanari and Meir (1990) corroborate this thesis: their simulations showed that as individuals arose with allelic combinations close to the optimum, the plastic alleles (representing the ability to learn) were selected out of the gene pool. In other words, the computational advantage of individual learning decreases over the course of an evolutionary optimization. Under these conditions, individual learning can only be maintained in a population subject to changing environmental conditions. A similar case has been made for phenotypic plasticity in general (West-Eberhard 1989, Stearns

1989, Scheiner 1993, Via 1993) as well as for sexual versus asexual reproduction (Maynard-Smith 1978).

34.1.3 Quantitative genetics models

In order to make some of these issues more explicit, it is useful to study the Baldwin effect under the general assumptions of quantitative genetics. A quantitative genetics methodology for modeling the effects of learning on evolution was developed by Anderson (1995a), and the primary results of this analysis are reviewed in this section. The limitations of this theoretical approach are well known. For example, quantitative genetics assumes infinite population sizes. Also, complete analysis is often limited to a single quantitative character. Nevertheless, such analyses can provide a baseline intuition regarding the effects of learning and evolution.

All essential elements of an evolutionary process subject to the Baldwin effect are readily incorporated into a quantitative genetics model. These elements include (i) a function for the generation of new genotypes through mutation and/or recombination, (ii) a mapping from genotype to phenotype, (iii) a model of the effects of learning on phenotype, and (iv) a selection function. In this section, this methodology is demonstrated for a simple, first-order model, where only the phenomenological *effects* of learning on selection are considered. More advanced models are discussed, which incorporate a model of the learning process, along with its associated costs and benefits. These analyses illustrate several underappreciated points: (i) learning generally slows genetic change, (ii) learning offers no long-term selective advantage in fixed environments, and (iii) the effects of learning are somewhat independent of the mechanisms underlying the learning process.

Learning as a phenotypic variance. For a first-order model, consider an individual whose genotype is a real-valued quantitative character subject to normal (Gaussian) selection:

$$w_s(g) \sim N(g_e, V_s) \tag{34.1}$$

where $w_s(g)$ represents selection as a function of genotype, g_e represents the optimal genotype, and $V_s(t)$ is variance of selection as a function of time. A direct mapping from genotype to phenotype is implicitly assumed.

What effect does learning have on this selection function? Learning allows an individual to modify its phenotype in response to its environment. Consider an individual whose genotype (g_i) is a given distance ($|g_i - g_e|$) from the environmental optimum (g_e). Regardless of the mechanisms underlying the learning process, the net effect of learning is to reduce the fitness penalty associated with this genetic distance. Because of its ability to learn, an individual with genotype g_i has a probability of modifying its phenotype to the environmental value g_e which is a function of the distance between these two values. A simple way to model this effect is to specify a phenotypic variance due to learning (V_l). This is equivalent to increasing the variance of selection. Thus, learning increases the width of the selection function such that V_s is replaced by $V_s^* = V_s + V_l$.

Fixed selection, constant learning. Consider a population subject to selection with a fixed environmental optimum. For simplicity, let $g_e = 0$. Assume an initial Gaussian distribution of genotypes, $f_p(g) = N(m(t), V_p(t))$, where $m(t)$ and $V_p(t)$ are the population mean and variance at time t. Each round of selection changes the distribution of genotypes according to

$$f_p^*(g) \sim f_p(g) w_s(g) \tag{34.2}$$

$$\sim \exp\left\{-\tfrac{1}{2}[(g - m(t))^2/V_p(t) + g^2/V_s^*]\right\} \Rightarrow N(m, V_p) * N(0, V_s) \tag{34.3}$$

$$\sim \exp\left\{-\tfrac{1}{2}[(g - m^*(t))^2/V_p^*(t)]\right\} \Rightarrow N(m^*(t), V_p^*(t)). \tag{34.4}$$

The population mean and variance after selection (m^*, V_p^*) can now be expressed in the form of dynamic equations:

$$m^*(t) = \frac{m(t) V_s^*}{V_p(t) + V_s^*} = m(t) - \frac{m(t) V_p(t)}{V_p(t) + V_s^*} \tag{34.5}$$

$$V_p^*(t) = \frac{V_p(t) V_s^*}{V_p(t) + V_s^*} = V_p(t) - \frac{V_p^2(t)}{V_p(t) + V_s^*}. \tag{34.6}$$

Lastly, mutations are introduced in the production of the next generation of trials. To model this process, assume a Gaussian mutation function with mean zero and variance V_μ. A convolution of the population distribution with the mutation distribution has the effect of increasing the population variance:

$$f_p^{**}(g) = \int_{-\infty}^{+\infty} f_p^*(s) f_m(g - s) \, ds = N(m^*(t), V_p^{**}(t)) \tag{34.7}$$

where

$$V_p^{**}(t) = V_p(t) - \frac{V_p^2(t)}{V_p(t) + V_s^*} + V_\mu. \tag{34.8}$$

Hence, in a fixed environment the population mean $m(t)$ will converge on the optimal genotype (Bulmer 1985), while a mutation–selection equilibrium variance occurs at

$$V_{peq}^{**} = \frac{V_\mu + (V_\mu^2 + 4 V_\mu V_s^*)^{1/2}}{2}. \tag{34.9}$$

Inspection of equations (34.5), (34.6), and (34.8) illustrates two important points. First, learning slows the convergence of both $m^*(t)$ and $V_p^*(t)$. Second, once convergence in the mean is complete, the utility of learning is lost, and learning only reduces fitness.

In a more elaborate version of this model, called the *critical learning period model* (Anderson 1995a), a second gene is introduced to regulate the fraction of an individual's life span devoted to learning (duration of the learning period). Specification of a critical learning period implicitly assigns cost associated with learning (the percent of life span not devoted to reproduction). Individuals are then selected for the optimal combination of genotype and learning investment. It is easily demonstrated that under these assumptions, learning ability is selected out of a population subject to fixed selection.

Constant-velocity environments.   Next, consider a simple case of changing selection—a constantly moving optimum, $g_e(t) = \delta t$, where δ is defined as the environmental velocity. Let the difference between the population mean and the environmental optimum be defined as $\phi = m(t) - g_e(t)$. The dynamic equation for ϕ is

$$\phi^*(t) = \phi(t) - \frac{\phi(t)V_p(t)}{V_p(t) + V_s^*} + \delta. \tag{34.10}$$

At equilibrium, $\phi^*(t) = \phi(t)$, hence

$$\phi_{eq} = \frac{V_p + V_s^*}{V_p}\delta \tag{34.11}$$

where the equilibrium is expressed as a distance from the optimum. A similar result can be found in the article by Charlesworth (1993), in his analysis of the evolution of sex in a variable environment. The equilibrium population variance remains the same as in the case of a fixed environment. Substituting (34.9) yields

$$\phi_{eq} = \frac{\delta}{2}\left[(1 + (1 + 4V_s^*/V_\mu)^{1/2}\right]. \tag{34.12}$$

Thus in an environment where the optimal phenotype is changing at constant rate, the population mean genotype converges on a constant 'phase lag' (ϕ_{eq}).

Learning actually increases the phase lag by protecting suboptimal genotypes from selection. But this model assumes $\bar{w}_s = 1$, so that only the relative magnitude of selection is accounted for. Strong selection without learning might actually lead to *extinction* in rapidly changing selection. Phenotypic variability (due to learning) has the effect of 'shielding' these marginal genotypes from selection (Wright 1931).

As environmental conditions change, so will the selective advantage of learning. The relations derived in this analysis show which equilibria will be reached for an assumed phenotypic variability, but the model does not yield information on what would represent the optimal investment in learning. Hence, a complete model of the benefits of the Baldwin effect must incorporate the costs associated with learning. The best way to estimate these costs is to develop a model of the underlying learning process.

Models of learning. A reasonable question to ask is how sensitive are the effects of learning to the mechanisms underlying the learning process. The most direct (and exhaustive) method for investigating this question would be to construct a computer simulation to compare the effects of two learning processes in an evolutionary program. However, estimates of comparative performance can also be obtained using quantitative genetics models according to the following methodology. First, one must develop a model of the learning process. Next the effects of the learning process must be mapped onto the selection function. A simple approximation is to construct a probabilistic or phenomenological model of the effect of learning on phenotype.

Under the critical learning period model (Anderson 1995a), learning consists of a series of independent trials conducted over a fraction of the individual's life span, or learning period. This simple model incorporates two important considerations: the sequential nature of learning and a model of the cost associated with learning. Despite the more complicated assumptions, the dynamical response of this model to various forms of selection (fixed, random variation, and constant velocity) were qualitatively comparable to those derived for the simple additive variance model analyzed here. Longer learning periods increase the investment in (and cost of) learning: consequently, the amount of learning investment generally only increased with increased environmental variability.

Other models of the learning process can be incorporated using the methodology outlined above. For example, under the critical learning period model, individuals were not allowed to benefit from successive trials within the learning period, nor were they allowed to begin exploitation of successful trials until after the learning period. Removing these two restrictions yields a *sequential* trial-and-error learning rule. Such a learning rule is a more appropriate model of the learning process in some systems, such as affinity maturation in the antibody immune system (Milstein 1990) or skill acquisition in neural systems (Bremermann and Anderson 1991, Anderson 1996b). For these initial models, including such details of individual learning was unwarranted, but any model of learning can be mapped onto a fitness function, although mapping a sequential trial-and-error learning rule onto a survival probability may be analytically more difficult. Often it turns out that this mapping masks the details of the underlying process (Anderson 1995a). This suggests that the *effects* of individual learning on evolution will be qualitatively the same.

34.1.4 Conclusions

Baldwin's essential insight was that if an organism has the ability to learn, it can exploit genes that only partially determine a structure—increasing the frequencies of useful genes in subsequent generations. The Baldwin effect has also been demonstrated to be operative in hybrid evolutionary algorithms. These empirical investigations can be used to quantify the benefits of incorporating individual learning into an evolutionary algorithm. Computation time is the obvious performance criterion; however, such comparisons are often limited to the particular application. Alternatively, phenomenological models can be used to generate reasonable estimates of performance expectations, deferring the arduous task of creating detailed computer simulations.

The introduction of individual learning can radically alter fitness landscapes. This is especially true if the learning algorithm operates on phenotypes according to a fundamentally different process. Clearly, if the learning algorithm is identical to the genetic algorithm, no computational savings are likely to be manifest.

Under certain conditions, learning slows genetic change by protecting suboptimal genotypes from selection. Thus, the benefits of individual learning will probably be accrued early in optimization, when the population is far from equilibrium, and learning will eventually impede algorithmic convergence. Accordingly, for optimizations on fixed fitness landscapes, a 'variable-learning-investment' strategy—where the computational resources applied toward learning are subject to change—should be considered (Saravanan *et al* 1995, Anderson 1995a).

34.2   Knowledge-augmented operators

David B Fogel

Evolutionary computation methods are broadly useful because they are general search procedures. The canonical forms of the evolutionary algorithms do not take advantage of knowledge concerning the problem at hand. For example, in the canonical genetic algorithm (Holland 1975), a one-point crossover operator is suggested with a crossover point to be chosen randomly across the parents' chromosomes. However it is generally accepted that the effectiveness of a particular search operator depends on at least three interrelated factors: (i) the chosen representation, (ii) the selection criterion, and (iii) the objective function to be minimized or maximized, subject to the given constraints if applicable. There is no single best search operator for all problems.

Rather than rely on simple operators that may generate unacceptably inefficient performance on a particular problem at hand, the search operators can be tailored for individual applications. For example, in evolution strategies and evolutionary programming, when searching for the minimum of a quadratic surface, Rechenberg (1973) showed that the best choice for the standard deviation σ when using a zero mean Gaussian mutation operator was

$$\sigma = 1.224 f(x)^{1/2}/n$$

where $f(x)$ is the quadratic function evaluated at the parent vector x, and n is the dimensionality of the function. This choice of σ incorporates knowledge about the function being searched in order to provide the greatest expected rate of convergence. In this particular case, however, knowledge that the function is a quadratic surface indicates the use of search algorithms that can take greater advantage of the available gradient information (e.g. Newton–Gauss).

There are other instances where incorporating domain-specific knowledge into a search operator can improve the performance of an evolutionary algorithm. In the traveling salesman problem, under the objective function of minimizing the Euclidean distance of the circuit of cities, and a representation of simply an ordered listing of cities to be visited, Fogel (1988) offered a mutation operator which selected a city at random and placed it in the list at another randomly chosen position. This operator was not based on any knowledge about the nature of the problem. In contrast, Fogel (1993) offered an operator that instead inverted a segment of the listing (i.e. like a 2-opt of Lin and Kernighan (1976)). The inversion operator in the traveling salesman problem is a knowledge-augmented operator because it was devised to take advantage of the Euclidean geometry present in the problem. In the case of a traveling salesman's tour, if the tour crosses over itself it is always possible to improve the tour by undoing the crossover (i.e. the diagonals of a quadrangle are always longer in sum than any two opposite sides). When the two cities just before and after the crossing point are selected and the listing of cities in between reversed, the crossing is removed and the tour is improved. Note that this use of inversion is appropriate in light of the traveling salesman problem, and no broader generality of its effectiveness as an operator is suggested, or can be defended.

Domain knowledge can also be applied in the use of recombination. For example, again when considering the traveling salesman problem, Grefenstette *et al* (1985) suggested a heuristic crossover operator that could perform a degree of local search. The operator constructed an offspring from two parents by (i) picking a random city as the starting point, (ii) comparing the two edges leaving the starting cities in the parents and choosing the shorter edge, then (iii) continuing to extend the partial tour by choosing the shorter of the two edges in the parents which extend the tour. If a cycle were introduced, a random edge would be selected. Grefenstette *et al* (1985) noted that offspring were on average about 10% better than the better parent when implementing this operator.

In many real-world applications, the physics governing the problem suggests settings for search parameters. For example, in the problem of docking small molecules into protein binding sites, the intermolecular potential can be precalculated on a grid. Gehlhaar *et al* (1995) used a grid of 0.2 Å, with each grid point containing the summed interaction energy between an atom at that point and all protein atoms within 6 Å. This suggests that under Gaussian perturbations following an evolutionary programming or evolution strategy approach, a standard deviation of several ångströms would be inappropriate (i.e. too large).

Whenever evolutionary algorithms are applied to specific problems with the intention of generating the best available optimization performance, knowledge about the domain of application should be considered in the design of the search operators (and the representation, selection procedures, and indeed the objective function itself).

34.3   Gene duplication and deletion

Martin Schütz

34.3.1   Historical review

The idea of using operators such as gene duplication and deletion in the context of evolutionary algorithms (EAs) is as old as the algorithms themselves.

Fogel *et al* (1966) seemed to be one of the first experimenting with variable-length genotypes. In their work they evolved finite-state machines of a varying number of states, therefore making use of operators such as addition and deletion. Typically, the 'add a state' operator was performed randomly, rather than a strict duplication. They also suggested a 'majority logic' operator that essentially created a machine in which each state was the composite of a state from each of the original machines; that is, this operator duplicated the majority logic vote at each state of multiple finite-state machines. Concerning engineering problems Schwefel (1968) was one of the first using gene duplication and deletion for solving the task of determining the internal shape of a two-phase jet nozzle with maximum thrust under constant starting conditions. Holland (1975, p 111) proposed the concepts of gene duplication and gene deletion in order to raise the computational power of EAs.

34.3.2   Basic motivations for the use of gene duplication and deletion

From these first attempts concerning variable-length genotypes until now many researchers have made use of gene duplication and deletion. Four different motivations may be classified.

(i) *Engineering applications.* Many difficult optimization tasks arise from engineering applications in which variable-dimensional mixed-integer problems have to be solved. Often these problems are of dynamic nature: the optimum is time dependent. Additionally, in order to obtain a reasonable model of the system under consideration, a large number of constraints has to be respected during the optimization. Solving the given task frequently assumes the integration of expert (engineer) knowledge into the problem solving strategy: into particular genetic operators in the case of EAs. Many such constrained, variable-dimensional, mixed-integer, time-varying engineering problems and their solutions can be found in the handbook by Davis (1991) and in the proceedings of several conferences, such as the International Conference on

Evolutionary Computation (ICEC), Conference on Genetic Algorithms and Their Applications (ICGA), Conference on Evolutionary Programming (EP) and Parallel Problem Solving from Nature (PPSN).

(ii) *Raising the computational power of EAs.* As Goldberg *et al* (1989, p 493; see also Goldberg *et al* 1990) state, 'nature has formed its genotypes by progressing from simple to more complex life forms', thereby using variable-length genotypes. He states that genetic algorithms (GAs) using variable-length genotypes, thus being able to use duplication and deletion operators,

> solve problems by combining relatively short, well-tested building blocks to form longer, more complex strings that increasingly cover all features of a problem. ... Specifically, and more positively, we assert that allowing more messy strings and operators permits genetic algorithms to form and exploit tighter, higher performance building blocks than is possible with random, fixed codings and relatively slow reordering operators such as inversion.

Transferring this idea to EAs in general hopefully leads to more efficient EAs.

(iii) *Extradimensional bypass.* One additional motivation underpinning the usefulness of variable-dimensional genotype lengths is given by the *extradimensional bypass thesis* of Conrad (1993) (given more formally by Conrad and Ebeling (1992)), which states (maximization):

> As the number of dimensions increases the chance of our sitting on top of an isolated peak decreases, assuming that the space has random topographic features. The peaks will be transformed to saddlepoints. The rate of evolution will then depend on how long it takes to discover an uphill running pathway that requires a series of short steps and not on how long it takes to make a long jump from one peak to another.

For example, imagine an alpinist walking in a two-dimensional environment standing in front of a crater whose top he would like to reach. Even if he cannot see the highest peak, climbing the crater and walking on the border of the crater (extradimensional bypass) will lead him to the top. Walking in a one-dimensional space would complicate the task. This time the surface consists of two separated peaks (cut through the crater). If the alpinist climbs the first peak (eventually the higher one) he sees the highest peak but since one dimension is lost the desired path along the borderline of the crater does not exist. This time the alpinist has to descend into the valley in order to solve his task. As one can recognize, introducing extra dimensions during the course of evolution may overcome the problem of becoming stuck in a local optimum or, to put it in other words, decrease the necessity of jumping from one basin of attraction to another.

(iv) *Artificial intelligence.* Another important field in which variable-dimensional techniques have also been used is the domain of artificial intelligence (AI), especially *machine learning* (ML) and *artificial life* (AL). Whereas in the field of ML (subordinated fields are, for example, genetic programming, classifier systems, and artificial neural networks) solving a possibly variable-dimensional optimization problem (depending on the actual subordinated field in mind) is one main objective, this aim plays a minor role in the AL field. AL research concentrates on computer simulations of simple hypothetical life forms and selforganizing properties emerging from local interactions within a large number of basic agents (life forms). A second objective of AL is the question of how to make the agents' behavior adaptive, thus often leading to agents equipped with internal rules or strategies determining their behavior. In order to learn/evolve such rule sets, learning strategies such as EAs are used. Since the number of necessary rules is not known *a priori*, a variable-dimensional problem instance arises. Despite the rule learning task, the modeling of the simple life forms itself makes use of variable-dimensional genotypes.

34.3.3 *Formal description of gene duplication and deletion*

From the preceding motivations one can see that solving variable-dimensional optimization problems with constraints forms one main task forcing the use of gene duplication and deletion. This sort of optimization (minimization) problem may be formalized as follows.

Definition 34.3.1 (variable-dimensional minimization problem with constraints).

Given $f : D \subseteq X = \cup_{i=1}^{\infty} G^i \to \mathbb{R}$, minimize $f(x)$ subject to

$$g_i(x) \geq 0 \qquad \forall i \in \{1, \ldots, m\}$$
$$h_j(x) = 0 \qquad \forall j \in \{1, \ldots, l\}$$
$$x = (x_1, \ldots, x_{n_x}) \in D \subseteq X \qquad n \in \mathbb{N} \text{ arbitrary but not fixed,}$$
$$f, g_i, h_j : X \to \mathbb{R}. \qquad \qquad \square$$

g_i are called *inequality constraints* and h_j *equality constraints*. The main difference concerning a non-variable-dimensional optimization (minimization) problem is the fact that the dimension of the objective vector x may vary; that is, it is not fixed. As a consequence the parameter space X has to be formulated as the union of all parameter spaces G^i of fixed sizes i. In the context of EAs the gene space G ought not to be a vector space as usual in classical optimization (most often a Banach space, e.g. $\mathbb{R}^n$), thereby omitting all the comfortable properties Banach spaces have with respect to analysis. Instead, G might be $\mathbb{B}, \mathbb{N}, \mathbb{Z}, \mathbb{Q}, \mathbb{C}, \mathbb{R}$ or any other complex space (not in the strict mathematical sense) representable by a complex data structure. The use of G is necessary because most duplication and deletion operators directly work on semantical entities represented by G. Davidor (1991a), for example, uses binary encoded vectors of triples (three angles) for representing a robot trajectory, thus G takes the form $G = \mathbb{B}^l \times \mathbb{B}^l \times \mathbb{B}^l$.

Until now we have presented motivations for the use of variable-length genotypes in the field of EAs. Unfortunately, nature gives no real hint at why using a variable-length genotype should be advantageous. A high degree of genepool diversity and a high flexibility to a changing environment may be one main benefit of non-fixed gene lengths, thus raising the evolutional power/adaptability of a population. One interesting fact nature offers is that gene duplication most often leads to viable individuals, whereas gene deletion does not (Futuyma 1986, p 78). (A brief and sufficient introduction into the concepts of *neo-Darwinism*, i.e. the *synthetic theory of evolution*, is given by Bäck (1996) and therefore omitted here. The more interested reader is referred to the book by Futuyma (1986).)

Although nature offers a variety of schemes one central idea of how these operators may be formalized can be extracted from nature as well as from several approaches in the context of EAs. Whereas the general working mechanism of both operators is very simple, the different achievements concerning distinct applications may vary. In order not to focus on a special construction a more abstract view of both operators is presented here (sufficing in the present context).

Imagine a genotype $x = (x_1, \ldots, x_n) \in X$ consisting of genes $x_i \in G, i \in 1, \ldots, n$ (n corresponds to the actual genotype length) from a gene space G. The deletion operator del may than be formalized as a function transforming a given individual $a = (x, s) \in I$ by deleting a gene x_i. If $I = X \times A_s$ is the individual space, where A_s is the strategy parameter space, which depends on the application and the EA, del has the form

$$\text{del} : I \to I, \text{ with del}(a) = \text{del}(x_1, \ldots, x_{i-1}, x_i, x_{i+1}, \ldots, x_n, s)$$
$$= (x_1, \ldots, x_{i-1}, x_{i+1}, \ldots, x_n, s) = a'.$$

In most cases an application-dependent probability $p_{\text{del}} \in (0, 1)$ is responsible for the decision of whether deletion should be applied or not. The position i fixing the gene which has to be deleted is usually uniformly chosen from the set $\{1, \ldots, n\}$. Since deletion and duplication produce genotypes of different length it is important to notice that the dimension n varies from individual to individual. Returning to our example (Davidor 1991a), deletions occur only after a recombination ($p_c = 1.0$) and typically have a probability of 0.05. A deleted gene x_i has the form $x_i \in \mathbb{B}^l \times \mathbb{B}^l \times \mathbb{B}^l$, where each bit vector of length l codes for an angle.

Similar to deletion, the duplication operator is simple. Generally, instead of deletion, a gene is added to the genotype, such that the operator may be formalized as follows:

$$\text{dup} : I \rightarrow I, \quad \text{with } \text{dup}(a) = \text{dup}(x_1, \ldots, x_i, \ldots, x_n, s)$$
$$= (x_1, \ldots, x_i, x_i', \ldots, x_n, s) = a'.$$

Analogously to deletion a duplication probability $p_{\text{dup}} \in (0, 1)$ is used and the index i is usually uniformly chosen. Concerning the policy for introducing the new gene x_i' several policies may be distinguished, such as:

- *Duplication.* The gene x_i' is a duplicate of x_i, such that a' has the form $a' = (x_1, \ldots, x_i, x_i, \ldots, x_n, s)$.
- *Related.* The initialization of the new gene x_i' is context dependent: x_i' is generated with help the of the actual values of x_i and x_{i+1}.
- *Addition.* x_i' is initialized at random.

For example, Davidor (1991a) performs a duplication with a probability of 0.06 only when a recombination takes place. Whereas the duplication and addition policy is intuitive the related policy may be further divided into two strategies. First, 'the added arm-configuration is such that either its end-effector is positioned at the mid distance between the two adjacent end-effector positions, or its links have a mid metric value between the corresponding link positions in the adjacent arm-configurations'.

Finally, both operators have to adapt the length of the parameter vector $s \in A_s$. Because this process depends on the form of A_s details are omitted here.

34.3.4 *Problems arising when using variable-length genotypes*

Despite the fact that variable-length genotypes may enhance the computational power of EAs (see motivations (ii) and (iii)), the introduction of this new concept borrowed from nature leads to several problems.

- The role of positions in a variable-length genotype is destroyed: the assignment of corresponding genes x_i on different 'homologous' chromosomes is not possible. In order to construct genetic operators which are able to generate interpretable individuals, thus being able to respect semantical blocks on the genotype, the 'assignment problem' has to be solved.
- In particular recombination is faced with the problem of finding the locus of corresponding genes.
- Whereas some authors introduced gene duplication and gene deletion operators 'in order to improve the stability of the strings' length' (Davidor 1991a, p 84) others waive these operators; that is, they believe that variable-dimensional recombination suffices for the stabilization of string lengths (see e.g. Harp and Samad 1991).

34.3.5 Solutions

The *evolution program approach* of Michalewicz (1992), i.e. combining the concept of evolutionary computation with problem-specific chromosome structures and genetic operators, may be seen as one main concept used to overcome the problems mentioned above. Although this concept is useful in practice, it prevents the conception of a more general and formalized view of variable-length EAs because there no longer exists 'the EA' using 'the representation' and 'the set of operators'. Instead, for each application problem a specialized EA exists. According to Lohmann (1992) and Kost (1993), for example, the formulation of operators such as gene duplication and deletion, used in their framework of *structural evolution*, is strongly application dependent, thus inhibiting a more formal, general concept of these operators. Davidor (1991a, b) expressed the need for revised and new genetic operators for his variable-length robot trajectory optimization problem. In contrast to the evolution program approach, Schütz (1994) formulated an application-independent, variable-dimensional mixed-integer evolution strategy (ES), thus following the course of constructing a more general sort of ES. This offered Schütz the possibility to be more formal than other researchers. Unfortunately, this approach is restricted to a class of problems which can easily be mapped onto the mixed-integer representation he used.

Because most work concerning variable-length genotypes uses the evolution program approach, a formal analysis of gene duplication and deletion is rarely found in the literature and is therefore omitted here. As a consequence, theoretical knowledge about the behavior of gene duplication and deletion is nearly unknown. Harvey (1993), for example, points out 'that gene-duplication, followed by mutation of one of the copies, is potentially a powerful method for evolutionary progress'. Most statements concerning nonstandard operators such as duplication and deletion have the same quality as Harvey's: they are far from being provable.

Because of the lack of theoretical knowledge we proceed by discussing some solutions used to circumvent the problems which arise when introducing variable-length genotypes. In the first place, we question how other researchers have solved the problem of noncomparable loci, i.e. the problem of respecting the semantics of loci. Mostly this 'gene assignment' problem is solved by explicitly marking semantical entities on the genotype. The form of the tagging varies from application to application and is carried out with the help of different representations.

- Davidor (1991a, b) used a binary encoded non-fixed-length vector of arm configurations, i.e. a vector of triples (three angles), for representing a robot trajectory, thus defining semantical blocks.
- The path of a mobile robot may be a variable-dimensional list of path nodes (triples consisting of the two Cartesian coordinates and a flag indicating whether a node is feasible or not).

- Harp and Samad (1991) implemented the tagging with the help of a special and more complex data structure representing the structure and actual weights of any feedforward net consisting of a variable number of hidden layers and a variable number of units.

- Goldberg *et al* (1989, 1990) extended the usual string representation of GAs by using a list of ordered pairs, with the first component of each tuple representing the position in the string and the second one denoting the actual bit value. Using genotypes of fixed length a variable dimension in the resulting messy GA was achieved by allowing strings not to contain full gene complement (underspecification) and redundant or even contradictionary genes (overspecification).

- Koza (1992, 1994) used rooted point-labeled trees with ordered branches (LISP expressions), thus having a genotype representing semantics very well.

Lohmann (1992) circumvented the 'assignment problem' using so-called *structural evolution*. The basic idea of structural evolution is the separation of structural and nonstructural parameters, thus leading to a 'two-level' ES: a multipopulation ES using isolation. While on the level of each population a parameter optimization, concerning a fixed structure, is carried out, on the population level several isolated structures compete with each other. In this way Lohmann was able to handle structural optimization problems with variable dimension: the dimension of the structural parameter space does not have to be constant. Since each ES itself worked on a fixed number of nonstructural parameters (here a vector of reals) no problem occurred on this level. On the structural level (population level) special genetic operators and a special selection criterion were formulated. The criticism concerning structural evolution definitively lies in the basic assumption that structural and nonstructural parameters can always be separated. Surely, many mixed-integer variable-dimensional problems are not separable. Secondly, on the structural level the well-known semantical problem exists, but was not discussed. Schütz (1994) totally omitted a discussion concerning semantical problems arising from variable-length genotypes.

If the genotype is sufficiently prepared, problems (especially) concerning recombination disappear, because the genetic operators may directly use the tagging in order to construct interpretable individuals. Another important idea when designing recombination operators for variable-length genotypes is pointed out by Davidor (1991a). He suggests a matching of parameters according to their genotypic character instead of to their genotypic position. Essentially, this leads to a matching on the phenotypic, instead of the genotypic level. Generally, Davidor points out:

> In a complex string structure where the number, size and position of the parameters has no rigid structure, it is important that the crossover

occurs between sites that control the same, or at least the most similar, function in the phenotypic space.

In case of the (two-point) segregation crossover used in his robot trajectory optimization problem, crossing sites were specified according to the proximity of the end effector positions.

34.3.6 Conclusions

One may remark that many ideas concerning the use of gene duplication and deletion exist. Unfortunately, most thoughts have been extremely application oriented, that is, not formulated generally enough. Probably the construction of a formal frame will be very complicated in the face of the diversity of problems and solutions.

References

Ackley D and Littman M 1991 Interactions between learning and evolution *Artificial Life II (Santa Fe, NM, February 1990)* ed C Langton, C Taylor, D Farmer and S Rasmussen (Redwood City, CA: Addison-Wesley) pp 487–509
——1994 A case for Lamarckian evolution *Artificial Life III* ed C Langton (Redwood City, CA: Addison-Wesley) pp 3–10
Anderson R W 1995a Learning and evolution: a quantitative genetics approach *J. Theor. Biol.* **175** 89–101
——1995b Genetic mechanisms underlying the Baldwin effect are evident in natural antibodies *Proc. 4th Ann. Conf. on Evolutionary Programming (San Diego, CA, March 1995)* ed J R McDonnell, R G Reynolds and D B Fogel (Cambridge, MA: MIT Press) pp 547–63
——1996a How adaptive antibodies facilitate the evolution of natural antibodies *Immunol. Cell Biology* **74** 286–91
——1996b Random-walk learning: a neurobiological correlate to trial-and-error *Prog. Neural Networks* at press
Bäck T 1996 *Evolutionary Algorithms in Theory and Practice* (New York: Oxford University Press)
Balakrishnan K and V Honavar 1995 *Evolutionary Design of Neural Architectures: a Preliminary Taxonomy and Guide to Literature* Artificial Intelligence Research Group, Department of Computer Science, Iowa State University, Technical Report CS TR 95-01
Baldwin J M 1896 A new factor in evolution *Am. Naturalist* **30** 441–51
Belew R K 1989 When both individuals and populations search: adding simple learning to the genetic algorithm *Proc. 3rd Int. Conf. on Genetic Algorithms (Fairfax, VA, June 1989)* ed J D Schaffer (San Mateo, CA: Morgan Kaufmann) pp 34–41
——1990 Evolution, learning and culture: computational metaphors for adaptive search *Complex Syst.* **4** 11–49
Bremermann H J and Anderson R W 1991 How the brain adjusts synapses—maybe *Automated Reasoning: Essays in Honor of Woody Bledsoe* ed R S Boyer (New York: Kluwer) pp 119–47

Bulmer M G 1985 *The Mathematical Theory of Quantitative Genetics* (Oxford: Oxford University Press) pp 150–2

Cecconi F, Menczer F and Belew R K 1995 *Maturation and the Evolution of Imitative Learning in Artificial Organisms* Technical Report CSE 506, University of California, San Diego; 1996 *Adaptive Behavior* **4** at press

Charlesworth B 1993 The evolution of sex and recombination in a varying environment *J. Heredity* **84** 345–50

Conrad M 1993 Structuring adaptive surfaces for effective evolution *Proc. 2nd Ann. Conf. on Evolutionary Programming (San Diego, CA)* ed D B Fogel and W Atmar (La Jolla, CA: Evolutionary Programming Society) pp 1–10

Conrad M and Ebeling W 1992 M V Volkenstein, evolutionary thinking and the structure of fitness landscapes *BioSystems* **27** 125–8

Davidor Y 1991a *Genetic Algorithms and Robotics. A Heuristic Strategy for Optimization (World Scientific Series in Robotics and Automated Systems 1)* (Singapore: World Scientific)

——1991b A genetic algorithm applied to robot trajectory generation *Handbook of Genetic Algorithms* ed L Davis (New York: Van Nostrand Reinhold) ch 12, pp 144–65

Davis L (ed) 1991 *Handbook of Genetic Algorithms* (New York: Van Nostrand Reinhold)

Fogel D B 1988 An evolutionary approach to the traveling salesman problem *Biological Cybern.* **60** 139–44

——1993 Applying evolutionary programming to selected traveling salesman problems *Cybern. Syst.* **24** 27–36

Fogel L J, Owens A J and Walsh M J 1966 *Artificial Intelligence through Simulated Evolution* (New York: Wiley)

Fontanari J F and Meir R 1990 The effect of learning on the evolution of asexual populations *Complex Syst.* **4** 401–14

French R and Messinger A 1994 Genes, phenes and the Baldwin effect *Artificial Life IV (Cambridge, MA, July 1994)* ed R A Brooks and P Maes (Cambridge, MA: MIT Press) pp 277–82

Futuyma D J 1986 *Evolutionary Biology* (Sunderland, MA: Sinauer)

Gehlhaar D K, Verkhivker G, Rejto P A, Fogel D B, Fogel L J and Freer S T 1995 Docking conformationally flexible small molecules into a protein binding site through evolutionary programming *Proc. 4th Ann. Conf. on Evolutionary Programming (San Diego, CA, March 1995)* ed J R McDonnell, R G Reynolds and D B Fogel (Cambridge, MA: MIT Press) pp 615–27

Goldberg D E, Deb K and Korb B 1990 Messy genetic algorithms revisited: studies in mixed size and scale *Complex Syst.* **4**

Goldberg D E, Korb B and Deb K 1989 Messy genetic algorithms: motivation, analysis, and first results *Complex Syst.* **3**

Grefenstette J J, Gopal R, Rosmaita B and Van Gucht D 1985 Genetic algorithms for the traveling salesman problem *Proc. Int. Conf. on Genetic Algorithms and their Applications* ed J J Grefenstette (Erlbaum) pp 160–6

Harp S A and Samad T 1991 Genetic synthesis of neural network architecture *Handbook of Genetic Algorithms* ed L Davis (New York: Van Nostrand Reinhold) ch 15, pp 202–21

Harvey I 1993 *The Artificial Evolution of Adaptive Behaviour* Master's Thesis, University of Sussex

Hightower R, Forrest S and Perelson A 1996 The Baldwin effect in the immune system: learning by somatic hypermutation *Adaptive Individuals in Evolving Populations: Models and Algorithms* ed R K Belew and M Mitchell (Reading, MA: Addison-Wesley) at press

Hinton G E and Nowlan S J 1987 How learning can guide evolution *Complex Syst.* **1** 495–502

Holland J H 1975 *Adaptation in Natural and Artificial Systems* (Ann Arbor, MI: University of Michigan Press)

Kost B 1993 *Structural Design via Evolution Strategies* Internal Report, Department of Bionics and Evolution Technique, Technical University of Berlin

Koza J R 1992 *Genetic Programming* (Cambridge, MA: MIT Press)

——1994 *Genetic Programming II* (Cambridge, MA: MIT Press)

Lin S and Kernighan B W 1976 An effective heuristic for the traveling salesman problem *Operat. Res.* **21** 498–516

Lohmann R 1992 Structure evolution and incomplete induction *Parallel Problem Solving from Nature, 2 (Proc. 2nd Int. Conf. on Parallel Problem Solving from Nature, Brussels, 1992)* ed R Männer and B Manderick (Amsterdam: Elsevier) pp 175–85

Männer R and Manderick B (eds) 1992 *Parallel Problem Solving from Nature, 2 (Proc. 2nd Int. Conf. on Parallel Problem Solving from Nature, Brussels, 1992)* ed R Männer and B Manderick (Amsterdam: Elsevier)

Maynard Smith J 1978 *The Evolution of Sex* (Cambridge: Cambridge University Press)

——1987 When learning guides evolution *Nature* **329** 761–2

Michalewicz Z 1992 *Genetic Algorithms + Data Structures = Evolution Programs* (Berlin: Springer)

Milstein C 1990 The Croonian lecture 1989 Antibodies: a paradigm for the biology of molecular recognition *Proc. R. Soc.* B **239** 1–16

Mitchell M and Belew R K 1995 Preface to G E Hinton and S J Nowlan How learning can guide evolution *Adaptive Individuals in Evolving Populations: Models and Algorithms* ed R K Belew and M Mitchell (Reading, MA: Addison-Wesley)

Morgan C L 1896 On modification and variation *Science* **4** 733–40

Osborn H F 1896 Ontogenic and phylogenic variation *Science* **4** 786–9

Nolfi S, Elman J L and Parisi D 1994 Learning and evolution in neural networks *Adaptive Behavior* **3** 5–28

Paechter B, Cumming A, Norman M and Luchian H 1995 Extensions to a memetic timetabling system *Proc. 1st Int. Conf. on the Practice and Theory of Automated Timetabling (ICPTAT 95) (Edinburgh, 1995)*

Parisi D, Nolfi S and Cecconi F 1991 Learning, behavior, and evolution *Toward a Practice of Autonomous Systems (Proc. 1st Eur. Conf. on Artificial Life (Paris, 1991))* ed F J Varela and P Bourgine (Cambridge, MA: MIT Press)

Rechenberg I 1973 *Evolutionsstrategie: Optimierung Technischer Systeme nach Prinzipien der Biologischen Evolution* (Stuttgart: Fromman-Holzboog)

Saravanan N, Fogel D B and Nelson K M 1995 A comparison of methods for self-adaptation in evolutionary algorithms *BioSystems* **36** 157–66

Scheiner S M 1993 Genetics and evolution of phenotypic plasticity *Ann. Rev. Ecol. Systemat.* **24** 35–68

Schütz M 1994 *Eine Evolutionsstrategie für gemischt-ganzzahlige Optimierungsprobleme mit variabler Dimension* Diploma Thesis, University of Dortmund

Schwefel H P 1968 *Projekt MHD-Staustrahlrohr: Experimentelle Optimierung einer Zweiphasendüse* Teil I Technischer Bericht 11.034/68, 35, AEG Forschungsinstitut, Berlin

Sober E 1994 The adaptive advantage of learning and *a priori* prejudice *From a Biological Point of View: Essays in Evolutionary Philosophy* (a collection of essays by E Sober) (Cambridge: Cambridge University Press) pp 50–70

Stearns S C 1989 The evolutionary significance of phenotypic plasticity—phenotypic sources of variation among organisms can be described by developmental switches and reaction norms *Bioscience* **39** 436–45

Stephens D W 1993 Learning and behavioral ecology: incomplete information and environmental predictability *Insect Learning. Ecological and Evolutionary Perspectives* ed D R Papaj and A C Lewis (New York: Chapman and Hall) ch 8, pp 195–218

Turney P D 1995 Cost-sensitive classification: empirical evaluation of a hybrid genetic decision tree induction algorithm *J. Artificial Intell. Res.* **2** 369–409

——1996 How to shift bias: lessons from the Baldwin effect *Evolutionary Comput.* at press

Turney P D, Whitley D and Anderson R W (eds) 1996 Special issue on evolution, learning, and instinct: 100 years of the Baldwin effect *Evolutionary Comput.* at press

Unemi T, Nagayoshi M, Hirayama N, Nade T, Yano K and Masujima Y 1994 Evolutionary differentiation of learning abilities—a case study on optimizing parameter values in Q-learning by a genetic algorithm *Artificial Life IV (July 1994)* ed R A Brooks and P Maes (Cambridge, MA: MIT Press) pp 331–6

Via S 1993 Adaptive phenotypic plasticity: target or by-product of selection in a variable environment? *Am. Naturalist* **142** 352–65

Waddington C H 1942 Canalization of development and the inheritance of acquired characters *Nature* **150** 563–5

Wcislo W T 1989 Behavioral environments and evolutionary change *Ann. Rev. Ecol. Systemat.* **20** 137–69

West-Eberhard M J 1989 Phenotypic plasticity and the origins of diversity *Ann. Rev. Ecol. Systemat.* **20** 249–78

Whitley D and Gruau F 1993 Adding learning to the cellular development of neural networks: evolution and the Baldwin effect *Evolutionary Comput.* **1** 213–33

Whitley D, Gordon S and Mathias K 1994 Lamarckian evolution, the Baldwin effect and function optimization *Parallel Problem Solving from Nature—PPSN III (Proc. Int. Conf. on Evolutionary Computation and 3rd Conf. on Parallel Problem Solving from Nature, Jerusalem, October 1994) (Lecture Notes in Computer Science 866)* ed Yu Davidor, H P Schwefel and R Männer (Berlin: Springer) pp 6–15

Wright S 1931 Evolution in Mendelian populations *Genetics* **16** 97–159

Further reading

More extensive treatments of issues related to the Baldwin effect can be found in the literature cited in section C3.4.1. The following are notable foundation and review papers.

1. Anderson R W 1995a Learning and evolution: a quantitative genetics approach *J. Theor. Biol.* **175** 89–101

2. Balakrishnan K and V Honavar 1995 *Evolutionary Design of Neural Architectures: a Preliminary Taxonomy and Guide to Literature* Artificial Intelligence Research Group, Department of Computer Science, Iowa State University, Technical Report CS TR 95-01

3. Baldwin J M 1896 A new factor in evolution *Am. Naturalist* **30** 441–51

4. Belew R K 1989 When both individuals and populations search: adding simple learning to the genetic algorithm *Proc. 3rd Int. Conf. on Genetic Algorithms (Fairfax, VA, June 1989)* ed J D Schaffer (San Mateo, CA: Morgan Kaufmann) pp 34–41

5. Hinton G E and Nowla Hinton G E and S J Nowlan 1987 How learning can guide evolution *Complex Syst.* **1** 495–502

6. Morgan C L 1896 On modification and variation *Science* **4** 733–40

7. Sober E 1994 The adaptive advantage of learning and *a priori* prejudice *From a Biological Point of View: Essays in Evolutionary Philosophy* (a collection of essays by E Sober) (Cambridge: Cambridge University Press) pp 50–70

8. Turney P D 1995 Cost-sensitive classification: empirical evaluation of a hybrid genetic decision tree induction algorithm *J. Artificial Intell. Res.* **2** 369–409

9. Turney P D, Whitley D and Anderson R W (eds) 1996 Special Issue on evolution, learning, and instinct: 100 years of the Baldwin effect *Evolutionary Comput.* at press

10. Waddington C H 1942 Canalization of development and the inheritance of acquired characters *Nature* **150** 563–5

11. Wcislo W T 1989 Behavioral environments and evolutionary change *Ann. Rev. Ecol. Systemat.* **20** 137–69

12. Whitley D and F Gruau 1993 Adding learning to the cellular development of neural networks: evolution and the Baldwin effect *Evolutionary Comput.* **1** 213–33

Index, Volume 1

A

Actuator placement on space structures 7
Adaptation 37
Adaptation in Natural and Artificial Systems (book) 46
Adaptive behavior 110
Adaptive landscape 36
Adaptive mutation 70
Adaptive Systems Workshop 46
Adaptive topography 24
Air combat maneuvering 9
Airborne pollution 8
Aircraft design 7
Alleles 64, 70, 164, 209, 263, 310, 311
Amino acids 33
Animats 110
ARGOT 77, 146
Arithmetic crossover 272
Artificial intelligence (AI) 90, 97, 189, 321
Artificial life (AL) 2, 321
Artificial neural networks. *See* Neural networks
Autoadaptation 43
Automatic control 43, 94
Automatic programming 40
Automatically defined functions (ADFs) 110, 158
Autonomous vehicle controller 8

B

Baldwin effect 308–17
    in evolutionary biology 309–13
Beam search 189
Bellman optimality 116
Bent pipe 49
Bias 66
Bidding procedures 119
Binary representation 60
Binary strings 64, 69, 75, 76, 127, 128, 131–5, 237, 238, 256–70
Binary tournament selection 168

Binary vectors. *See* Binary strings
Biological systems 2, 256
Biology 7, 9
Biomorphs 228
Bit strings. *See* Binary strings
Bitwise crossover 73
Bitwise simulated crossover (BSC) 70
Bitwise uniform crossover 73
Boltzmann selection 174, 195–200, 202
    mechanisms 195
Boltzmann selection operator 169
Boltzmann tournament selection (BTS) 196
Boltzmann trials 195
    double acceptance/rejection 197
    single acceptance/rejection 197
Boolean parse tree 249
Branch-and-bound techniques 130
Breeding strategies 45
Bucket brigade algorithm 117, 118
Building block hypothesis (BBH) 70, 90

C

Canonical genetic algorithms 132
Cellular automata (CAs) 7
CHC algorithm 67
Chemistry 7
Chromatids 32
Chromosomes 27, 32, 33, 35, 64
Classification, applications 9, 10
Classifier systems (CFS) 2, 46
Clonal selection theory 37
Combinatorial problems (CES) 76
Combined representations 130, 131
Competitive selection 203
Compress mutation operation 158
Computer-generated force (CGF) 99
Computer programs 103, 109, 110
Computer simulation 1
Constant learning 314, 315
Constant-velocity environments 315
Control applications 8
Control systems 98

Convergence-controlled variation (CCV) 70, 75
Convergence-controlled variation hypothesis (CCVH) 70, 71
Convergence rate 240, 241
  theory 49
Corridor model 241
Creeping random search method 50
Criminal suspects 8
Critical learning period model 315
Crossover 64, 65, 68–75, 235
  bias 268, 269
  in tetrad 32
  mathematical characterizations 261, 262
  mechanisms 256–61
  one-point 69
  points 257
  probability 260
  rate 257, 258
  two-point 69
  uniform 69
Crossover operators 76, 132, 238, 257, 258
  characterizations 361
Cyanide production 31
Cyanogenic hybrid 31
Cycle crossover 145

D

Darwinian evolution 89, 309
Darwinism 27
Deception 72
Deceptive functions 147, 149
Deceptive problems 72, 73
Decision making, multicriterion. *See* Multicriterion decision making
Decision variables 127
Defining length 265
Delta coding 77
Deoxyribonucleic acid (DNA) 33
Derivative methods 103–13
Design applications 6, 7
Deterministic hill climbing 1
Dihybrid ratio 31
Dimensionality 21
Diplodic representation 251
Diploid 27, 35
Diploid representations 164

Discount parameter 116
Discrete recombination 270
Disruption analysis 264
Disruptive selection 202
Distribution bias 269
Diversity 192
DNA (deoxyribonucleic acid) 33
Document retrieval 9
Domain-specific knowledge 318
Double helix 33
Drift 36, 46, 209
Drug design 98
Dynamic programming (DP) 116

E

Economics 7, 9
  interaction modeling 7
Electromagnetics 8
Elitist strategy 66, 210
Embryonic development 110
Encapsulate operator 158
Endosymbiotic systems 37
Engineering applications 7
Enzymes 33
Epistasis 31, 32
Equivalence 214–18
Euclidean search spaces 81
Eukaryotic cell 33
Evaluations 89
  noise 222–4
Evolution and learning 308, 309
Evolutionary algorithms (EAs) 6, 7, 20–2, 318
  admissible 191
  basic 59
  Boltzmann 195
  common properties 59
  computational power 320
  development 41
  general outline 59
  mainstream instances 59
  strict 191
  theory 40, 41
  see also specific types and applications
Evolutionary computation (EC)
  advantages (and disadvantages) 20–2
  applications 4–19
  consensus for name 41

discussion 3
history 40–58
use of term 1
Evolutionary Computation (journal) 47
Evolutionary game theory 37
Evolutionary operation (EVOP) 40
Evolutionary processes 37
  overview 23–6
  principles of 23–6
Evolutionary programming (EP) 1, 60,
    136, 163, 167, 217, 218
  basic concepts 89–102
  basic paradigm 94
  continuous 95
  convergence properties 100
  current directions 97–100
  diversification 44
  early foundations 92–4
  early versions 95
  extensions 94–7
  future research 100
  genesis 90
  history 40, 41, 90–7
  main components 89
  main variants of basic paradigm 95
  medical applications 98
  original 95, 96
  original definition 91
  overview 40, 41
  self-adaptive 95, 96
  standard form 89
  v. GAs 90
Evolutionary robotics, *see also* Robots
Evolutionary strategies (ESs) 1, 48–51,
    60, 64, 81–8, 136, 163
  $(1 + 1)$ 48, 83
  $(1 + \lambda)$ 48
  $(\mu + 1)$ 83
  $(\mu + \lambda)$ 48, 67, 83, 86, 167, 169,
    189, 206, 210, 217, 220, 224, 230
  (μ, λ) 189, 206, 210, 220, 222, 231
  $(\mu + \mu)$ 170
  alternative method to control internal
    parameters 86
  archetype 81, 82
  contemporary 83–6, 85
  development 40
  multimembered $(\mu > 1)$ 48
  nested 86, 87
  overview 48–51

population-based 83
steady-state 83
two-membered 48, 50
Exons 34
Expected, infinite-horizon discounted
    cost 116
Expression process 231
Extradimensional bypass thesis 320

F

Fault diagnosis 7
Feedback networks 97
Feedforward networks 97
Fertility factor 192
Fertility rate 192
Filters, design 6
Financial decision making 9
Finite impulse response (FIR) filters 6
Finite-length alphabet 91
Finite-state machines 44, 60, 91, 92, 95,
    129, 134, 152, 153, 162, 246–8
Finite-state representations 151–4
  applications 152–4
Fitness-based scan 273
Fitness criterion 228
Fitness evaluation 108
Fitness function 178
Fitness functions 172–5
  monotonic 190, 191
  scaling 66
  strictly monotonic 190, 191
Fitness landscapes 229, 308, 311
Fitness measure 235
Fitness proportional selection (FPS) 218
Fitness scaling 174, 175, 187
Fitness values 59, 64, 66
Fixed selection 314, 315
Flat plate 48
*Foundations of Genetic Algorithms
    (FOGA)* (workshop) 47
Functions 104, 105
Fundamental theorem of genetic
    algorithms 177
Fuzzy logic systems 163
Fuzzy neural networks 97
Fuzzy systems 44

G

Game playing 9
  programs 45
Game theory 98
Gametes 27
Gametogenesis 27
Gaming 43
Gauß–Seidel-like optimization strategy 87
Gaussian distribution 242
Gaussian mutations 241
Gaussian selection 313
Geiringer's theorem II 263, 264
Gene duplication and deletion 319
  basic motivations 319
  engineering applications 319, 320
  formal description 321–3
  historical review 319–21
Gene flow 36
Gene frequencies 36
Generation gap methods 205–11
  historical perspective 206
Generational EAs 207
Generational models 216
Generic control problem 114
Genes 30, 33, 34, 64, 310
  segregating independently 30
GENESIS 311
Genetic algorithms (GAs) 1, 2, 59, 60, 64–80, 103, 136, 155, 167
  basics 65–8
  breeder 67
  canonical 64
  generational 67
  history 40–58
  implementation 46, 65
  messy 72, 73, 164
  operation 70
  overview 44–8, 64–80
  pseudocode 65
  steady-state 67
  v. EP 90
  see also specific applications
Genetic drift 36, 46, 209
Genetic operators 65, 106–8, 110
Genetic Program Builder (GLiB) 158
Genetic programming (GP) 23, 60, 156, 167
  defined 103–8

development 108, 109
functions 106
fundamental concepts 103–8
initialization 106
specialized representation as executable program 104–6
value 109, 110
Genetics
  fundamental concepts 27–33
  principles of 27–39
GENITOR system 67, 189, 209
Genomic DNA 35
Genotypes 23, 64, 231, 232
Geometrical crossover 273
Global convergence 199, 200
Gradient descent 1
Gray code 76, 133
Gray-coded strings 128
Grow mutation operator 248

H

Hamiltonian circuit 139
Hamming cliffs 128
Hamming distance 75, 133, 148
Haploid 27, 35
Hardy–Weinberg theorem 35
Heating, ventilation and air conditioning (HVAC) controllers 98
Heterozygote 30
Heuristic crossover 273
H-infinity optimal controllers 8
Hinton and Nowlan's model 310, 311, 312
Hitchhiking effects 71
Homozygotes 30
Hybrid algorithms 308–11
Hybridizations 60
Hydrodynamics 81
Hyerplanes 72, 177, 178, 179
  analysis 69

I

Identification applications 7, 8
IEEE World Congress on Computational Intelligence (WCCI) 41
Image processing 9
  applications 44
Implicit parallelism 134, 190, 191
Incremental models 217, 220–2

Inductive bias 268
Infinite impulse response (IIR) filters 6
Information retrieval (IR) systems 9
Information storage 9
Inheritance systems 28, 29, 37
Initialization 74, 89
Insert operator 245, 246
Intelligent behavior 42
Interactive evolution (IE) 228–34
    application areas 231, 232
    approach 229–31
    difficulties encountered 231
    formulation of algorithm 230
    further developments and
        perspectives 232, 233
    history and prospects 228, 229
    minimum requirement 228
    overview 231
    problem definition 229
    samples of evolved objects 233
    selection 229
    standard system 229
Interceptor 94
*Interdisciplinary Workshop in Adaptive
    Systems* 46
Intermediate recombination 85
*International Conference on Genetic
    Algorithms (ICGA)* 47
International Society for Genetic
    Algorithms (ISGA) 47
Introns 163, 164
Inverse permutation 141, 142
Inversion 145
Inversion operator 77
Island models 36, 77

J

Job shop scheduling (JSS) 5
Joint plate 48
Juxtapositional phase 74

K

k-ary alphabet 134
Knapsack problems (KP) 6, 132
Knowledge-augmented operators 317–19
Kohonen feature map design 6

L

Lamarckian inheritance 308
Languages 60
Laplace-distributed mutation 240
Learning
    and evolution 308, 309
    as phenotypic variance 313, 314
Learning algorithms 308
Learning classifier systems (LCSs)
    114–23
    introduction 117, 118
    Michigan approach 118
    operational cycle 118
    Pitt approach 118
    stimulus response 118
    structure 117
Learning models 316
Learning problems 114–18
Life cycle model 28
LINAC. *See* Linear accelerator
Linear accelerator, design 7
Linear-quadratic-Gaussian controllers
    (LQG) 8
Linear ranking 188, 215
Linguistics 9
Linkage, equilibrium 264
LISP 75, 108, 109, 128, 129
Local search (LS) 149

M

Machine intelligence 2
Machine learning (ML) problems 321
Mapping function 142
Markov decision problem 115
Mask 257
Mathematical analyses 44
Matrix representations 143–5
Maximum independent set problem 132
Medical applications 98
Medicine 44
Meiosis 27, 32
Memory cache replacement policies 5
Mendel, Gregor 28
Mendelian experiment 28, 29, 30
Mendelian inheritance 261
Mendelian ratios 31
Messenger RNA (mRNA) 33, 34
Messy GA 251

Messy genetic algorithms (mGAs) 72, 73, 164
Metropolis algorithm 196
Military applications 9
Minimax optimization 87
Mixed-integer optimization 87
Mixed-integer representation 163
Mixing events 268
Monohybrid ratio 30
Monte Carlo (MC) generators 7
Multicriterion problems 50
Multimodal objective functions 84
Multiple criterion decision making (MCDM) 22
Multiple-input, multiple-output (MIMO) model 98, 99
Multipoint crossover 266
Multiprocessor system 5
Multivariate zero-mean Guassian random variable 137
Mutation 23, 35, 42, 59, 60, 61, 68–75, 89, 108, 152, 228, 237–55
2-opt, 3-opt and k-opt 244, 245
  function 239
  optimum standard deviations 241
  successful 241
Mutation function 314
Mutation operators 84, 92, 93, 125, 132, 158, 237–55
Mutation-selection equilibrium variance 314
Mutations 36

N

Near misses 308
Neighborhood, model 77
Neo-Darwinism 23, 24, 37
Nesting technique 86
Network design problems 6
Neural networks 6, 43, 44, 99, 163
  design 97
  training 43, 44
Neutral molecular evolution theory 37
Niching methods 50
No-free-lunch (NFL) theorem 20, 21
Nodes 104
Nondeterministic-polynomial-time (NP) complete problems 97
Nondisruptive crossovers 267

Nonlinear optimization problems 1
Nonlinear ranking 188, 189, 215
Nonlinear ranking selection 202, 203
Nonoverlapping populations 205
Nonoverlapping systems 208
Nonregressive evolution 43
Nonuniform mutation 243
n-point crossover 259, 266

O

Object parameters 136
Object variables 136–8
Objective functions 172, 173, 178
Objective values 192
Offspring machines 42, 152
One-point crossover 258, 265, 266, 271
Online planning/navigating 5
Operator descriptions 149
Operon system 34
Optical character recognition (OCR) 9
Optimization methods 1, 2, 89, 160
Optimum convergence 241
Order-based mutation 246
Ordering schemata (o-schemata) 146, 147
Oren–Luenberger class 85
Overlapping populations 205, 208, 209, 210
Ovum 27

P

Packing problems 6
Pairwise implementation 70
Pairwise mating 70, 71, 75
Parallel genetic algorithms (PGAs) 77
Parallel Problem Solving from Nature (PPSN) (workshop) 41
Parallel recombinative simulated annealing (PRSA) 196, 197
  parameters and their settings 197
  pseudocode for common variation 197
  working mechanism 196, 197
Parameter optimization 133
Parameter settings 22
Parasites 120, 121
Parse tree representation 155–9
  complex numerical function 157
  primitive language 156, 157

Parse trees 134, 248–50
Partially matched crossover (PMX)
      operators 147
Pattern recognition 45
Payoff matrix 99
Penalty functions 95, 130
Permutations 75, 128, 129, 134, 139–50,
      243–6
   inverse 141, 142
   mapping integers to 140
Pharmaceutical fermentation process
      data 8
Phenotype table 31
Phenotypes 23, 31, 232, 313, 314
Pitt approach to rule learning 47
Planning applications 4–6
Pleiotropy 23
Ploidy 35
Point mutation 35
Polygeny 23
Poolwise CCV algorithm 75
Poolwise mating 75
Poolwise methods 71
Poolwise schemes 70
Population models 214
Population parameters 228
Population size 198
Populations 35–7
Position-based mutation 246
Positional bias 269
Primordial phase 74
Prisoner's dilemma 98, 99, 153, 154
Probability density function (PDF) 239,
      240
Proportional selection 64, 90, 167,
      172–82
   theory 176–80
Protein secondary-structure
      determination 9
Protein segments 9
PRSA. See Parallel recombinative
      simulated annealing
Pseudo-Boolean optimization problems
      132, 134
Pseudocode 166–70
Punctuated crossover 260

Q

Q-learning 116

Q-values 116
Quantitative genetics models 313–16
Query construction 10

R

Random decisions 48
Random keys 146
Random mutation hill climbing
      (RMHC) 243
Random program trees 106
Random search 1
Randomness 1
Rank-based selection 66, 169, 187–94,
      209
   overview 187, 188
   theory 190–4
Ranking 187, 188, 215, 216, 218, 219
Real-valued parameters 75
Real-valued vectors 60, 128, 134,
      136–8, 239–43, 270–4
Recombination 59, 60, 85, 106, 107,
      152, 228, 256–307
   dynamics 262, 263
   events 257
   formal analysis 261
Recombination bias 269
Recombination distributions 261, 264,
      265, 270
Recombination events 265, 266, 269
Recombination-mutation-selection loop
      61
Recurrence relations 263
Reduced surrogate recombination 261
Reinforcement learning problem 115,
      117
Replacement
   biased v. unbiased strategy 67
   selection 67
Representation 75–7, 127, 128, 228, 235
   alternative 145, 146
   guidelines for suitable encoding
      160–2
   importance of 128–30
   nonstandard 163, 164
   see also specific representations
Representations 127–31
Reproduction 66
Reproductive plans 45
Reproductive rate 192

Resource scheduling 139
REVOP 40
Ribonucleic acid (RNA) 33
Robbins equilibrium 264
Robots
    applications 9
    control systems 8
    optimization applications of EP 43
    path planning 5
    see also Evolutionary robotics
Robustness 1, 2, 20
Roulette wheel, sampling algorithm 175,
    176
Route planning 43
Routing problems 4, 5, 97
Rule-based learning, Pitt approach to 47

S

Sampling algorithms 175–7
Scaling 66
Scheduling 5
Scheduling problems 5
Schema analysis 45
Schema bias 270
Schema processing 44, 130, 134
Schema theorem 134, 177, 178
Schemata 69, 134, 265, 266, 269
Scramble mutation operator 245, 246
Search operators 228
    introduction 235, 236
Second-order schemata, transmission
    probabilities for 268
Segmented crossover 73
Selection 24, 25, 37, 45, 59, 89
    biasing 66, 67
    introduction 166–71
    operators 166–71
    working mechanisms 166, 167
Selection algorithm
    monotonic 191
    strictly monotonic 191
Selection differential 179, 192, 193
Selection intensity 184, 192, 213, 224
Selection mechanisms, comparison
    212–27
Selection methods 201–4
    analytic comparison 224, 225
Selection operators 60, 167
Selection parameters 225

Selection pressure 213, 215, 219, 220,
    224
Selection probabilities 175
Selection process
    steps involved 172, 187
    see also specific processes
Selectionist theories 37
Selective pressure 170, 171
Self-adaptation 43, 83, 84, 238, 241,
    242, 248
    by collective learning 50
    procedures 138
Sequence prediction 43
Set covering problem (SCP) 132
Sex chromosomes 28
S-expressions 75
Shifting balance theory 36
Shrink mutation operator 248
Shuffle crossover 261
Sigma scaling 174, 175
SIMD (small instruction, multiple data)
    parallel machines 50
Simplex crossover 273
Simplex design 40
Simulated annealing (SA) 196, 197
Simulated evolution (SE) 43
Simulation applications 7
Simulations 213, 214, 218–22
Small instruction, multiple data (SIMD)
    parallel machines 50
Soft brood selection 202
Solutions 127, 128
Source apportionment problem 8
Speciation 36, 37
Sperm 27
Sphere model 241
Stack genetic programming 109
Staff rota 5
Steady-state genetic algorithms 83, 207,
    208
Steady-state selection 95
Stepping stone model 36
Stimulus-response LCS 118
Stochastic universal sampling (SUS) 176
Strategy parameters 59, 137, 138
Strength 118
Strict building block hypothesis (SBBH)
    70, 71, 74, 75
Structural gene 34
Substitution theorem 193, 194

Supersolution 168
Superstrings 199
Supervised learning 114, 115
Swap mutation 245, 246
Switch mutation operator 248
System identification problems 8

T

Takeover time 179, 180, 182, 193
Target sampling rates 177, 190
Terminals 104
Termination criterion 61
Testing applications 7
Tetraploid 35
Three-dimensional convergent-divergent
        nozzle 49
Threshold 189
Threshold selection 189, 190
Timetables 5
Tournament competition 97
Tournament selection 66, 181–6, 201,
        202, 215
    binary 168
    concatenation of tournaments 183
    formal description 182
    loss of diversity 184
    operator 168
    parameter settings 182
    properties 183–5
    working mechanism 181, 182
Tournament size 181, 184
Transcription process 33, 34
Transfer RNA (tRNA) 34
Translation process 34
Transmission probabilities for
        second-order schemata 268

Transportation problem 4, 5
Traveling salesman problem (TSP) 4,
        43, 76, 97, 139, 143, 146, 147,
        161, 162, 240, 245, 318
Tree-like hierarchies 75
Tree-structured genetic material 107
Tree-structured representation 104
Truncation selection 189, 217, 218
Two-dimensional foldable plate 81
Two-dimensional packing problems 6
Two-membered ES 48, 50
Two-phase nozzle optimization 49
Two-point crossover 267

U

Uncertainty 3
Unequal-area facility layout problem 6
Uniform crossover 73, 85, 259, 267
Uniform scan operator 271

V

Variable lifespan 203
Variable-length genotypes 319, 320
Variance 208, 209
Variation 89
Vehicle routing problem 4, 5
Virtual machine design task 105

W

Watson–Crick model 33

Z

Zero-one knapsack problems 6
Zygote 35